AF360724

Recent Advances in Applied Microbiology and Food Sciences

Recent Advances in Applied Microbiology and Food Sciences

Editors

Marek Kieliszek
Przemyslaw Lukasz Kowalczewski

MDPI • Basel • Beijing • Wuhan • Barcelona • Belgrade • Manchester • Tokyo • Cluj • Tianjin

Editors
Marek Kieliszek
Warsaw University of Life
Sciences—SGGW
Poland

Przemyslaw Lukasz
Kowalczewski
Poznan University of Life
Sciences
Poland

Editorial Office
MDPI
St. Alban-Anlage 66
4052 Basel, Switzerland

This is a reprint of articles from the Special Issue published online in the open access journal *Applied Sciences* (ISSN 2076-3417) (available at: https://www.mdpi.com/journal/applsci/special_issues/Applied_Microbiology_Food_Sciences).

For citation purposes, cite each article independently as indicated on the article page online and as indicated below:

LastName, A.A.; LastName, B.B.; LastName, C.C. Article Title. *Journal Name* **Year**, *Volume Number*, Page Range.

ISBN 978-3-0365-6638-2 (Hbk)
ISBN 978-3-0365-6639-9 (PDF)

Contents

About the Editors

Marek Kieliszek

Marek Kieliszek is Professor at the Department of Food Biotechnology and Microbiology of the Institute of Food Sciences of the Warsaw University of Life Sciences—SGGW. His is author and co-author of several dozen scientific articles from the JCR list and has delivered lectures at numerous international universities. In addition, he is a regular reviewer in many journals listed in the Journal Citation Reports (JCR) database. He serves as Manager of individual research projects financed by the National Science Centre, Contractor of two POIG projects and internal grants of the Institute of Food Sciences of the Warsaw University of Life Sciences (SGGW), and is the winner of many scientific awards, such as in 2019, when he was awarded by the Minister of Science and Higher Education of Poland as an outstanding young scientist. He is a co-author of international and national patents and has deposited fragments of nucleotide sequences of microorganism strains in the database of the National Center for Biotechnology Information. His primary area of research is on the bioaccumulation of selenium in the cellular biomass of yeasts. An important aspect of his scientific work is also related to the identification of new microorganisms using molecular biology methods. He is also interested in the possibility of obtaining various metabolites produced by yeasts and bacteria through microbiological processes, which can be widely used in various branches of food and pharmaceutical industry (e.g., bioplexes, enzymes, and β-glucan). Moreover, he actively partakes in research works conducted not only at the Institute of Food Sciences of the Warsaw University of Life Sciences—SGGW but also outside his home research unit.

Przemyslaw Lukasz Kowalczewski

Przemysław Łukasz Kowalczewski is Assistant Professor at the Faculty of Food Science and Nutrition, Poznań University of Life Sciences, Poland. He received his Ph.D. in Food Science and Nutrition in 2016, focusing on research tasks linked to health-promoting food, especially design and testing of the biological activity of such products. As a result of his scientific activity, a number of valuable scientific papers were published in well-renowned journals. These works describe the biological activity of potato juice (cytotoxicity and genotoxicity) in relation to tumor growth in gastrointestinal tract as well as anti-inflammatory activity. A number of health-promoting technologies for food products containing potato juice and a method for the production of bioactive components of potato juice have been developed. Several production technologies of health-promoting food products containing bioactive ingredients of potato juice have been patented. He has also led several research activities and projects dealing with design and testing of biological activity of health-promoting food products, such as on plant-based meat analogues and edible insects, among others. In 2020, he was awarded by the Minister of Science and Higher Education of Poland as an outstanding young scientist. He has authored more than 100 research articles published in peer-reviewed international journals and has presented many communications in national and international seminars and conferences within the areas of Food Sciences and Technology, Nutrition and Dietetics, multidisciplinary Chemistry, and Agronomy.

Editorial

Recent Advances in Applied Microbiology and Food Sciences

Marek Kieliszek [1],* and Przemysław Łukasz Kowalczewski [2],*

[1] Department of Food Biotechnology and Microbiology, Institute of Food Sciences,
Warsaw University of Life Sciences—SGGW, 02-776 Warsaw, Poland
[2] Department of Food Technology of Plant Origin, Poznań University of Life Sciences, 60-624 Poznań, Poland
* Correspondence: marek_kieliszek@sggw.edu.pl (M.K.); przemyslaw.kowalczewski@up.poznan.pl (P.Ł.K.)

Citation: Kieliszek, M.;
Kowalczewski, P.Ł. Recent Advances
in Applied Microbiology and Food
Sciences. *Appl. Sci.* **2022**, *12*, 10786.
https://doi.org/10.3390/
app122110786

Received: 14 October 2022
Accepted: 21 October 2022
Published: 25 October 2022

Publisher's Note: MDPI stays neutral
with regard to jurisdictional claims in
published maps and institutional affiliations.

1. Introduction

Changes are taking place on many fronts, including socio-demographic changes, economic changes, and technological progress. These changes lead to both contributions and inspirations for the work of many scientists around the world. Food science deserves special attention. Newer and newer food ingredients, production methods, and technologies are being sought. Microbiological safety is an inevitable element that affects the quality of the obtained products.

With this mission in mind, the scope of this Special Issue of *Applied Sciences*, entitled "Recent Advances in Applied Microbiology and Food Sciences", is devoted to the latest achievements in analytics and to the application of new methods and recent advanced technologies in food science and microbiology, especially modern methods of the production and testing of food as well as the sustainable development of agriculture and the food industry. In this SI, nine original research manuscripts and two review articles were collected and published. The articles included in this collection are briefly described below.

2. Microbiology and Food Biotechnology

Kot et al. [1] conducted studies with the use of selected cations and B vitamins for the biosynthesis of carotenoids by the yeast *Rhodotorula mucilaginosa* MK1. At the same time, yeast was grown on waste substrates from the agri-food industry (glycerol and potato wastewater). As a result of the conducted analyses, it was found that the presence of appropriate microelements (including barium and aluminum) in the culture medium is necessary to obtain high yields of carotenoids. Moreover, the B vitamin (niacin) stimulated the biosynthesis of torularhodin by yeast. Its content increased from 22.79% (control culture) to 33.79% (vitamin medium). It is worth noting that this is the first study to describe the effect of niacin on the biosynthesis of carotenoids by *Rhodotorula* red yeast cells.

Olea-Rodríguez et al. [2] investigated the properties and behavior of *Staphylococcus aureus*, *Salmonella* spp., *Listeria monocytogenes*, and staphylococcal toxin during the development and maturation of Cotija cheese. Cotija cheese is a semi-hard Mexican cheese that comes from a town called Cotija de la Paz in the state of Michoacan. The authors showed that during the production of this cheese there are interactions between the microorganisms that make up the complex ecosystem of the microbiota of the end product. *Staphylococcus* and *Salmonella* showed a gradual reduction in the number of bacteria due to an increase in acidity and a decrease in pH. To Summarize, it should be noted that the cheese ripening process influences if the obtained final product is safe from a microbiological point of view for the end consumer.

Isakova et al. [3] conducted research aimed at understanding the key role of mitochondria in the regulation of aging in the yeast *Endomyces magnusii*. At the same time, the authors showed that the survival of yeasts during long-term cultivation and their high metabolic activity may be related to the active function of mitochondria. Moreover, at the stage of logarithmic growth of yeast cells, the triacylglycerol (TAG) content was a minority in the mitochondria. In the case of diacylglycerols (DAG), their content was highest in the logarithmic phase. It is worth noting that the lipid content and composition of the mitochondrial membrane also changed significantly during the growth and aging of yeast cultures.

Ranjha et al. [4] in their review, entitled "Nutritional and Health Potential of Probiotics: A Review", discussed selected probiotics and considered their potential health benefits. In addition, they reviewed the health-promoting role of probiotics and their mechanisms of action in the prevention of several common diseases, including reduction in obesity, minimizing diabetes (type 1 and 2), and mechanisms against gestational diabetes mellitus and chronic kidney disease.

A second review article, entitled "Antimicrobials from Medicinal Plants: An Emergent Strategy to Control Oral Biofilms", was written by Milho et al. [5]. The authors presented the characteristics of medicinal plants and their importance in the control of oral biofilms. They also presented evidence of the effectiveness of the use of essential oils from *Cymbopogon citratus* and *Lippia alba*, which can be a great alternative to antibiotics in the treatment of oral diseases.

3. Food Sciences and Technology

Marciniak-Lukasiak et al. [6] conducted research on the use of chestnut flour and the method of packaging for the quality of gluten-free bread. The obtained breads were packed in a PA/PE barrier foil with air and under vacuum before being stored. Then, the water content, texture, and color parameters were determined, and sensory evaluation and microbiological analyses were performed. Results show that the use of chestnut flour in an amount below 10% and vacuum packaging allows one to obtain sensory-attractive bread with appropriate properties.

The study of Miedzianka et al. [7] focused on the hidden potential of plant seeds and microalgae rich in protein that can be used in food production. They investigated antioxidant and antimicrobial properties and amino acid profiles of selected spice plants and microalgae, which were also digested in vitro to determine the potential bioavailability of the polyphenols. They found that the seeds of black cumin, milk thistle, white mustard, eggfruit, and chlorella did not contain any limiting amino acids. Moreover, they suggested that fenugreek seeds appeared to be a good candidate as an attractive food ingredient due to their amino acid profile and high antioxidant activity.

A group of researchers led by Siejak [8] analyzed the impact of using various extractants (water, methanol, ethanol, and acetone) against lipid peroxidation in soybean L-α-phosphatidylcholine liposomes. Various spectroscopic and chromatographic methods were used to evaluate the effect of different extracts of *Prunus padus* on the stability of a biomembrane. Their results showed that *P. padus* is a rich source of antioxidant compounds that can effectively inhibit the degradation of the lipid membrane.

Starch is one of the most abundant biopolymers in nature. In its native form, it often does not meet the expectations of food technologists, and therefore it is subject to various modifications. One of the more interesting of these is octenyl succinic anhydride (OSA)-modified starch. Ali et al. [9] analyzed the stability of a nanoemulsion prepared using vitamin E and β-carotene co-entrapped within oil-in-water nanoemulsions of carrier oils, including tuna fish oil and medium-chain triglycerides that were stabilized by OSA-modified starch and Tween-80. The emulsion with OSA-modified starch was shown to have higher stability due to the production of denser interfacial coatings that can protect the trapped compounds from the water phase. The presented results may be useful in the design of food containing lipophilic bioactive compounds in products containing significant amounts of water.

The study of Woźniak et al. [10] focused on the extraction of galactolipids from waste by-products by supercritical fluid extraction (SFE) and ultrasound-assisted extraction (UAE). This work was focused on the application of these two green extraction technologies for the recovery of galactolipids from rosehip pomace. SFE using pure CO_2 was not an effective method for the extraction of galactolipids such as monogalactosyldiacylglycerols (MGDGs) and digalactosyldiacylglycerols (DGDGs), although the use of an additional co-solvent such as ethanol significantly improved the separation. The use of such a low solid–liquid ratio improved the extraction efficiency. Nevertheless, about a 36% improvement in efficiency was observed after the use of sonication. SFE yields were approximately 5-fold higher than the control sample, thus proving the superiority of this technique. The results of UAE were also very promising and allow for the improvement of the extraction yield by up to 74%. The green chemistry approaches used for galactolipid isolation were compared with a conventional processing method and proved to be an interesting alternative for traditional techniques.

In turn, Hou et al. [11] presented the results of studies assessing the possibility of increasing the growth rate of plants using non-thermal plasma and reducing the accumulation of heavy metals in water spinach. The accumulation of heavy metals in water spinach depended on the type and concentration of heavy metals in the soil. It was shown that the accumulation of cadmium in spinach leaves was significantly reduced after plasma treatment of the seeds or water used for irrigation. However, no similar effect was found in the case of lead accumulation, the level of which is also important for the safety of consumed food.

Author Contributions: All authors have contributed to the conceptualization, writing, review, and editing of this manuscript. All authors have read and agreed to the published version of the manuscript.

Funding: This research received no external funding.

Acknowledgments: This Special Issue has been made possible by the contributions of several authors, reviewers, and editorial team members. Their efforts are acknowledged, and thanks are due to all of them.

Conflicts of Interest: The authors declare no conflict of interest.

References

1. Kot, A.M.; Błażejak, S.; Brzezińska, R.; Sęk, W.; Kieliszek, M. Effect of Selected Cations and B Vitamins on the Biosynthesis of Carotenoids by *Rhodotorula mucilaginosa* Yeast in the Media with Agro-Industrial Wastes. *Appl. Sci.* **2021**, *11*, 11886. [CrossRef]
2. Olea-Rodríguez, M.d.l.Á.; Chombo-Morales, P.; Nuño, K.; Vázquez-Paulino, O.; Villagrán-de la Mora, Z.; Garay-Martínez, L.E.; Castro-Rosas, J.; Villarruel-López, A.; Torres-Vitela, M.R. Microbiological Characteristics and Behavior of *Staphylococcus aureus*, *Salmonella* spp., *Listeria monocytogenes* and Staphylococcal Toxin during Making and Maturing Cotija Cheese. *Appl. Sci.* **2021**, *11*, 8154. [CrossRef]
3. Isakova, E.P.; Gessler, N.N.; Dergacheva, D.I.; Tereshina, V.M.; Deryabina, Y.I.; Kieliszek, M. Lipid Remodeling in the Mitochondria upon Ageing during the Long-Lasting Cultivation of *Endomyces magnusii*. *Appl. Sci.* **2021**, *11*, 4069. [CrossRef]
4. Ranjha, M.M.A.N.; Shafique, B.; Batool, M.; Kowalczewski, P.Ł.; Shehzad, Q.; Usman, M.; Manzoor, M.F.; Zahra, S.M.; Yaqub, S.; Aadil, R.M. Nutritional and Health Potential of Probiotics: A Review. *Appl. Sci.* **2021**, *11*, 11204. [CrossRef]
5. Milho, C.; Silva, J.; Guimarães, R.; Ferreira, I.C.F.R.; Barros, L.; Alves, M.J. Antimicrobials from Medicinal Plants: An Emergent Strategy to Control Oral Biofilms. *Appl. Sci.* **2021**, *11*, 4020. [CrossRef]
6. Marciniak-Lukasiak, K.; Lesniewska, P.; Zielińska, D.; Sowinski, M.; Zbikowska, K.; Lukasiak, P.; Zbikowska, A. The Influence of Chestnut Flour on the Quality of Gluten-Free Bread. *Appl. Sci.* **2022**, *12*, 8340. [CrossRef]
7. Miedzianka, J.; Lachowicz-Wiśniewska, S.; Nemś, A.; Kowalczewski, P.Ł.; Kita, A. Comparative Evaluation of the Antioxidative and Antimicrobial Nutritive Properties and Potential Bioaccessibility of Plant Seeds and Algae Rich in Protein and Polyphenolic Compounds. *Appl. Sci.* **2022**, *12*, 8136. [CrossRef]
8. Siejak, P.; Smułek, W.; Nowak-Karnowska, J.; Dembska, A.; Neunert, G.; Polewski, K. Bird Cherry (*Prunus padus*) Fruit Extracts Inhibit Lipid Peroxidation in PC Liposomes: Spectroscopic, HPLC, and GC–MS Studies. *Appl. Sci.* **2022**, *12*, 7820. [CrossRef]
9. Ali, A.; Rehman, A.; Jafari, S.M.; Ranjha, M.M.A.N.; Shehzad, Q.; Shahbaz, H.M.; Khan, S.; Usman, M.; Kowalczewski, P.Ł.; Jarzębski, M.; et al. Effect of Co-Encapsulated Natural Antioxidants with Modified Starch on the Oxidative Stability of β-Carotene Loaded within Nanoemulsions. *Appl. Sci.* **2022**, *12*, 1070. [CrossRef]
10. Woźniak, Ł.; Wojciechowska, M.; Marszałek, K.; Skąpska, S. Extraction of Galactolipids from Waste By-Products: The Feasibility of Green Chemistry Methods. *Appl. Sci.* **2021**, *11*, 12088. [CrossRef]
11. Hou, C.-Y.; Kong, T.-K.; Lin, C.-M.; Chen, H.-L. The Effects of Plasma-Activated Water on Heavy Metals Accumulation in Water Spinach. *Appl. Sci.* **2021**, *11*, 5304. [CrossRef]

applied sciences

Article

Effect of Selected Cations and B Vitamins on the Biosynthesis of Carotenoids by *Rhodotorula mucilaginosa* Yeast in the Media with Agro-Industrial Wastes

Anna Maria Kot [1,*], Stanisław Błażejak [1], Rita Brzezińska [2], Wioletta Sęk [1] and Marek Kieliszek [1]

1   Department of Food Biotechnology and Microbiology, Institute of Food Sciences, Warsaw University of Life Sciences, Nowoursynowska 159 C, 02-776 Warsaw, Poland; stanislaw_blazejak@sggw.edu.pl (S.B.); wioletta_sek@sggw.edu.pl (W.S.); marek_kieliszek@sggw.edu.pl (M.K.)
2   Department of Chemistry, Institute of Food Sciences, Warsaw University of Life Sciences, Nowoursynowska 159C, 02-776 Warsaw, Poland; rita_brzezinska@sggw.edu.pl
*   Correspondence: anna_kot@sggw.edu.pl

Citation: Kot, A.M.; Błażejak, S.; Brzezińska, R.; Sęk, W.; Kieliszek, M. Effect of Selected Cations and B Vitamins on the Biosynthesis of Carotenoids by *Rhodotorula mucilaginosa* Yeast in the Media with Agro-Industrial Wastes. *Appl. Sci.* **2021**, *11*, 11886. https://doi.org/10.3390/app112411886

Academic Editor: Francisco Artés-Hernández

Received: 15 November 2021
Accepted: 12 December 2021
Published: 14 December 2021

Publisher's Note: MDPI stays neutral with regard to jurisdictional claims in published maps and institutional affiliations.

Abstract: In recent years, there has been an increase in the search for novel raw materials for the production of natural carotenoids. Among yeasts, *Rhodotorula* species have the ability to synthesize carotenoids, mainly β-carotene, torulene, and torularhodin, depending on the culture conditions. This study aimed to determine the effect of selected cations (barium, zinc, aluminum, manganese) and B vitamins (biotin, riboflavin, niacin, pantothenic acid) on the biosynthesis of carotenoids by *Rhodotorula mucilaginosa* MK1 and estimate the percentages of β-carotene, torulene, and torularhodin synthesized by the yeast. The cultivation was carried out in a medium containing glycerol (waste resulting from biodiesel production) as a carbon source and potato wastewater (waste resulting from potato starch production) as a nitrogen source. Carotenoid biosynthesis was stimulated by the addition of aluminum (300 mg/L) or aluminum (300 mg/L) and niacin (100 μg/L) to the medium. The number of carotenoids produced by *R. mucilaginosa* MK1 in the medium containing only aluminum and in the medium with aluminum and niacin was 146.7 and 180.5 $\mu g/g_{d.m.}$, respectively. This content was 101% and 147% higher compared to the content of carotenoids produced by yeast grown in the control medium (73.0 $\mu g/g_{d.m.}$). The addition of aluminum and barium seemed to have a positive effect on the biosynthesis of torulene, and the percentage of this compound increased from 31.86% to 75.20% and 68.24%, respectively. Niacin supplementation to the medium increased the percentage of torularhodin produced by the yeast from 23.31% to 31.59–33.79%. The conducted study showed that there is a possibility of intensifying carotenoid biosynthesis by red yeast and changing the percentages of individual carotenoids fractions by adding cations or B vitamins to the medium. Further research is needed to explain the mechanism of action of niacin on the stimulation of torularhodin biosynthesis.

Keywords: red yeast; carotenoids; metals; vitamins; glycerol; potato wastewater

1. Introduction

Carotenoids are a group of natural pigments that are commonly found in nature. Because of their well-documented health effects, researchers are continually looking for new methods to obtain these compounds. Chemical synthesis processes provide a substantial proportion of carotenoids for industrial applications. However, as the need for natural dyes continues to rise, the production of these compounds with the use of microorganisms is becoming increasingly important [1,2]. Among the microbes that can produce carotenoids, yeasts of the *Rhodotorula* genus are particularly interesting. The major carotenoids synthesized by these organisms are β-carotene, torulene, and torularhodin [3]. Beta-carotene is the most potent precursor of vitamin A and possesses strong antioxidant properties, and hence has a number of health benefits, such as reducing the risk of diseases and certain types of cancer, protecting against macular degeneration, and strengthening the immune system. It is used as a dye and an ingredient of dietary supplements in the food industry, while in the cosmetic industry, it is applied as an ingredient in sun creams [4]. Torulene, a carotene containing 13 double bonds, exhibits pro-vitamin A and anticancer properties. It has an intense orange color and is indicated as a potential food additive [5]. However, torulene production is not carried out on a large scale because the exact pathway of its biosynthesis and the optimal cultivation conditions are unknown [5,6]. Torularhodin belongs to the group of xanthophylls, and, depending on the concentration, its color ranges from pink to orange [5]. It can be used in medicine, for example, for the prevention of prostate cancer or in pharmaceuticals, which is associated with the apoptosis of neoplastic cells (reduction of Bcl-2 proteins and increased expression of Bax protein and specific caspases [7]. Furthermore, it possesses strong antimicrobial properties and, therefore, can be used as a natural antibiotic [8].

The selection of appropriate strains that can synthesize higher amounts of these compounds, as well as optimization of culture conditions, is essential for the profitable production of carotenoids using microorganisms. Several factors influence the process of carotenoid biosynthesis, of which the most important are carbon and nitrogen source, pH, oxygenation level, and availability of light [2]. Various agri-food wastes have been used for production to reduce the costs of microbiological media. It was shown that glycerol, a by-product of the biodiesel production process, can be successfully used as a carbon source, and potato wastewater, obtained in large amounts from potato starch production, as a nitrogen source [9]. The presence of appropriate micronutrients in the culture medium is also necessary to achieve high yields of carotenoids [10], as these compounds act as cofactors of cellular enzymes involved in their biosynthesis. For example, Mg^{2+} or Mn^{2+} ions act as cofactors of phytoene synthase, and Fe^{2+} ions ensure the proper activity of β-lycopene cyclase [3]. The presence of some metals may also be an environmental stress factor for yeast cells, inducing increased carotenogenesis [11]. Another factor that influences carotenoid biosynthesis in red yeast cells is the presence of vitamins. Among vitamins, B group plays an important role in yeast metabolism [12–15]. However, no studies have analyzed the effect of these vitamins on the biosynthesis of carotenoids by *Rhodotorula* yeasts so far. Therefore, this study aimed to investigate the effect of supplementation of the culture medium prepared from two types of industrial wastes (glycerol and potato wastewater) with selected cations and B vitamins on the biosynthetic efficiency of *Rhodotorula mucilaginosa* as well as the profile of carotenoids produced by this yeast species.

2. Materials and Methods

2.1. Biological Material

The study used the *R. mucilaginosa* MK1 strain, which was isolated from bee bread and identified by Kieliszek et al. [16]. It is deposited under GenBank Accession Number LC527461.1 (National Center of Biotechnology Information).

2.2. Wastes Used as Components of Media

The culture media used for yeast cultivation were prepared from two industrial wastes, glycerol and potato wastewater. Technical glycerol (ORLEN Południe, Trzebinia, Poland) was used as a carbon source, while potato wastewater, which was prepared in laboratory conditions as described previously [17], was used as a nitrogen source.

2.3. Media and Culture Conditions

Two types of culture media were used in the study. The control medium was prepared from potato wastewater and technical glycerol, in which the final concentration of glycerol was 3% (*w/v*). One type of experimental media was prepared with zinc ($ZnSO_4 \cdot 7H_2O$), aluminum ($Al(NO_3)_3 \cdot 9H_2O$), barium ($BaCl_2$), and manganese ($MnSO_4 \cdot H_2O$) cations, which were added at the concentration range of 50–300 mg/L. Another type of experimental media was prepared with selected B vitamins (niacin, biotin, riboflavin, pantothenic acid), the concentrations of which ranged from 100 to 1000 µg/L. The vitamin solutions prepared for supplementation were sterilized through a sterile membrane filter (0.22 µm) and added at an appropriate amount to the cooled, sterile medium containing potato wastewater and 3% glycerol.

2.4. Bioscreen C Cultures

In the first stage, yeast cultivation was carried out in a Bioscreen C apparatus (Oy Growth Curves Ab Ltd., Finland). For this purpose, a 24-h yeast inoculum was prepared in yeast extract–peptone–dextrose (YPD) medium (200 rpm, 25 °C) and diluted in sterile physiological saline to achieve a cell density of 1×10^6 CFU/mL. Then, 270 µL of the control medium and experimental media were transferred to the microwells on a plate, and 30 µL of the inoculum was added. Cultivation was carried out under intensive shaking at 25 °C. The apparatus monitored changes in the optical density (OD) of the cultures every hour (broadband filter λ = 420–580 nm). Based on the OD values observed after 120 h of cultivation, the duration of the adaptation phase (Δt_{lag}) and logarithmic phase (Δt_{log}), the minimum and maximum OD values in the logarithmic growth phase, the specific growth rate coefficient ($\mu_{max} = \ln OD_{max} - \ln OD_{min})/\Delta t_{log}$), and the total increase in OD were determined.

2.5. Cultures on the Shaker

The control and experimental media were inoculated with a 24-h yeast culture such that the initial cell density was 1×10^5 CFU/mL. Cultivation was performed for 120 h at 25 °C on a rotary shaker (200 rpm; SM-30 Control, Edmund Bühler, Germany).

2.6. Biomass Yield

The cellular biomass yield was determined by weighing. After 120 h, 10 mL of culture was removed and centrifuged at $5000\times g$ for 10 min (Centrifuge 5804R, Eppendorf, Germany). The wet biomass was washed twice with sterile deionized water and dried at 105 °C until a constant weight was reached. The results are expressed in grams of dry matter ($g_{d.m.}$) per liter of the medium.

2.7. Content and Profile of Carotenoids in Yeast Biomass

To disintegrate the cellular biomass and extract carotenoids from the yeast cells, 2 mL of dimethyl sulfoxide and 0.5 g of glass beads (diameter: 500 µm) were added to approximately 100 mg of wet biomass. The samples were shaken in a rotator (Multi Bio RS-24, Biosan) at 70 rpm for 1 h. Then, 2 mL of petroleum ether, 2 mL of acetone, and 2 mL of 20% sodium chloride were added, and the samples were shaken again for 1 h. Next, the samples were centrifuged at $3500\times g$ for 5 min (Centrifuge 5804R, Eppendorf, Germany) to separate the phases. The absorbance of the ether fraction was measured at 490 nm (UV1800, Rayleigh). The total carotenoid content ($\mu g/g_{d.m.}$) was calculated from the formula $= A_{max} \times D \times V/(E \times W)$, where A_{max} is the absorbance at 490 nm, D is the sample dilution factor, V is the volume of the ether fraction, E is the extinction coefficient of carotenes (0.16), and W is the yeast dry matter ($g_{d.m.}$) [6].

2.8. Carotenoids Profile

The individual carotenoids were identified by high-performance liquid chromatography (HPLC). Petroleum ether was evaporated from the ethereal fractions under a nitrogen atmosphere. One milliliter of the HPLC phase consisting of a mixture of acetonitrile, isopropanol, and ethyl acetate (4:4:2, *v/v/v*) was added to the tube and filtered through a 0.45-µm nylon filter. The samples were analyzed on a C18 analytical column (Bionacom; 250 mm $\times$ 4.6 mm, 5 µm). The flow rate of the mobile phase was set to 0.7 mL/min (isocratic). A UV–VIS detector operating at 490 nm was used for detection. β- and γ-carotene were identified based on the retention time of the standards (Sigma-Aldrich), while torulene and torularhodin were based on the retention time of the products separated using thin-layer chromatography [17]. The percentages of these four compounds in the carotenoid mixture were determined based on the area of their peaks.

2.9. Statistical Analysis of the Results

All cultures and analyses were performed in triplicate. Statistical analysis of the results was performed in R program (version i386 2.15.3). The normal distribution of data was verified by the Shapiro–Wilk test and the homogeneity of variance by the Levene test. The significance of differences was determined by one-way analysis of variance and Tukey's test at a level of $\alpha = 0.05$.

3. Results

3.1. First Stage—Growth in Bioscreen C

The changes observed in the OD values during cultivation in the Bioscreen C apparatus indicated that *R. mucilaginosa* MK1 yeast was able to grow in the control medium and the experimental medium enriched with selected cations (Table 1).

Table 1. Growth characteristics of *Rhodotorula mucilaginosa* MK1 yeast observed in Bioscreen C during cultivation in media containing different cations.

Type of Medium	Concentration [mg/L]	t_{lag} [h]	t_{log} [h]	μ_{max} [h^{-1}]	ΔOD
Barium	0	6.0	18.0	0.0341 ± 0.0033 *	1.542 ± 0.204 *
	50	4.0	22.0	0.0334 ± 0.0020 *	1.442 ± 0.060 *
	100	4.0	22.0	0.0313 ± 0.0011 *	1.447 ± 0.210 *
	150	6.0	20.0	0.0342 ± 0.0017 *	1.319 ± 0.098 *
	200	6.0	22.0	0.0314 ± 0.0005 *	1.442 ± 0.082 *
	250	6.0	20.0	0.0338 ± 0.0006 *	1.381 ± 0.044 *
	300	6.0	22.0	0.0298 ± 0.0020 *	1.339 ± 0.147 *
Zinc	0	6.0	18.0	0.0341 ± 0.0033 [ab]	1.542 ± 0.204 *
	50	6.0	18.0	0.0366 ± 0.0010 [a]	1.605 ± 0.155 *
	100	6.0	20.0	0.0325 ± 0.0018 [ab]	1.645 ± 0.096 *
	150	6.0	20.0.	0.0358 ± 0.0019 [ab]	1.554 ± 0.170 *
	200	6.0	20.0	0.0318 ± 0.0012 [ab]	1.653 ± 0.320 *
	250	6.0	20.0	0.0324 ± 0.0028 [ab]	1.600 ± 0.057 *
	300	6.0	20.0.	0.0306 ± 0.0022 [b]	1.535 ± 0.295 *
Aluminum	0	6.0	18.0	0.0341 ± 0.0033 [a]	1.542 ± 0.204 *
	50	6.0	20.0	0.0321 ± 0.041 [abc]	1.344 ± 0.077 *
	100	6.0	20.0	0.0322 ± 0.0030 [ab]	1.337 ± 0.169 *
	150	6.0	24.0	0.0286 ± 0.0032 [abc]	1.370 ± 0.053 *
	200	6.0	24.0	0.0265 ± 0.0024 [bc]	1.306 ± 0.058 *
	250	8.0	26.0	0.0251 ± 0.0021 [c]	1.339 ± 0.096 *
	300	8.0	28.0	0.0234 ± 0.0022 [c]	1.305 ± 0.052 *
Manganese	0	6.0	18.0	0.0341 ± 0.0033 *	1.542 ± 0.204 *
	50	6.0	18.0	0.0367 ± 0.0037 *	1.412 ± 0.047 *
	100	6.0	18.0	0.0385 ± 0.0024 *	1.411 ± 0.018 *
	150	6.0	18.0	0.0371 ± 0.0058 *	1.383 ± 0.212 *
	200	6.0	18.0	0.0378 ± 0.0007 *	1.440 ± 0.113 *
	250	6.0	18.0	0.0397 ± 0.0024 *	1.398 ± 0.117 *
	300	6.0	18.0	0.0366 ± 0.0043 *	1.345 ± 0.091 *

Indexes [a,b] … denote homogeneous groups determined by Tukey's test. * means no significant difference. A one-way analysis of variance was performed for the results obtained for each cation separately and compared with the control sample.

Statistical analysis showed that the specific growth rate (μ_{max}) in the media containing barium (50–300 mg/L), zinc (50–300 mg/L), aluminum (50–150 mg/L), and manganese (50–300 mg/L) was comparable to that obtained in the control medium (0.0341 h^{-1}) (Table 1). A decreased growth rate was noted in the remaining media variants. In particular, a significant reduction in μ_{max}, as well as an extended logarithmic phase, was observed in cultures supplemented with aluminum at the concentrations of 200–300 mg/L, which proves that this element has a toxic effect on cells and slows down the reproductive process.

The growth characteristics of the studied yeast strain differed in the experimental media supplemented with B vitamins (Table 2). Specific growth rate values above 0.03 h^{-1} were estimated in cultures containing biotin at the concentrations of 100–700 µg/L, riboflavin at the lowest concentration of 100 µg/L, and pantothenic acid at the concentrations of 100–500 µg/L. Importantly, growth was found to be delayed in the media supplemented with 100 µg/L niacin. In all other media supplemented with vitamins, the total increase in OD was significantly lower (0.694–1.202) compared to the control media (1.542).

Table 2. Characteristics of *Rhodotorula mucilaginosa* MK1 yeast observed in Bioscreen C during cultivation in media supplemented with B vitamins.

Type of Medium	Concentration [µg/L]	t_{lag} [h]	t_{log} [h]	μ_{max} [h^{-1}]	ΔOD
Control	0	6.0	18.0	0.0341 ± 0.0033 *	1.542 ± 0.204 [a]
Biotin	100	8.0	14.0	0.0323 ± 0.0056 *	1.116 ± 0.143 [b]
	300	8.0	14.0	0.0339 ± 0.0064 *	1.113 ± 0.071 [b]
	500	8.0	16.0	0.0335 ± 0.0022 *	1.202 ± 0.038 [b]
	700	8.0	16.0	0.0301 ± 0.0014 *	1.179 ± 0.072 [b]
	850	8.0	16.0	0.0279 ± 0.0014 *	1.098 ± 0.097 [b]
	1000	10.0	16.0	0.0293 ± 0.0020 *	1.210 ± 0.193 [b]
Riboflavin	0	6.0	18.0	0.0341 ± 0.0033 [a]	1.542 ± 0.204 [a]
	100	6.0	18.0	0.0306 ± 0.0021 [ab]	1.173 ± 0.178 [b]
	300	6.0	18.0	0.0289 ± 0.0017 [ab]	1.109 ± 0.097 [b]
	500	6.0	20.0	0.0291 ± 0.0020 [ab]	1.129 ± 0.137 [b]
	700	6.0	20.0	0.0283 ± 0.0011 [b]	1.108 ± 0.079 [b]
	850	6.0	20.0	0.0268 ± 0.0010 [b]	1.140 ± 0.040 [b]
	1000	8.0	20.0	0.0269 ± 0.0022 [b]	1.153 ± 0.053 [b]
Niacin	0	6.0	18.0	0.0341 ± 0.0033 [a]	1.542 ± 0.204 [a]
	100	8.0	20.0	0.0210 ± 0.0075 [b]	1.030 ± 0.152 [bc]
	300	8.0	20.0	0.0221 ± 0.0027 [b]	1.083 ± 0.138 [bc]
	500	8.0	20.0	0.0226 ± 0.0022 [b]	1.163 ± 0.126 [b]
	700	10.0	20.0	0.0214 ± 0.0007 [b]	0.868 ± 0.110 [bc]
	850	10.0	24.0	0.0169 ± 0.0017 [b]	0.762 ± 0.123 [bc]
	1000	12.0	26.0	0.0150 ± 0.0020 [b]	0.694 ± 0.069 [c]
Pantothenic acid	0	6.0	18.0	0.0341 ± 0.0033 [a]	1.542 ± 0.204 [a]
	100	6.0	16.0	0.0324 ± 0.0039 [a]	1.041 ± 0.154 [b]
	300	6.0	16.0	0.0318 ± 0.0044 [ab]	1.064 ± 0.180 [b]
	500	6.0	16.0	0.0334 ± 0.0031 [a]	1.142 ± 0.191 [ab]
	700	8.0	20.0	0.0270 ± 0.0033 [bc]	0.986 ± 0.037 [b]
	850	10.0	20.0	0.0224 ± 0.0043 [bc]	0.941 ± 0.204 [b]
	1000	10.0	20.0	0.0199 ± 0.0023 [c]	0.850 ± 0.097 [b]

Indexes [a,b] ... denote homogeneous groups determined by Tukey's test. * means no significant difference. A one-way analysis of variance was performed for the results obtained for each vitamin separately and compared with the control sample.

3.2. Second Stage—Growth and Biosynthesis of Carotenoids in Shaker Culture in Media Supplemented with Cations or B Vitamins

After cultivation in the Bioscreen C apparatus, the yeast cultures were transferred to shaker flasks. At the beginning of cultivation, the biomass yield, which is defined as the amount of dry yeast cell biomass in 1 L of the medium, was determined (Table 3). In the control medium, the biomass yield after cultivation was estimated at 21.48 $g_{d.m.}$/L. Statistical analysis indicated that the addition of zinc at a concentration of 50–300 mg/L did not influence the biomass yield. On the other hand, the addition of other cations caused a difference in the values of this index. The biomass yield determined after cultivation in the media containing manganese at doses from 50 to 300 mg/L was found to be significantly higher than that in the control medium and ranged from 25.15 to 26.57 $g_{d.m.}$/L. In the media variants containing barium, a decrease in the growth of cellular biomass was found with an increase in the dose of the cation. The determined cellular biomass yields ranged from 15.65 to 9.10 $g_{d.m.}$/L. The addition of aluminum at a concentration of 300 mg Al^{3+}/L caused a significant reduction in the biomass yield (15.71 $g_{d.m.}$/L). Among the tested vitamins, supplementation with biotin and riboflavin in the entire concentration range did not cause any significant difference in the determined values of the biomass yield, which ranged from 19.52 to 23.23 $g_{d.m.}$/L. The addition of niacin at the doses of 700, 850, and 1000 µg/L caused a significant reduction in biomass yield, which ranged from 15.35 to 12.12 $g_{d.m.}$/L. The same phenomenon was observed with pantothenic acid at concentrations of 850 and 1000 µg/L, and the values of biomass yield were 16.55 and 14.68 $g_{d.m.}$/L, respectively.

Table 3. Biomass yield and content and profile of carotenoids after 120 h of cultivation in media enriched with selected cations.

Type of Medium	Concentration [mg/L]	Biomass Yield [$g_{d.m.}$/L]	Total Carotenoid Content in Biomass [$\mu g/g_{d.m.}$]	Volumetric Yield of Carotenoids [mg/L]	Percentages of Carotenoids			
					Torularhodin	Torulene	β-Carotene	Others
Barium	0	21.48 ± 1.24 [a]	73.06 ± 10.08 *	1.56 ± 0.13 [a]	22.79 ± 2.03 [a]	31.85 ± 0.71 [b]	43.08 ± 1.52 [a]	2.29 ± 0.20 *
	50	15.65 ± 0.07 [b]	71.96 ± 8.85 *	1.13 ± 0.14 [ab]	21.29 ± 2.60 [a]	34.34 ± 2.81 [b]	41.38 ± 0.43 [a]	3.00 ± 0.23 *
	100	15.65 ± 1.56 [b]	83.71 ± 1.14 *	1.31 ± 0.15 [ab]	19.95 ± 2.42 [a]	36.39 ± 1.51 [b]	41.30 ± 3.64 [a]	2.37 ± 0.28 *
	150	14.80 ± 0.21 [b]	73.70 ± 12.83 *	1.09 ± 0.21 [ab]	19.37 ± 1.55 [a]	35.60 ± 1.92 [b]	43.28 ± 3.58 [a]	1.76 ± 0.11 *
	200	15.10 ± 0.85 [b]	77.28 ± 14.15 *	1.16 ± 0.15 [ab]	16.79 ± 0.60 [a]	37.56 ± 4.51 [b]	43.03 ± 4.56 [a]	2.63 ± 0.65 *
	250	11.70 ± 1.48 [bc]	83.07 ± 16.96 *	0.96 ± 0.08 [b]	8.89 ± 0.35 [b]	63.77 ± 4.04 [a]	24.94 ± 3.37 [b]	2.41 ± 0.33 *
	300	9.10 ± 1.27 [c]	81.98 ± 3.83 *	0.75 ± 0.14 [b]	7.32 ±1.32 [b]	68.24 ± 6.00 [a]	21.74 ± 3.39 [b]	2.71 ± 1.29 *
Zinc	0	21.48 ± 1.24 *	73.06 ± 10.08 *	1.56 ± 0.13 *	22.79 ± 2.03 *	31.85 ± 0.71 *	43.08 ± 1.52 *	2.29 ± 0.20 *
	50	22.60 ± 2.62 *	64.65 ± 6.12 *	1.45 ± 0.03 *	22.07 ± 2.03 *	36.77 ± 2.16 *	38.92 ± 3.63 *	2.25 ± 0.56 *
	100	24.80 ± 2.90 *	61.46 ± 10.21 *	1.51 ± 0.08 *	20.02 ± 2.04 *	35.57 ± 5.57 *	42.01 ± 2.50 *	2.41 ± 1.03 *
	150	21.18 ± 1.59 *	75.90 ± 13.01 *	1.60 ± 0.15 *	18.93 ± 0.80 *	34.75 ± 2.06 *	44.17 ± 2.43 *	2.16 ± 0.42 *
	200	24.07 ± 2.09 *	68.29 ± 15.28 *	1.63 ± 0.23 *	22.32 ± 2.04 *	32.99 ± 3.15 *	40.74 ± 0.87 *	3.96 ± 0.24 *
	250	24.53 ± 1.73 *	64.82 ± 13.41 *	1.58 ± 0.22 *	20.62 ± 2.45 *	32.80 ± 3.08 *	43.63 ± 0.93 *	2.96 ± 0.30 *
	300	20.17 ± 0.53 *	68.80 ± 9.80 *	1.39 ± 0.23 *	20.78 ± 0.83 *	32.98 ± 3.65 *	44.10 ± 3.25 *	2.15 ± 0.42 *
Aluminum	0	21.48 ± 1.24 [a]	73.06 ± 10.08 [b]	1.56 ± 0.13 *	22.79 ± 2.03 [a]	31.85 ± 0.71 [d]	43.08 ± 1.52 [a]	2.29 ± 0.20 *
	50	22.42 ± 1.59 [ab]	72.58 ± 23.83 [b]	1.61 ± 0.42 *	10.74 ± 0.71 [bc]	58.57 ± 2.40 [b]	28.36 ± 1.41 [bc]	2.34 ± 0.28 *
	100	21.83 ± 2.58 [ab]	66.73 ± 10.28 [b]	1.47 ± 0.40 *	13.54 ± 1.06 [b]	50.78 ± 0.41 [c]	33.58 ± 1.56 [b]	2.10 ± 0.91 *
	150	20.48 ± 1.03 [ab]	61.89 ± 6.08 [b]	1.27 ± 0.19 *	9.26 ± 1.37 [bcd]	62.75 ± 3.17 [b]	26.24 ± 1.65 [c]	1.76 ± 0.15 *
	200	17.83 ± 1.10 [ab]	64.19 ± 12.47 [b]	1.14 ± 0.15 *	8.27 ± 1.24 [cd]	64.15 ± 1.55 [b]	25.08 ± 1.03 [cd]	2.51 ± 1.33 *
	250	16.85 ± 0.99 [ab]	92.74 ± 23.34 [ab]	1.57 ± 0.49 *	5.94 ± 0.45 [d]	71.36 ± 1.58 [a]	20.27 ± 1.62 [de]	2.45 ± 0.40 *
	300	15.71 ± 2.46 [b]	146.73 ± 14.12 [a]	2.32 ± 0.58 *	4.50 ± 0.90 [d]	75.20 ± 0.64 [a]	17.80 ± 0.64 [e]	2.51 ± 0.90 *
Manganese	0	21.48 ± 1.24 [b]	73.06 ± 10.08 *	1.56 ± 0.13 *	22.79 ± 2.03 [a]	31.85 ± 0.71 *	43.08 ± 1.52 *	2.29 ± 0.20 [ab]
	50	25.90 ± 1.27 [a]	76.96 ± 13.85 *	1.98 ± 0.26 *	12.85 ± 3.39 [b]	35.09 ± 6.70 *	48.91 ± 3.74 *	3.17 ± 0.43 [a]
	100	26.57 ± 0.46 [a]	73.93 ± 18.68 *	2.02 ± 0.59 *	16.31 ± 2.06 [ab]	28.64 ± 5.31 *	53.25 ± 2.82 *	1.81 ± 0.42 [b]
	150	26.08 ± 1.10 [a]	78.63 ± 5.34 *	2.05 ± 0.23 *	13.82 ± 1.96 [b]	29.61 ± 0.72 *	53.86 ± 1.56 *	2.72 ± 0.33 [ab]
	200	25.18 ± 2.79 [a]	69.41 ± 4.06 *	1.75 ± 0.30 *	15.81 ± 2.18 [ab]	31.11 ± 5.18 *	50.74 ± 3.54 *	2.35 ± 0.54 [ab]
	250	25.15 ± 2.12 [a]	69.38 ± 9.89 *	1.73 ± 0.10 *	14.35 ± 1.26 [b]	32.35 ± 4.41 *	51.13 ± 2.67 *	2.18 ± 0.48 [ab]
	300	25.47 ± 0.67 [a]	80.43 ± 3.08 *	2.05 ± 0.13 *	14.10 ± 3.46 [b]	34.30 ± 3.46 *	49.30 ± 2.89 *	2.32 ± 0.91 [ab]

Indexes [a, b] … denote homogeneous groups determined by Tukey's test. * means no significant difference. A one-way analysis of variance was performed for the results obtained for each cation separately and compared with the control sample.

The content of total carotenoids in the yeast cell biomass obtained after cultivation in the medium with potato wastewater and 3% glycerol was, on average, 73.06 $\mu g/g_{d.m.}$ (Table 3). A significant increase in the carotenoid content (146.73 $\mu g/g_{d.m.}$) was noted only in the medium containing 300 mg Al^{3+}/L. Supplementation of medium with barium, zinc, and manganese, as well as B vitamins, did not cause an increase in the overall carotenoid content in the cellular biomass. Based on the values of cellular biomass yield and the total content of carotenoids, the volumetric efficiency of biosynthesis (defined as the number of carotenoids obtained from 1 L of the culture medium) was determined. After cultivation in the control medium, the volumetric efficiency amounted to an average of 1.56 mg/L, while after cultivation in the medium containing 300 mg/L aluminum, it increased to 2.32 mg/L.

Supplementation of the culture medium with barium, aluminum, and manganese cations or niacin caused changes in the content of synthesized carotenoids (Tables 3 and 4). The proportions of torularhodin, torulene, and β-carotene determined after cultivation in the control medium were 22.79%, 31.85%, and 43.08%, respectively. With an increase in the concentration of barium cations in the medium from 50 to 300 mg/L, a decrease was observed in the proportion of torularhodin from 21.29% to 7.32%, respectively. However, an increase in torulene was noted from 34.34% to 68.24%. The proportion of β-carotene in the biomass obtained after cultivation in the medium supplemented with 300 mg/L barium cations was 21.74%, which was twofold lower than that in the control sample. The addition of aluminum cations to the media containing potato wastewater and glycerol caused similar changes in the percentages of carotenoids, such as barium. With an increase in the concentration of Al^{3+} ions in the medium, the proportion of torulene increased from 31.85% to 75.20%. This was due to a decrease in the biosynthesis of torulene (from 22.79% to 4.50%) and β-carotene (from 43.08% to 17.80%). Similarly, the addition of manganese at doses of 250 and 300 mg/L resulted in a decrease in torularhodin to 14.20–14.35%, while the content of β-carotene increased (49.30–51.13%). Among the examined B group vitamins, only the addition of niacin to the medium caused significant changes in the composition of carotenoids. This vitamin stimulated the biosynthesis of torularhodin, the percentage of which increased from 22.79% (control culture) to 29.79–33.79% in the media containing niacin at concentrations from 100 to 1000 µg/L. At the same time, an increase in torulene content was also noted in the niacin-supplemented media (from 31.85% in the control culture to 40.63–45.89%).

Table 4. Biomass yield and content and profile of carotenoids after 120 h of cultivation in media enriched with selected B vitamins.

Type of Medium	Concentration [mg/L]	Biomass Yield [$g_{d.m.}$/L]	Total Carotenoid Content in Biomass [$\mu g/g_{d.m.}$]	Volumetric Yield of Carotenoids [mg/L]	Percentages of Carotenoids			
					Torularhodin	Torulene	β-Carotene	Others
Biotin	0	21.48 ± 1.24 *	73.06 ± 10.08 *	1.56 ± 0.13 *	22.79 ± 2.03 *	31.85 ± 0.71 *	43.08 ± 1.52 *	2.29 ± 0.20 *
	100	20.32 ± 3.29 *	75.02 ± 13.22 *	1.50 ± 0.02 *	17.91 ± 0.63 *	41.49 ± 0.11 *	38.72 ± 0.81 *	1.89 ± 0.07 *
	300	23.23 ± 3.50 *	79.48 ± 7.62 *	1.86 ± 0.46 *	20.84 ± 0.28 *	37.36 ± 4.22 *	39.32 ± 2.87 *	2.49 ± 1.63 *
	500	22.90 ± 2.12 *	77.40 ± 9.49 *	1.76 ± 0.05 *	19.02 ± 0.64 *	35.64 ± 1.49 *	41.91 ± 0.33 *	3.45 ± 1.80 *
	700	20.38 ± 2.51 *	84.18 ± 7.34 *	1.71 ± 0.06 *	21.70 ± 0.92 *	34.51 ± 0.30 *	41.75 ± 2.06 *	2.05 ± 0.83 *
	850	21.13 ± 0.39 *	82.60 ± 5.70 *	1.74 ± 0.09 *	21.80 ± 1.27 *	32.57 ± 1.27 *	43.00 ± 0.35 *	2.65 ± 0.29 *
	1000	21.80 ± 1.34 *	70.32 ± 6.61 *	1.54 ± 0.24 *	21.39 ± 2.58 *	37.20 ± 5.28 *	39.76 ± 2.45 *	1.66 ± 0.25 *
Riboflavin	0	21.48 ± 1.24 *	73.06 ± 10.08 *	1.56 ± 0.13 *	22.79 ± 2.03 *	31.85 ± 0.71 *	43.08 ± 1.52 *	2.29 ± 0.20 *
	100	22.72 ± 2.65 *	78.66 ± 14.24 *	1.77 ± 0.11 *	20.40 ± 1.34 *	36.04 ± 1.91 *	41.80 ± 0.47 *	1.77 ± 0.09 *
	300	21.67 ± 3.08 *	77.57 ± 16.31 *	1.66 ± 0.11 *	22.90 ± 1.98 *	35.23 ± 2.83 *	39.53 ± 0.01 *	2.34 ± 0.85 *
	500	21.30 ± 1.48 *	84.52 ± 4.44 *	1.80 ± 0.03 *	19.31 ± 0.78 *	36.30 ± 1.37 *	41.80 ± 0.78 *	2.60 ± 0.20 *
	700	21.00 ± 2.83 *	80.33 ± 10.61 *	1.67 ± 0.01 *	22.61 ± 0.95 *	35.49 ± 1.56 *	39.60 ± 1.35 *	2.32 ± 0.74 *
	850	20.52 ± 2.58 *	79.08 ± 8.22 *	1.61 ± 0.04 *	22.29 ± 0.49 *	32.99 ± 1.18 *	43.00 ± 1.82 *	1.73 ± 0.14 *
	1000	19.52 ± 1.45 *	85.55 ± 10.62 *	1.68 ± 0.33 *	22.08 ± 1.42 *	33.78 ± 0.66 *	41.68 ± 2.17 *	2.47 ± 0.08 *
Niacin	0	21.48 ± 1.24 ab	73.06 ± 10.08 *	1.56 ± 0.13 ab	22.79 ± 2.03 b	31.85 ± 0.71 b	43.08 ± 1.52 a	2.29 ± 0.20 *
	100	20.85 ± 0.28 ab	83.28 ± 12.64 *	1.73 ± 0.24 a	33.79 ± 2.04 a	42.51 ± 0.81 a	21.74 ± 1.97 b	1.98 ± 0.88 *
	300	24.50 ± 0.85 a	70.47 ± 3.52 *	1.73 ± 0.03 a	31.83 ± 3.25 a	43.70 ± 2.99 a	21.70 ± 0.62 b	2.78 ± 0.35 *
	500	23.25 ± 3.96 a	75.61 ± 7.18 *	1.74 ± 0.13 a	31.59 ± 0.78 a	43.90 ± 1.85 a	22.19 ± 1.34 b	2.34 ± 0.28 *
	700	15.35 ± 1.27 bc	87.99 ± 10.75 *	1.34 ± 0.05 ab	33.34 ± 2.67 a	44.03 ± 0.78 a	20.75 ± 1.97 b	1.89 ± 0.08 *
	850	13.73 ± 0.32 c	83.51 ± 15.46 *	1.14 ± 0.19 ab	29.79 ± 2.04 a	45.89 ± 1.50 a	22.52 ± 0.87 b	1.81 ± 0.33 *
	1000	12.12 ± 1.10 c	79.84 ± 7.00 *	0.97 ± 0.17 b	33.18 ± 1.48 a	40.63 ± 1.41 a	23.97 ± 2.84 b	2.23 ± 0.06 *
Pantothenic acid	0	21.48 ± 1.24 ab	73.06 ± 10.08 *	1.56 ± 0.13 *	22.08 ± 1.42 *	33.78 ± 0.66 *	41.68 ± 2.17 *	2.47 ± 0.08 *
	100	22.77 ± 3.01 a	77.96 ± 4.47 *	1.77 ± 0.13 *	25.80 ± 1.53 *	27.35 ± 1.62 *	44.35 ± 2.68 *	2.51 ± 0.47 *
	300	21.35 ± 2.40 a	75.96 ± 7.73 *	1.61 ± 0.02 *	26.92 ± 1.24 *	30.05 ± 1.39 *	41.20 ± 2.77 *	1.84 ± 0.15 *
	500	22.58 ± 1.45 a	79.02 ± 8.01 *	1.78 ± 0.07 *	25.43 ± 7.14 *	31.35 ± 4.09 *	40.72 ± 2.88 *	2.51 ± 0.17 *
	700	18.78 ± 0.60 ab	73.25 ± 16.32 *	1.37 ± 0.26 *	21.33 ± 2.43 *	32.39 ± 0.65 *	44.62 ± 0.90 *	1.67 ± 0.88 *
	850	16.55 ± 0.71 b	86.91 ± 12.43 *	1.44 ± 0.27 *	24.10 ± 0.65 *	32.44 ± 0.16 *	41.42 ± 0.78 *	2.05 ± 0.29 *
	1000	14.68 ± 1.38 b	81.50 ± 16.10 *	1.21 ± 0.35 *	25.14 ± 4.74 *	32.84 ± 3.35 *	40.02 ± 1.48 *	2.01 ± 0.08 *

Indexes [a,b] … denote homogeneous groups determined by Tukey's test. * means no significant difference. A one-way analysis of variance was performed for the results obtained for each vitamin separately and compared with the control sample.

3.3. Third Stage—Growth and Biosynthesis of Carotenoids in Shaker Culture in Media Supplemented with Both Cations and B Vitamins

In the third stage, the cultivation was carried out in the media simultaneously supplemented with selected cations and vitamins. Barium cations (300 mg/L) and niacin (100 µg/L) were added to the first experimental medium, and aluminum cations (300 mg/L) and niacin (100 µg/L) to the second. After cultivation, the cellular biomass yield obtained in the medium supplemented with barium and niacin was 9.59 $g_{d.m.}$/L, while in the medium containing aluminum and niacin, the yield was 16.46 $g_{d.m.}$/L. In both cases, the mean carotenoid content in the cell biomass was found to be higher than that in the control medium and in the media supplemented with single compounds and amounted to 106.44 $µg/g_{d.m.}$ (Ba^{2+} + niacin) and 180.50 $µg/g_{d.m.}$ (Al^{3+} + niacin). Based on the cellular yield, the mean volumetric efficiency of carotenoid biosynthesis was estimated at 1.02 and 2.97 mg/L, respectively. Simultaneous enrichment of media with cations and vitamins also resulted in changes in the percentages of the synthesized carotenoids. In the medium supplemented with barium and niacin, a significant reduction in the percentage of β-carotene was found (from 43.08% in the control sample to 7.30%), which resulted in an increase in the proportions of torularhodin (26.69%) and torulene (63.58%).

4. Discussion

Of the four cations tested, barium, which belongs to the group of alkaline earths, was found to have the most toxic effect on the cells of *R. mucilaginosa* MK1. Unfortunately, the literature lacks data concerning its influence on the growth and biosynthesis of carotenoids by yeasts. Among the few studies focusing on the effect of barium on yeast growth, the study by Alvino et al. [18] showed that barium added in the form of barium ferrite nanoparticles (BaFeNP) at doses ranging from 15 to 500 µg/mL did not cause any significant effect on the growth of *Candida albicans* ATCC 10,231 yeast strain.

Zinc, which is a chemical element from the zinc group, is a component of many enzymes, transcription factors, and regulatory proteins. It ensures optimal metabolism of nucleic acids and proteins and plays an important role in the growth, division, and functioning of cells. In yeast cells, zinc is present in both free and bound forms, with the latter being dominant [19]. In the present study, the addition of this element did not influence the growth and biosynthesis of carotenoids by *R. mucilaginosa* MK1 yeast at the studied concentrations. In a study by El-Banna et al. [20], the addition of $ZnSO_4$ salt at a dose of 0.1 g/L stimulated carotenoid synthesis by *Rhodotorula glutinis* yeast which produced 638 $µg/g_{d.m.}$ of carotenoids (volume yield: 2.81 mg/L), while the content and volume yield determined in the medium lacking this salt after cultivation were, respectively, 292 $µg/g_{d.m.}$ and 1.13 mg/L. A study by Rusinov-Videva et al. also confirmed that carotenoid biosynthesis was stimulated in a medium supplemented with zinc ions in the form of $ZnSO_4$ [21]. The yeast *Sporobolomyces salmonicolor* AL_1 was found to produce 16-fold more β-carotene in the medium containing 112 ppm of Zn^{2+} ions (from 11.2 to 189.2 $µg/g_{d.m.}$) compared to the medium lacking these ions. Interestingly, in the presence of Zn^{2+}, the biosynthesis of coenzyme Q_{10} was completely inhibited. It was assumed that zinc ions activated the enzyme catalytic centers responsible for isoprenoid synthesis and thus contributed to the biosynthesis of β-carotene. Some authors analyzed whether supplementation of zinc in the form of nanoparticles, rather than inorganic salt, caused an increase in the efficiency of biosynthesis of these compounds. Ibrahim and Mahmoud [22] reported that the addition of zinc in the form of ZnO nanoparticles to the culture medium resulted in a high content of carotenoids in the yeast cell biomass. After 72 h of cultivation, the volumetric yield of biosynthesis by *Rhodotorula toruloides* MH023518 yeast strain increased from 124 mg/L (control medium) to 264 mg/L (medium containing 50 ppm ZnO). It is worth noting, however, that cultivation was carried out in the YPD model medium, which contains optimal nutrients. Nanoparticles generate cellular stress leading to the formation of reactive oxygen species (ROS). This, in turn, increases the biosynthesis of carotenoids to neutralize ROS [22].

The above-presented results prove that it is necessary to select the appropriate doses of zinc, as well as the ideal composition of the propagating medium, for each strain. If higher doses of zinc ions were used in this study, they would have stimulated carotenoid biosynthesis by the *R. mucilaginosa* MK1 yeast strain. However, a high concentration of this element in the medium may significantly reduce the yeast growth rate and inhibit carotenoid biosynthesis. This was demonstrated in the study by Rovinaru et al. [23], who showed that after 120 h of cultivation of the yeast *R. glutinis* CCY 020-002-033 in the medium supplemented with 500 mg/L zinc, the volumetric efficiency of β-carotene biosynthesis and the cellular biomass yield were 0.92 mg/L and 0.16 $g_{d.m.}$/L, respectively, while after cultivation in the control medium the values were 2.4 mg/L and 1.94 $g_{d.m.}$/L, respectively. The authors reported that the synthesis of β-carotene (7.1 mg/L) was highest in the medium containing 250 mg Zn^{2+}/L.

At acidic pH, aluminum ions are toxic to both plant and animal cells. Many studies suggest that they increase the peroxidation of phospholipids and proteins in cell membranes and induce oxidative stress in cells [11,24]. It was also indicated that the addition of Al^{3+} to the medium at doses from 200 to 300 mg/L resulted in a significant reduction of the biomass yield (Table 3). Among the cations tested in the present study, only aluminum ions increased the content of carotenoids in the cell biomass of the *R. mucilaginosa* MK1 strain and changed the proportions of β-carotene, torulene, and torularhodin synthesized by the yeast. The share of torulene increased significantly from 31.85% in the control sample to 75.20% in the biomass obtained after cultivation in the medium supplemented with 300 mg Al^{3+}/L. A study by Elfeky et al. [11] showed that the enrichment of the culture medium with aluminum sulfate (0.7 mM) led to an increase in the synthesis of carotenoid pigments and torulene, which confirms the role of aluminum ions in the induction of carotenoid biosynthesis in yeast cells. Wang et al. [25] found that the addition of aluminum ions in the medium resulted in an increase in the activity of malate dehydrogenase and the intracellular accumulation of citric acid by *Rhodotorula taiwanensis* RS1 yeast strain.

Manganese is an important trace element that belongs to the group of transition metals. As a cell component, it plays a key role among enzymes involved in various metabolic processes and is also part of an antioxidant—superoxide dismutase, which protects cells by inactivating free radicals. Manganese also acts as an electron store, from where electrons are transferred to reaction sites, and as an activator of many enzymes involved in oxidation, carboxylation, carbohydrate metabolism, and the series of phosphorus and citric acid reactions. Similar to magnesium, this element is involved in the binding of ATP with enzyme complexes (phosphotransferase and phosphokinase) and activates the enzymes taking part in the Krebs cycle, such as dehydrogenase and decarboxylase, as well as RNA polymerase [26]. Hence, this study tested the influence of magnesium on carotenoids biosynthesis by yeast. The results showed no increase in the number of carotenoids in cellular biomass in the control medium, but after cultivation in the medium supplemented with manganese at a dose of 300 mg/L, the proportion of torularhodin in the total pool of carotenoids decreased from 22.8% (control) to 14.1%. Interestingly, Buzzini et al. [27] observed the opposite effect in their study on carotenoid biosynthesis by *Rhodotorula graminis* DBVPG 7021 strain, which showed that the presence of manganese ions inhibited the biosynthesis of torulene and torularhodin. The authors hypothesized that manganese cations had a possible negative influence on the activity of desaturases involved in carotenoid biosynthesis by yeast.

Similar to micronutrients, B vitamins are critical for the proper functioning and metabolism of yeast cells. For example, biotin is the prosthetic group of acetyl-CoA carboxylase, an enzyme that catalyzes the formation of malonyl-CoA, a substrate in de novo lipid biosynthesis [28]. Riboflavin is essential for the activity of enzymes involved in the TCA cycle, as well as for the electron transport chain, β-oxidation, and various biosynthetic processes [15]. Pantothenic acid (vitamin B_3) is a metabolic precursor of coenzyme A and many acyl carrier proteins, which are cofactors of many enzymes [12,13]. Niacin is involved in the biosynthesis of nicotinamide adenine dinucleotide [14]. Due to the significant role played by B vitamins in yeast cell metabolism, this study investigated their influence on carotenoid biosynthesis. To the best of the authors' knowledge, there are no data on this subject in the scientific literature. It was found that among the tested vitamins, only niacin caused a change in the profile of carotenoids synthesized by *R. mucilaginosa* MK1. A significant increase in the content of torulene and torularhodin was observed, while the proportion of β-carotene was decreased. It was also noted that supplementation of the medium with a concentration of niacin significantly slowed down the growth of yeast, indicating that the vitamin inhibits certain metabolic processes. Perhaps the high concentration of niacin decreased the activity of lycopene cyclase, which catalyzes the conversion of γ-carotene to β-carotene, resulting in an increase in the content of torulene and torularhodin in the total carotenoid pool. This is the first study to have examined the effect of niacin on the biosynthesis of carotenoids, and therefore, further studies are needed to understand the mechanism underlying the effect of niacin on the enzymes involved in the carotenoid biosynthesis pathway in yeast cells.

5. Conclusions

The study proved that the profile of carotenoids synthesized by the yeast *R. mucilaginosa* MK1 in media containing potato juice water and glycerol could be changed by supplementing the medium with selected cations and B vitamins. The total carotenoid content increased significantly after yeast cultivation in the medium supplemented with aluminum at a dose of 300 mg/L. In addition, the amount of torulene increased to 72.2% (from 31.8% for the control medium) in the presence of aluminum. Torulene biosynthesis was also stimulated by enriching the medium with barium at an amount of 300 mg/L (percentage: 68.24%). The addition of niacin to the medium increased the production of torularhodin from 23.31% to 31.59–33.79%. The highest content of carotenoids (180.50 $\mu g/g_{d.m.}$) was observed after cultivation in the medium supplemented with both aluminum ions (300 mg/L) and niacin (100 $\mu g/L$). This is the first study to show the effect of niacin on the biosynthesis of carotenoids by red yeast cells, and therefore, further research is needed in this area to understand the mechanism underlying the influence of niacin.

Author Contributions: The individual contributions are as follows: Conceptualization, A.M.K. and S.B.; methodology, A.M.K. and M.K.; performance of experiments and data analysis, A.M.K., M.K., R.B. and W.S.; writing—original draft preparation, A.M.K.; writing—review and editing, A.M.K. and S.B.; project administration, A.M.K.; funding acquisition, A.M.K. All authors have read and agreed to the published version of the manuscript.

Funding: This work was supported by the National Science Centre, Poland "Miniatura" (no. 2019/03/X/NZ9/00163). The research was carried out using the equipment purchased as part of the "Food and Nutrition Centre—modernization of the WULS campus to create a Food and Nutrition Research and Development Centre (CŻiŻ)" co-financed by the European Union from the European Regional Development Fund under the Regional Operational Programme of the Mazowieckie Voivodeship for 2014–2020 (Project No. RPMA.01.01.00-14-8276/17).

Institutional Review Board Statement: Not applicable.

Informed Consent Statement: Not applicable.

Data Availability Statement: Not applicable.

Conflicts of Interest: The authors declare no conflict of interest.

References

1. Saini, R.K.; Keum, Y.S. Microbial platforms to produce commercially vital carotenoids at industrial scale: An updated review of critical issues. *J. Ind. Microbiol. Biotechnol.* **2019**, *46*, 657–674. [CrossRef] [PubMed]
2. Mata-Gómez, L.C.; Montañez, J.C.; Méndez-Zavala, A.; Aguilar, C.N. Biotechnological production of carotenoids by yeasts: An overview. *Microb. Cell Factories* **2014**, *13*, 12. [CrossRef] [PubMed]

3.   Frengova, G.I.; Beshkova, D.M. Carotenoids from *Rhodotorula* and *Phaffia*: Yeasts of biotechnological importance. *J. Ind. Microbiol. Biotechnol.* **2009**, *36*, 163–180. [CrossRef] [PubMed]

4.   Gul, K.; Tak, A.; Singh, A.K. Chemistry, encapsulation, and health benefits of β-carotene- A review. *Cogent Food Agric.* **2015**, *1*, 1018696. [CrossRef]

5.   Zoz, L.; Carvalho, J.C.; Soccol, V.T.; Casagrande, T.C.; Cardoso, L. Torularhodin and torulene: Bioproduction, properties and prospective applications in food and cosmetics—A review. *Braz. Arch. Biol. Technol.* **2015**, *58*, 278–288. [CrossRef]

6.   Cheng, Y.T.; Yang, C.F. Using strain *Rhodotorula mucilaginosa* to produce carotenoids using food wastes. *J. Taiwan Inst. Chem. Eng.* **2016**, *61*, 270–275. [CrossRef]

7.   Du, C.; Li, Y.; Guo, Y.; Han, M.; Zhang, W.; Qian, H. The suppression of torulene and torularhodin treatment on the growth of PC-3 xenograft prostate tumors. *Biochem. Biophys. Res. Commun.* **2016**, *469*, 1146–1152. [CrossRef]

8.   Ungureanu, C.; Dumitriu, C.; Popescu, S.; Enculescu, M.; Tofan, V.; Popescu, M.; Pirvua, C. Enhancing antimicrobial activity of TiO$_2$/Ti by torularhodin bioinspired surface modification. *Bioelectrochemistry* **2016**, *107*, 14–24. [CrossRef]

9.   Kot, A.M.; Błażejak, S.; Kieliszek, M.; Gientka, I.; Bryś, J. Simultaneous production of lipids and carotenoids by the red yeast *Rhodotorula* from waste glycerol fraction and potato wastewater. *Appl. Biochem. Biotechnol.* **2019**, *189*, 589–607. [CrossRef] [PubMed]

10. Bhosale, P.B.; Gadre, R.V. Production of β-carotene by a mutant of *Rhodotorula glutinis*. *Appl. Microbiol. Biotechnol.* **2001**, *55*, 423–427. [CrossRef]

11. Elfeky, N.; Elmahmoudy, M.; Zhang, Y.; Guo, J.; Bao, Y. Lipid and carotenoid production by *Rhodotorula glutinis* with a combined cultivation mode of nitrogen, sulfur, and aluminium stress. *Appl. Sci.* **2019**, *9*, 2444. [CrossRef]

12. White, W.H.; Gunyuzlu, P.L.; Toyn, J.H. *Saccharomyces cerevisiae* is capable of de Novo pantothenic acid biosynthesis involving a novel pathway of beta-alanine production from spermine. *J. Biol. Chem.* **2001**, *6*, 10794–10800. [CrossRef] [PubMed]

13. White, W.H.; Skatrud, P.L.; Xue, Z.; Toyn, J.H. Specialization of function among aldehyde dehydrogenases: The ALD2 and ALD3 genes are required for beta-alanine biosynthesis in *Saccharomyces cerevisiae*. *Genetics* **2003**, *163*, 69–77. [CrossRef]

14. Li, Y.F.; Bao, W.G. Why do some yeast species require niacin for growth? Different modes of NAD synthesis. *FEMS Yeast Res.* **2007**, *7*, 657–664. [CrossRef] [PubMed]

15. Hucker, B.; Wakeling, L.; Vriesekoop, F. Vitamins in brewing: Presence and influence of thiamine and riboflavin on wort fermentation. *J. Inst. Brew.* **2016**, *122*, 126–137. [CrossRef]

16. Kieliszek, M.; Kolotylo, V.; Mikołajczuk-Szczyrba, A.; Giurgiulescu, L.; Kot, A.M.; Kalisz, S.; Pobiega, K.; Cendrowski, A. Isolation and identification of new yeast strains from bee bread. *Carpathian J. Food Sci. Technol.* **2021**, *13*, 207–213.

17. Kot, A.M.; Błażejak, S.; Kurcz, A.; Bryś, J.; Gientka, I.; Bzducha-Wróbel, A.; Maliszewska, M.; Reczek, L. Effect of initial pH of medium with potato wastewater and glycerol on protein, lipid and carotenoid biosynthesis by *Rhodotorula glutinis*. *Electron. J. Biotechnol.* **2017**, *27*, 25–31. [CrossRef]

18. Alvino, L.; Pacheco-Herrero, M.; López-Lorente, A.I.; Quiñones, Z.; Cárdenas, S.; González-Sánchez, Z.I. Toxicity evaluation of barium ferrite nanoparticles in bacteria, yeast and nematode. *Chemosphere* **2010**, *254*, 126786. [CrossRef]

19. Plum, L.M.; Rink, L.; Haase, H. The essential toxin: Impact of zinc on human health. *Int. J. Environ. Res.* **2010**, *7*, 1342–1365. [CrossRef]

20. El-Banna, A.; El-Razek, A.; El-Mahdy, A. Some Factors Affecting the Production of Carotenoids by *Rhodotorula glutinis* var. *glutinis*. *Food Sci. Nutr.* **2012**, *3*, 64–71.

21. Rusinova-Videva, S.; Dimitrova, S.; Georgieva, K.; Katsarova, M.; Pavlova, K. Effect of Zn^{2+}, Cu^{2+} and Fe^{2+} ions for accumulation of ergosterol, β–carotene and coenzyme Q10 by antarctic yeast strain *Sporobolomyces salmonicolor* AL1. *Comptes Rendus L'Acad. Bulg. Des Sci.* **2016**, *69*, 1005–1012.

22. Ibrahim, A.B.M.; Mahmoud, G.A.E. Chemical- vs. sonochemical-assisted synthesis of ZnO nanoparticles from a new zinc complex for improvement of carotene biosynthesis from *Rhodotorula toruloides* MH023518. *Appl. Organomet. Chem.* **2021**, *35*, e6086. [CrossRef]

23. Rovinaru, C.; Pasarin, D.; Capra, L.; Stoica, R. The effect of ZnSO$_4$ in the cultivation medium on *Rhodotorula glutinis* CCY 020-002-033 yeast biomass growth, β-carotene production and zinc accumulation. *J. Microbiol. Biotechnol. Food Sci.* **2018**, *8*, 931–935.

24. Ezaki, B.; Sivaguru, M.; Ezaki, Y.; Matsumoto, H.; Gardner, R.C. Acquisition of aluminum tolerance in *Saccharomyces cerevisiae* by expression of the BCB or NtGDI1 gene derived from plants. *FEMS Microbiol. Lett.* **1999**, *171*, 81–87. [CrossRef] [PubMed]

25. Wang, C.; Wang, C.Y.; Zhao, X.Q.; Chen, R.F.; Lan, P.; Shen, R.F. Proteomic analysis of a high aluminum tolerant yeast *Rhodotorula taiwanensis* RS1 in response to aluminum stress. *Biochim. Biophys. Acta* **2013**, *1834*, 1969–1975. [CrossRef] [PubMed]

26. Mousavi, S.; Shahsavari, M.; Rezaei, M. A general overview on manganese importance for crops production. *Aust. J. Basic Appl. Sci.* **2011**, *5*, 1799–1803.

27. Buzzini, P.; Martini, A.; Gaetani, M.; Turchetti, B.; Pagnoni, U.A.; Davoli, P. Optimization of carotenoid production by Rhodotorula graminis DBVPG 7021 as a function of trace element concentration by means of response surface analysis. *Enzym. Microb. Technol.* **2005**, *36*, 687–692. [CrossRef]

28. Sitepu, I.R.; Sestric, R.; Ignatia, L.; Levin, D.; Bruce German, J.; Gillies, L.A.; Almada, L.A.; Boundy-Mills, K.L. Manipulation of culture conditions alters lipid content and fatty acid profiles of a wide variety of known and new oleaginous yeasts species. *Bioresour. Technol.* **2013**, *144*, 360–369. [CrossRef]

Article

Microbiological Characteristics and Behavior of *Staphylococcus aureus, Salmonella* spp., *Listeria monocytogenes* and Staphylococcal Toxin during Making and Maturing Cotija Cheese

María de los Ángeles Olea-Rodríguez [1], Patricia Chombo-Morales [2], Karla Nuño [3], Olga Vázquez-Paulino [1], Zuamí Villagrán-de la Mora [4], Luz E. Garay-Martínez [1], Javier Castro-Rosas [5], Angélica Villarruel-López [1,*] and Ma. Refugio Torres-Vitela [1,*]

[1] Departamento de Farmacobiología, Centro Universitario de Ciencias Exactas e Ingenierías, Universidad de Guadalajara, Blvd. Gral. Marcelino García Barragán 1421, Olímpica, Guadalajara 44430, Mexico; mariangeles_olea@hotmail.com (M.d.l.Á.O.-R.); olga.vazquez@academicos.udg.mx (O.V.-P.); eduviges.garay@academicos.udg.mx (L.E.G.-M.)

[2] Food Technology Department, Centro de Investigacion y Asistencia en Tecnología y Diseño del Estado de Jalisco (CIATEJ), Av. Normalistas 800, Colinas de la Normal, Guadalajara 45019, Mexico; pchombo@ciatej.mx

[3] Departamento de Ciencias Biomédicas, Centro Universitario de Tonalá, Universidad de Guadalajara, Nuevo Perif. Ote. 555, Ejido San José, Tateposco, Tonalá 45425, Mexico; karlajanette.nuno@cutonala.udg.mx

[4] Departamento de Ciencias de la Salud, Centro Universitario de Los Altos, Universidad de Guadalajara, Av. Rafael Casillas Aceves 1200, Tepatitlán de Morelos 47620, Mexico; blanca.villagran@academicos.udg.mx

[5] Instituto de Ciencias Básicas e Ingeniería, Universidad Autónoma del Estado de Hidalgo, Carretera Pachuca-Tulancingo Km 4.5, Pachuca, 42183, Mexico; jcastro@uaeh.edu.mx

[*] Correspondence: angelica.vlopez@academicos.udg.mx (A.V.-L.); ma.torres@academicos.udg.mx (M.R.T.-V.)

Citation: Olea-Rodríguez, M.d.l.Á.; Chombo-Morales, P.; Nuño, K.; Vázquez-Paulino, O.;Villagrán-de la Mora, Z.; Garay-Martínez, L.E.; Castro-Rosas, J.; Villarruel-López, A.; Torres-Vitela, M.R. Microbiological Characteristics and Behavior of *Staphylococcus aureus, Salmonella* spp., *Listeria monocytogenes* and Staphylococcal Toxin during Making and Maturing Cotija Cheese. *Appl. Sci.* **2021**, *11*, 8154. https://doi.org/10.3390/app11178154

Academic Editor: Marek Kieliszek

Received: 19 July 2021
Accepted: 30 August 2021
Published: 2 September 2021

Abstract: Cotija cheese is an artisanal matured Mexican cheese from unpasteurized milk. This work determined the microbiological characteristics and behavior of *Staphylococcus aureus, Salmonella* spp., *Listeria monocytogenes* and staphylococcal toxin during cheese elaboration and ripening. Sixteen 20-kg cheeses were used, eight inoculated with 6 log CFU/mL of each pathogen, and eight uninoculated, and samples were taken at each stage of the process. In the uninoculated samples, the survival of *S. aureus* and *L. monocytogenes* decreased after 30 days of ripening. The average counts of *S. aureus* in milk, curd, and serum were 7 log MPN /mL, and 8.7 log MPN /g in cheese, decreasing from day 15. *Salmonella* spp. counts (initial inoculum: 7.2 log MPN /mL) decreased after 24 h, and *L. monocytogenes* counts (8.7 log MPN/g at 24 h) decreased from day 15 in the cheese. *Salmonella* spp. and *L. monocytogenes* were not detected in any sample after 60 days of ripening, unlike *S. aureus*, which was detected at the end of the study. Lactic acid bacteria counts were 9 log CFU/mL in milk and whey and 7.2 log CFU/g in cheese. Pathogens behavior during the ripening process reduces the health risks by consuming products made with unpasteurized milk when subjected to ripening.

Keywords: Cotija; Staphylococcus aureus; Salmonella spp.; Listeria monocytogenes; maturing cheese

1. Introduction

The prevalence of pathogens in milk is influenced by numerous factors, such as the size of the farm, the number of animals, hygiene, variability in the sampling, and the techniques used for sample analysis, geographic location, and the season of the year. However, within the variability that these factors may have, milk is one of the most important sources of pathogens that cause foodborne diseases [1]. In Mexico, most fresh cheeses are handmade using rustic methods in micro-industries located in small cities. These industries generally have little or no quality control and high composition and sensory parameters variability, leading to a limited shelf life [2–4]. However, artisanal matured cheeses show a different reality due to the fermentative processes during the maturation time, giving rise to the formation of inhibitory compounds for certain microorganisms [5]. The microbiota is an essential component that plays an essential role during cheese maturation [6,7]. Interaction within the milk microbiota is vital for the characteristics of the final product. It participates from the acidification of the curd to the formation of compounds that determine the aroma, flavor, and texture during ripening [8,9]. Even raw milk is used in the elaboration of some matured cheeses. Therefore, the diversity of the native microbiota contributes to the final sensory properties and biopreservation [10,11].

Cotija cheese is a typical matured Mexican cheese that is artisanal-made and produced exclusively with milk from crosses of zebu cattle (*Bos taurus indicus*). It is manufactured from July to October in the Cotija area, located between the states of Jalisco and Michoacán. The salt added during the manufacturing process contributes to its flavor and firm consistency. During the aging process, salt concentration and other solutes produce a reduction in the water activity, promoting a selective medium in which halo-tolerant microorganisms like various genera of lactic acid bacteria dominate others, transform the physicochemical properties and composition of the cheese. Although the action of the different microbial populations from the milk used, the environment of the cheese factory and the production practices, the climatic conditions of the geographical area where the cheese is made and matured, favor the dynamics of the microbial populations, characteristics and, consequently, the distinctive characteristics of this cheese [5,12,13].

During the ripening process, adverse conditions are created for the development of pathogens. In the case of Cotija cheese, salt is added. However, some pathogenic bacteria may survive, causing health problems. An example occurred in Illinois due to the consumption of cheese produced in the USA, marketed as Cotija, contaminated with *Salmonella enterica* serotype Newport [14]. Another pathogen frequently recovered in Cotija cheese is *Staphylococcus aureus*, which can originate from the mammary glands of infected cows [15].

The objective of this work was to investigate microbiological characteristics and behavior of *Staphylococcus aureus*, *Salmonella* spp., *Listeria monocytogenes* and staphylococcal toxin during elaboration and maturation of Cotija cheese.

2. Results

Each 20-kg cheese was made using 180 L of unpasteurized milk. Cheeses were made immediately after the milk was obtained, and the main matrix was salted within the first 9 h. Sixteen cheeses were made for this work. Eight were left uninoculated (negative controls). Eight were inoculated with 6 log CFU/mL of *S. aureus* ATCC 13565, *S. Typhimurium* ATCC 14028, *L. monocytogenes* ATCC BAA-751, respectively, and staphylococcal toxin. The analysis of the cheese already molded and in the process of maturing began at 24 h. It is worth mentioning that at each sampling time, a complete cheese was taken. For sampling, each cheese was divided into three uniform vertical: (A) Core, (B) Main matrix, and (C) Surface (Figure 1).

2.1. Microbiological Parameters in Uninoculated Samples

Milk, Curd, and Whey Analysis

Before the beginning of the cheese production, the natural presence of *Staphylococcus aureus*, *Salmonella* spp., *Listeria monocytogenes* and staphylococcal toxin was analyzed, as well as lactic acid bacteria (LAB), aerobic mesophilic bacteria (AMB), coliform organisms (CO), molds and yeasts (M/Y) indicators in milk, uninoculated curd and whey. Only *S. aureus* and *Listeria* were recovered in milk, curd, and whey (Table 1), and two species were identified, *L. moncytogenes* and *L. ivanovii*. Enterotoxin (<1 ng/mL) and *Salmonella* were not detected in any non-inoculated samples.

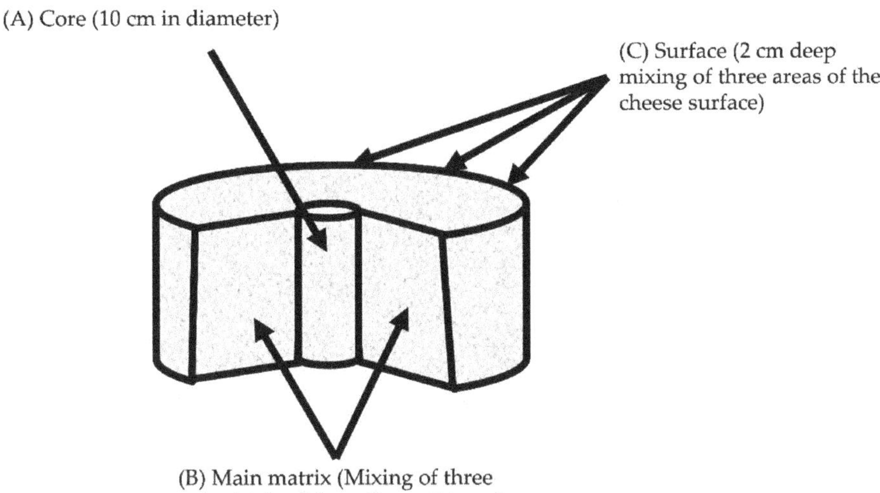

Figure 1. Revaluated sections of Cotija cheese (From Chombo-Morales et al., 2016 [12]).

Table 1. Behavior of pathogens in Cotija cheese made with uninoculated milk.

	Staphylococcus aureus			*Listeria monocytogenes*		
	(Log MPN/mL or g)			**(Log MPN/mL or g)**		
	Milk	**Curd**	**Whey**	**Milk**	**Curd**	**Whey**
0 *	5.9	6.4	5.9	5.9	7.5	7
Cheese	Surface	Main matrix	Core	Surface	Main matrix	Core
24 h		7.4			7.2	
7 d	6.5 [a]	6.1 [b]	4.9 [b]	3.4 [A]	4.3 [B]	4.1 [B]
15 d	6.7 [a]	6.5 [b]	5.6 [b]	3.1 [A]	2.7 [B]	2.9 [B]
30 d	4.5 [a]	3.9 [b]	ND	1 [A]	0.3 [B]	ND
45 d	1.1 [a]	ND	ND	ND	ND	ND
60 d	ND	ND	ND	ND	ND	ND
75 d	ND	ND	ND	ND	ND	ND
90 d	ND	ND	ND	ND	ND	ND

d = days of maturation; ND = not detected; * *Salmonella* and staphylococci enterotoxin were not detected. Values with different superscripts ([a,b] or [A,B]) in the rows, are significantly different between adjacent columns of the same pathogen ($p < 0.05$).

Staphylococcus aureus. Once the cheese was molded, it was left in the ripening chamber for 24 h, and then the native *S. aureus* was enumerated, finding values of 7.4 log MPN/g in the non-inoculated cheese. Table 1 shows the behavior of *S. aureus* in the different sections and sampling times of the non-inoculated cheese.

The statistical analysis of the non-inoculated cheese did not show a significant difference in the distribution of *S. aureus* regarding the analyzed sampled area up to 45 days ($p = 0.343$), the last sample obtained in which the bacteria was recovered. In addition, during the ripening process, no traces of staphylococcal enterotoxin were detected (<1 ng/mL).

Listeria monocytogenes. In cheese made with non-inoculated milk, native *L. monocytogenes* counts increased at 24 h, reaching a population of 7.2 log MPN/g and then drastically decreased on day 15, when samples from the core had 2.9 log MPN/g, and later, no *L. monocytogenes* was detected. On the other hand, the main matrix and surface samples allowed the pathogen to survive until day 30, with a recovery of 0.3 and 1 log MPN/g, respectively (Table 1).

2.2. Microbiological Parameters in Inoculated Samples

Staphylococcus aureus. In milk inoculated with 7 log CFU of *S. aureus* ATCC 13565 *per* mL, pathogen counts in curd and serum were 7.8 log MPN/mL and 7.1 log MPN/mL, respectively, while in the 24 h cheese, 8.7 log MPN/mL were recovered.

After seven days of maturation of the inoculated cheese, microorganisms from samples taken from the core survived until day 45 (0.15 log MPN/g), unlike the main matrix samples where it was only possible to detect them after 15 days (2.55 log MPN/g). The area where the longest survival time of *S. aureus* was observed was the surface, since counts were detected until day 90 with 3.6 log MPN/g. The statistical analysis of the inoculated cheese showed no significant differences regarding the recovery of the microorganisms on day seven ($p = 0.326$). From the second sampling time (15 days), a statistically significant difference was observed in the distribution of *S. aureus* ($p = 0.001$) in cheese.

All the samples, regardless of the sampling area, were negative for staphylococcal enterotoxin ($\geq$1 ng/mL) in inoculated cheese after 24 h. Initially was detected in milk, rennet, and whey (Table 2).

Table 2. Behavior of pathogens in Cotija cheese made with inoculated milk.

	Staphylococcus aureus			*Salmonella* **Typhimurium**			*Listeria monocytogenes*		
	(Log MPN/mL or g)			**(Log MPN/mL or g)**			**(Log MPN/mL or g)**		
Milk Inoculum	7.3			7.2			7.3		
Cheese	**Surface**	**Main Matrix**	**Core**	**Surface**	**Main Matrix**	**Core**	**Surface**	**Main Matrix**	**Core**
24 h		8.7			4.8			8.7	
7 d	6.7 [a]	6.42 [b]	5.45 [b]	3.55 [A]	2.82 [B]	3 [B]	7.4 [a]	7.4 [b]	6.9 [b]
15 d	7.2 [a]	3.77 [b]	5.05 [b]	3.85 [A]	2.4 [B]	4 [B]	7.1 [a]	5.6 [b]	5.3 [b]
30 d	6.4 [a]	ND	1.8 [b]	4.51 [A]	0.67 [B]	3.6 [B]	6.1 [a]	ND	ND
45 d	5.3 [a]	ND	0.15	1.85 [A]	0.72 [B]	ND	4.5 [a]	ND	ND
60 d	4.7 [a]	ND	ND	ND	ND	ND	ND	ND	ND
75 d	4 [a]	ND	ND	ND	ND	ND	ND	ND	ND
90 d	3.6 [a]	ND	ND	ND	ND	ND	ND	ND	ND

d = days of maturation; ND = not detected. Values with different superscripts ([a,b] or [A,B]) in the rows, are significantly different between adjacent columns of the same pathogen ($p < 0.05$).

Salmonella Typhimurium. In the inoculated milk, *S. Typhimurium* ATCC 14028 showed a gradual and continuous decrease in all samples from the 7.2 log MPN/mL initial inoculum. After 24 h of storage, cheese samples had a reduction of 2.5 log (4.8 log MPN/g of cheese). In the following samplings, the concentration of the pathogen gradually decreased until it stopped recovering on day 45 (core) and 60 (main matrix and surface) (Table 2).

Listeria monocytogenes. Regarding cheese made with inoculated milk (7.3 log of MPN/mL), a similar behavior was observed compared to the uninoculated milk. In cheese, a pathogen concentration increase was observed (8.7 log of MPN/g) after 24 h of storage, then its presence decreased. In the core and the main matrix, only 5.3 and 5.9 log of MPN/g were recovered until day 15, respectively, while on the surface, *L. monocytogenes* ATCC BAA-751 survived longer since its last recovery was on day 45 with a 4.5 log of MPN/g. Table 2 shows the behavior of each pathogen in the different parts of Cotija cheese that were analyzed during its elaboration and maturation process.

Indicator groups. Lactic acid bacteria (LAB), aerobic mesophilic bacteria (AMB), coliform organisms (CO), molds and yeasts (M/Y) were quantified. Table 3 shows the microbiota content in milk, curd, whey, and cheese with 24 h of maturation, without inoculation. No significant differences were observed in LAB, CO, M/Y of milk, curd, and whey ($p > 0.05$). Only the AMB had a higher content in milk and cheese ($p < 0.05$). The four types of samples showed <10 Log CFU/ g or mL of molds and yeasts (Figure 2).

Table 3. Indicator groups in milk, curd and whey not inoculated with pathogens.

	Lactic Acid Bacteria	Aerobic Mesophilic Bacteria	Coliforms	Molds and Yeast
Milk *	9.1 [a]	8.0 [a]	6.5 [a]	3.0 [b]
Curd **	9.4 [a]	6.9 [b]	6.1 [a]	3.1 [b]
Whey *	8.9 [a]	6.9 [b]	5.8 [a]	3.0 [b]
Cheese **	7.2 [a]	8.2 [a]	7.2 [a]	4.2 [b]

* Log of CFU/mL; ** Log of CFU/g. Values with different superscripts ([a,b]) in the columns, are significantly different between rows ($p < 0.05$).

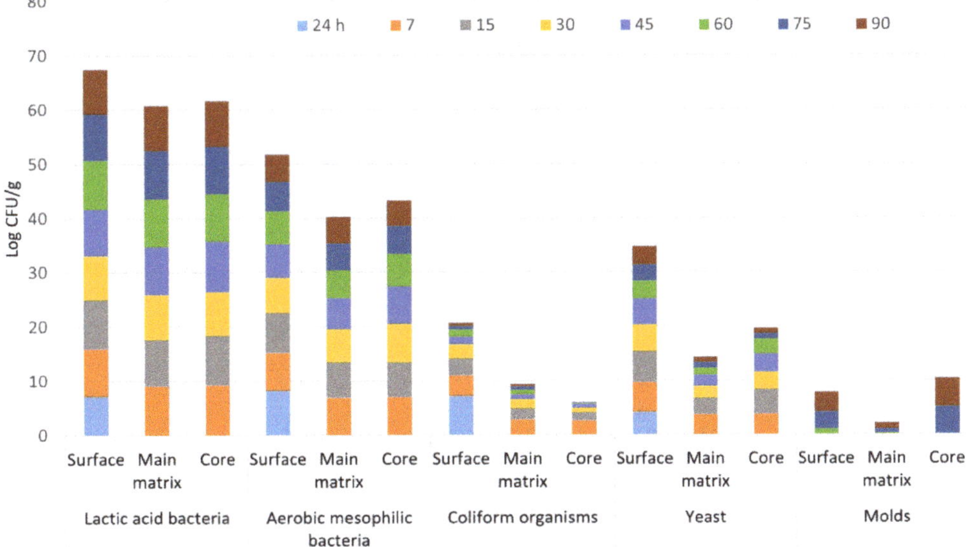

Figure 2. Behavior of indicator groups in Cotija cheese during its maturation process.

The indicator groups in cheese showed that after seven days of maturation, LAB continued to show a homogeneous behavior in the microbial content of the different areas analyzed (core, main matrix, and surface) and between the non-inoculated and inoculated cheese. In non-inoculated cheese, indicator groups maintained an average of around 9.0 log of CFU/g until day 45, then decreased to 1 log CFU/g from day 60, maintaining those count values until the end of this work. In the inoculated Cotija cheese, the LAB behavior was similar. However, it was maintained with an 8.8 log CFU/g until the end of the study.

The statistical analysis showed a significant difference between the main matrix and the core in the indicator groups at 15, 75 and 90 days ($p < 0.05$). There was no difference in either of the two groups, inoculated and not inoculated ($p > 0.05$).

From cheeses with seven days of maturation, main matrix and surface samples from days 7 and 15 had the highest content of indicator groups with an average of 5.9 and 6.4 log of CFU/g, respectively, while the core had 5.0 log of CFU/g. After 30 days, a gradual decrease in the microbial load was observed, more critical in the main matrix and the core. At 90 days, the AMB content was 2.4 log CFU/g in the core, 2.2 log CFU/g in the main matrix and 3.4 log CFU/g in the surface. Statistically, there was a difference in the AMB content in the surface regarding the main matrix and the core on days 7, 30, 45, 60,

75 and 90, while on day 15; the ABM content in the core, main matrix and surface was different ($p < 0.05$). The coliform content gradually decreased in the non-inoculated and the inoculated cheeses until it practically disappeared. In the non-inoculated cheese, the main matrix and the core had 3 log CFU/g at 7 days, ceasing to recover at 90 and 60 days in the main matrix and core, respectively, while in the surface, the CO content at 7 days was 5.8 log CFU/g, decreasing to 0.2 log CFU/g on day 90. In the inoculated cheese, the initial count of the core, main matrix and surface was 3 log CFU/g on day 7, and at the end of the study, the counts decreased to 0.3, 0.5, and <1 log CFU/g in the main matrix, surface, and core. A significant difference was observed on days 7, 30 and 60 between the surface, core and main matrix of non-inoculated cheese. Inoculated cheese showed a significant difference in surface compared to main matrix and core on days 7, 15 and 45; and between the core compared to the surface and main matrix on days 30 and 60 ($p < 0.05$).

In the case of yeasts, the content was heterogeneous in both types of cheese, non-inoculated and inoculated. The initial load (7 days) in the non-inoculated cheese was 1.4, 6.2 and 7.5 log CFU/g for the main matrix, surface, and core. At the end of the study, 1.4 log CFU/g were found in the main matrix, 4.4 log CFU/g in the surface and 1.2 log of CFU/g in the core. Yeast content in inoculated cheese was higher on the surface with 5.5 log CFU/g and lower in the main matrix and core (3.7 log CFU/g in both).

At 90 days, the content was higher in the surface (4.4 log CFU/g) and lower in the main matrix and core (1 log CFU/g). Statistically, all sampling times showed differences between the non-inoculated and the inoculated cheeses ($p < 0.05$). Finally, molds appeared after 60 days in not inoculated and inoculated cheeses. In non-inoculated cheese, no molds were recovered from the core in any sample (<1 log CFU/g). The main matrix had a mold content of 2.8, 2.1, and 0.67 log CFU/g on days 60, 75 and 90, respectively, while the surface presented 1.3, 2.3 and 2.9 log CFU/g on days 60, 75 and 90, respectively. In inoculated cheese, the main matrix had an increase from 0.3 log CFU/g at 60 days to 1 log CFU/g on days 75 and 90, while in the surface, mold increased by 1 log CFU/g on day 60 at 3.2 log CFU/g on day 75, to 3.5 log CFU/g on day 90. In the core, molds were only detected on days 75 and 90 (5.1 and 5.2 log CFU/g, respectively). The statistical analysis showed a significant difference between the three types of samples (main matrix, surface, and core) and the three sampling times in which the presence of molds in the non-inoculated cheese was detected. In contrast, in the inoculated cheese, there was only a significant difference on days 75 and 90 between the core, the main matrix, and the surface ($p < 0.05$).

2.3. Physicochemical Parameters

Changes were observed during the maturation time in titratable acidity, pH, water activity (Wa), and sodium chloride percentage (%) within the cheese's physicochemical parameters analyzed in different sections. The changes of each variable had a different dynamic and gave rise to a distinguishing composition between each cheese section: (A) Core (10 cm in diameter); (B) Main matrix (Section between the core and the surface), and (C) Surface (Obtained 2-cm deep from the surface) (Figure 1).

The surface had a constant increase of 0.22% of lactic acid, increasing to 1.1% at the end of the study (90 days); pH dropped from 5.9 after seven days of maturation to 5.5 at day 90; Wa decreased from 0.96 to 0.86, and sodium chloride increased from 3.2 to 4.06 due to moisture loss on the surface. In the cheese main matrix, the initial lactic acid was 0.4% and gradually increased to 0.9% at day 90, while the pH dropped slightly from 5.5 to 5.01; the Wa decreased from 0.91 to 0.89 and sodium chloride increased from 3 to 3.4. The samples from the core of the cheese showed the most contrasting changes compared to the surface; the acidity content increased from 0.32% (day 7) to 1.52% on day 90; the pH decreased from 5.9 at day 7 of maturation to 5.0 at day 90; Wa also decreased from 0.92 on day 7 to 0.89, and the sodium chloride content increased from 2.9 to 3.5 (Table 4).

Table 4. Physicochemical parameters during the cheese maturation process.

	Surface				Main Matrix				Core			
	Acidity	Wa	pH	NaCl (%)	Acidity	Wa	pH	NaCl (%)	Acidity	Wa	pH	NaCl (%)
7	0.22 [a]	0.92 [a]	5.9 [a]	3.2 [a]	0.32 [A]	0.921 [A]	5.5 [A]	2.89 [A]	0.4 [a]	0.916 [a]	5.01 [a]	3 [a]
15	0.48 [a]	0.9 [a]	6.2 [a]	3.21 [a]	0.43 [A]	0.918 [A]	5.6 [A]	3.04 [A]	0.42 [a]	0.916 [a]	5.2 [a]	3 [a]
30	0.76 [a]	0.87 [a]	6.2 [a]	3.23 [a]	0.92 [B]	0.899 [A]	5.6 [A]	3.12 [A]	0.47 [a]	0.910 [a]	5.09 [a]	3 [a]
45	0.9 [b]	0.87 [a]	5.5 [b]	3.87 [a]	1.21 [B]	0.899 [A]	5.1 [B]	3.32 [A]	0.64 [a]	0.907 [a]	5 [a]	3.26 [a]
60	1 [b]	0.83 [b]	5.5 [b]	4 [a]	1.32 [B]	0.866 [A]	5.0 [B]	3.45 [A]	0.85 [b]	0.897 [a]	4.9 [a]	3.29 [a]
75	1.19 [b]	0.84 [b]	5.5 [b]	3.85 [a]	1.4 [B]	0.870 [A]	5.0 [B]	3.49 [A]	0.89 [b]	0.891 [a]	5.0 [a]	3.07 [a]
90	1.1 [b]	0.86 [b]	5.5 [b]	4.06 [a]	1.52 [B]	0.885 [A]	5.0 [B]	3.52 [A]	0.9 [b]	0.895 [a]	5.01 [a]	3.4 [a]

Wa: water activity Values with different superscripts ([a,b] or [A,B]) in the cheese section columns, are significantly different between rows ($p < 0.05$).

3. Discussion

Dairy production follows several stages, ranging from obtaining the milk on the farm to further processing in the dairy company or, in many cases, on the farm itself. During this process, the risk of contamination can enter at various points along the supply chain, compromising food safety [16]. During cheese making, microbial contamination can occur at any stage from cheese production to consumption, making cheese a significant vehicle for foodborne illness [4,17]. In the case of Cotija cheese, it is a Mexican artisan product made from raw cow milk without any heat treatment. The ripening process occurs spontaneously and is influenced by environmental conditions [12,18].

The fermentation of Cotija cheese is carried out by the native biota present in the milk; since starter cultures are not added, the initial microbiota is diverse and will influence its sensory characteristics and the safety of the final product. Because of the above, we decided to carry out this project in the region of Cotija. It was used the milk produced in the area. There was a space to work with inoculated milk, and the cheeses were stored for 90 days for their maturation, such as the traditional process of Cotija cheesemaking.

Therefore, the native biota as control (not inoculated) and inoculated cheeses had the same diversity of microorganisms. This allowed us to know the natural behavior that the microbiota has throughout the cheese maturation process and the microbial diversity in the milk obtained in the Cotija region, Michoacán, to the final product generated after 90 days of maturation.

The use of unpasteurized milk has represented a debate in the dairy industry due to the risk posed by the possibility that it is contaminated with pathogens, causing health problems for consumers. There are multiple reports of products made with raw milk in various outbreaks [19–21]. However, this does not mean that all products, particularly cheeses made with raw milk, pose a health risk to consumers. The presence of pathogens in cheese depends on several factors: the type of cheese, the microbiological quality of the milk used, and the storage conditions. Fresh cheeses made with unpasteurized milk represent a greater risk because nothing inhibits the pathogen that may be present in the raw milk, favoring the development and/or survival of pathogens [2,4,20]. The difference with a fermented cheese is that the same fermentation process causes the production of compounds and changes in some parameters that can inhibit the development of the pathogens, like pH, lactic acid and the bacteriocins produced by some of the lactic acid bacteria that participate in the fermentation and cheese ripening [18]. It is important to mention that this study was done in the Cotija region using milk sourced from the dairy herd, which is regularly used for cheesemaking under the same environmental conditions as Cotija ripening.

In the present work, the inoculated pathogens: *S. aureus*, *S. Typhimurium* and *L. monocytogenes* were no longer detected in the different sections of the cheese after day 45 of ripening. However, *S. aureus*, was detected until the end of the maturation process (90 days) in surface samples, probably because of their halotolerant characteristics [22]. Cheese native *S. aureus* from uninoculated cheese was not detected at 45 days. Meanwhile, a different behavior was observed in *S. aureus* (ATCC 13565) from inoculated cheese which was still being detected until the end of the study (90 days). In both cases, they remained longer on the surface of the cheese. It may be due to the increased sensitivity of the native strain to the metabolites of cheese microbiota or the lower halotolerance in comparison to ATCC strain. *S. aureus* was recovered in the surface in concentrations significantly higher than in the core and the main matrix ($p < 0.05$). The behavior of the distribution of *S. aureus* in the inoculated and non-inoculated cheese did not show differences ($p > 0.05$); both were decreasing as time passed. On the surface of both cheeses, inoculated and non-inoculated, always remained in higher concentration (2 log MPN/g). Behavior of native *L. monocytogenes* of no inoculated cheese was like inoculated strain; they were not detected to 30 and 40 days, respectively. On the other hand, the distribution of *L. monocytogenes* in the inoculated cheese showed significant differences ($p = 0.0045$). The surface showed 3 log more with respect to the core and the main matrix. The distribution of *L. monocytogenes* in the inoculated and non-inoculated cheese did not show significant differences ($p > 0.05$). In both cheeses, inoculated and non-inoculated, the surface had the highest concentration of this pathogen. While the distribution of *S.* Typhimurium in the inoculated cheese, showed significant differences ($p = 0.0005$). It was observed that it was similar in the main matrix and the core, meanwhile, the surface showed 1.5 log more than the main matrix.

The inhibition of pathogens in ripened cheeses made with raw milk, like Cotija, has also been reported by other authors [23] who have shown the gradual demise of pathogens such as *L. monocytogenes* [24], and the inhibition of other pathogens in fermented cheeses [25,26]. The absence of inoculated pathogens is associated with the metabolic processes that occurred during cheese maturation and the inhibitory compounds that the other microorganisms have, such as lactic acid bacteria [18,26,27]. García-Cano et al. [25], reported production of bacteriocins from *Enterococcus faecium* and *E. faecalis* recovered from Cotija cheese. This author showed inhibition of *S. aureus*, *Yersinia enterocolitica*, *S.* Typhimurium and *Pseudomonas aeruginosa*. Escobar-Zepeda et al. [18] did not detect pathogenic bacteria like *Salmonella*, *Listeria monocytogenes*, *Brucella*, or *Mycobacterium* in a metagenomic analysis performed on Cotija cheese. They reported three predominant phyla (*Lactobacillus*, *Leuconostoc* and *Weissella*) associated with the production of various metabolites with antimicrobial activity. Within the complex microbiota in Cotija cheese, the presence of lactic acid bacteria can inhibit the inoculated pathogens, and the yeast present could have participated in this process [28–30]. Chombo et al. [12] observed that surface of Cotija cheese showed higher counts of yeasts, which is like our study. These microorganisms could contribute to the inactivation of pathogens observed in cheese's surface, in conjunction with physicochemical factors. Also, yeasts can be present in different cheeses [28,31].

As previously described, the metabolic activity of the microbiota present during the fermentation process generates many compounds ranging from peptides, like bacteriocins, to peroxide and organic acids, among other metabolites that determine the predominance of some microbial groups such as lactic acid bacteria, responsible for the synthesis of these compounds, as well as the typical organoleptic characteristics of Cotija cheese [7,18,32]. Even though only the production of lactic acid was quantified in the present investigation, it should be noted that the absence of pathogens coincided with the increase of lactic acid and the consequent drop in pH. Water activity and its decrease also influenced, although to a lesser degree, to the inactivation of the inoculated pathogens. Lactic acid concentration and pH had the highest impact on the decrease of *S.* Typhimurium and *S. aureus*. Both pathogens showed a gradual decrease associated with increasing acidity and decreasing pH. On the other hand, the inactivation of *L. monocytogenes* showed no relationship with the parameters determined since it occurred drastically.

The present study shows the importance of interaction between microorganisms that constitute the complex microbiota ecosystem of cheese made with no pasteurized milk. The same microbiota present in cheese is used for its production. In Cotija cheese, the robustness of its native ecology is decisive to ensure its quality. Quality control considers the maturity stage, which ensures the absence of pathogens and takes advantage of the characteristics *per se* of Cotija cheese, including its functional properties provided by its characteristic microbiota [12]. Likewise, the quality of raw material in Cotija cheese is important because it is the one that will determine the type of microorganisms that will carry out the fermentation process and give the desired sensory characteristics. However, the raw material and milk must be free of pathogens like *Brucella* and *Mycobacterium* [18].

4. Materials and Methods

4.1. Bacterial Strains

Strains used in inoculated cheese were typified: *Staphylococcus aureus* ATCC 13565, *Salmonella* Typhimurium ATCC 14028 and *Listeria monocytogenes* ATCC BAA-751. In uninoculated cheese, native bacteria were isolated from fresh cheese and biochemically characterized.

4.2. Inoculum Preparation

All strains were separately reactivated in trypticase soy broth with yeast extract (CS-TEL) twice a row. Subsequently, each bacterium was inoculated in Roux bottles containing 200 mL of trypticase soy agar with yeast extract and 20 mL of CSTEL, then incubated at 35 °C for 18 h. The biomass obtained was separated from the agar using sterile glass beads and extracted with a pipette. Physiological saline solution was added to the biomass obtained to a final volume of 40 mL. From this suspension, the count was carried out in selective agar for each microorganism (Bright Green Agar, Oxford Agar and Baird Parker agar, respectively) to know the bacterial concentration. Additionally, in the culture of *S. aureus*, the presence of staphylococcal toxin was verified by visual immunoassay (Tecra 3M).

4.3. Cheesemaking

The fresh raw milk (180 L) was filtered through a clean gauze cloth to remove foreign matter, and once it reached a temperature between 30 to 33 °C, rennet was added, leaving it to rest until it formed a firm curd, which was cut using lira to obtain approximately 1 cm^3 cube.

After a few minutes of standing, the whey was drained, and the curd was transferred to a stainless-steel table and mixed with 5–10% NaCl until forming a homogeneous mass. The molding of the dough was done in previously washed and disinfected cylindrical hoop-shaped stainless-steel molds, provided with an ixtle or cotton blanket. The mass was pressed for 24 h, and the main matrix was removed from the press. The mold and the blanket with which the main matrix was wrapped allow it to give the traditional cylindrical shape of the cheese and print the characteristic mark on the rind. The blanket was removed, and the hoop was left for the next 15 days. Every 18–24 h, the cheese was removed from the mold, turned over, and the surface cleaned with a clean cotton cloth. This process was carried out daily for the first eight days and then alternates every third day until 15 days were completed; the goal was to have complete drainage and a sturdy piece that could be handled without the shaping ring. Finally, the cheese was stored to begin its ripening process. Ripening was carried out for three months within the same geographical area of Cotija, Michoacán, Mexico, to provide the environmental conditions (Temperature < 28 °C, Relative humidity: 60 to 95% and Altitude: 700 to 1700 m above the level of the sea) that give the cheese its specific organoleptic characteristics (Figure 3).

Figure 3. Cotija region in the State of Michoacan, Mexico.

4.4. Microbiological Parameters

During Cotija cheese production, samples were taken in each process: milk, curd, whey, and cheese during the aging process (NMX-F-735-COFOCALEC-2011). For the cheese analysis during aging, samples were taken directly in the ripening chambers, and seven sampling times were considered (1, 7, 15, 30, 45, 60, 75 and 90 days).

4.5. Milk and Cheese Sampling

4.5.1. Milk

The uninoculated milk was analyzed for *S. aureus*, staphylococcal toxin, lactic acid bacteria, aerobic mesophilic bacteria, coliforms, molds, yeasts, pH, and titratable acidity. In the case of the previously inoculated milk, each pathogen was quantified, and the presence of staphylococcal toxin was confirmed.

4.5.2. Cheese

For the first sampling (24 h of maturation), only one sample was taken that ranged from the core to the cheese surface. From the 7th ripening sampling: each cheese was divided into three uniform vertical sections to take a sample from each section: (A) core (10 cm in diameter); (B) main matrix (Section between the core and the surface), and (C) surface (obtained 2-cm deep from the surface) (Figure 1). The samples were placed in separate sterile bags (A, B and C, respectively) where they were homogenized manually.

4.6. Evaluation of Microbiological Parameters

4.6.1. Pathogen Count in Cheese

Each pathogen was counted on days 1, 7, 15, 30, 45, 60, 75 and 90, employing traditional counting per culture. In addition, at time 7, lactic bacteria, aerobic mesophilic, coliforms, molds, yeasts, pH, and titratable acidity were quantified in both cheeses, inoculated and non-inoculated. In all cases, the samples were analyzed in duplicate.

The most probable number technique (MPN) was used to enumerate the pathogens in the cheese. It was weighed 25 g of each sample, then 225 mL of CSTEL were added, homogenized for 1 min in a Stomacher (Lab Blender 400), and subsequently seven dilutions were made. Tubes were inoculated with 9 mL of CSTEL and incubated at 35 °C for 24 h from each dilution. Baird Parker agar (ABP) plates for *S. aureus* were inoculated from the tubes that showed bacterial growth; Bright green agar (AVB) for *Salmonella* and Oxford agar (OXA) for *Listeria monocytogenes* to confirm the development of each pathogen.

The growth of each pathogen on selective agar was first confirmed by Gram staining and the biochemical tests corresponding to each microorganism through the API system. Additionally, specific tests were carried out for each bacterium. For *Salmonella* strains, serological tests were performed with antiserum O Anti-serum Poly A–I & Vi (BD Difco TM), and the hemolysis test and CAMP (Cristie-Atkins-Munch-Peterson Test) for *L. monocytogenes*.

4.6.2. Staphylococcal Toxin

The presence of staphylococcal toxin was investigated by the visual immunoassay method (Tecra 3M). For which 10 mL of milk or 10 g of cheese were taken and homogenized with 20 mL of sterile water for 3 min, a 3 mL aliquot was taken. The pH was adjusted to 4; next, the aliquot was centrifuged at $1000\times g$, and then, the liquid was passed through a 45 µm filter, and the pH was adjusted between 6–7. Subsequently, 1 mL was placed in an Eppendorf tube, and 50 µL of the additive was added and shaken manually 200 µL of this mixture were plated in 96-well plates, incubated at 35 °C for 2 h. The liquid was discarded, and four washes were performed with the washing solution. Plates were allowed to dry, and 200 µL of the conjugate was added to each well and incubated for 1 h at room temperature (20–25 °C). The liquid was poured out, and they were carefully washed. Finally, 200 µL of substrate were added to each well and incubated at room temperature (20–25 °C) for 30 min and then the samples were read. The support was placed on a white background, and the reading was done visually by comparing it with the color card.

4.6.3. Lactic Bacteria Count

To 25 g of sample, 225 mL Man-Rogosa-Sharpe (MRS) broth were added twice the concentration with aerobic lactic acid bacteria broth (3M) supplement and homogenized for 1 min in a Stomacher (Lab Blender 400). Subsequently, seven decimal dilutions were made. The Petri film plates (3M) for anaerobic bacteria were inoculated with 1 mL of each dilution and incubated at 37 °C for 48–72 h in anaerobic jars provided with a gas pack.

4.6.4. Indicator Groups

The determination of indicator groups (aerobic mesophilic bacteria, coliforms, molds, and yeasts) was made from 25 g of each sample and homogenized with 225 mL of peptone diluent for 2 min in a Stomacher (Lab Blender 400). From here, seven decimal dilutions were made. From each dilution, Petri-film plates (3M) were inoculated with Aerobic Count and *E. coli*/Coliform Count, both incubated at 35 °C for 24 h, and Yeast and Mold Count incubated at 20 °C for 5 to 7 days.

4.7. Physicochemical Parameters

In addition to the microbiological determinations, pH, titratable acidity, water activity and chlorides were quantified from non-inoculated cheese samples. The pH was measured in 10 g of sample, placing them in a 100 mL volumetric flask with distilled water; subsequently, it was filtered, and pH was determined with an Orion3 Star pH Benchtop meter (Thermo Fisher Scientific, Waltham, MA, USA. In acidity, this was determined by titration from 10 g of sample, using 0.1 N NaOH and phenolphthalein as indicators. The test was performed by titration in triplicate. Water activity determination was carried out using 3TE Series AquaLab equipment (Decagon Device Inc., Pullman, Washington, USA). To determine sodium chloride, the AOAC 983.14 "Chloride in Cheese" method was used, weighing 2 g of ground cheese, and titrating the samples with a standard solution of 0.0856 M $AgNO_3$.

4.8. Statistical Analysis

A one-factor analysis of variance (ANOVA) with an alpha level of $\alpha = 0.05$ was performed on the results of the indicator groups, behavior, and distribution of pathogen in cheese. LSD Fisher method (Least significant difference) was performed for homogeneous groups. Statistical analysis was performed with Statgraphics Centurion XIX version 19.2.01 software.

5. Conclusions

This work shows important results on the behavior of pathogens and the dynamics that follow during the ripening process directly in the place where Cotija cheese is traditionally made. In addition, it shows the reduction of the risk represented by the consumption of products made with unpasteurized milk by being subjected to a maturation process, without neglecting good hygiene practices from obtaining the milk to the end of the cheese fermentation process and its subsequent marketing.

Author Contributions: Conceptualization, A.V.-L. and M.R.T.-V.; methodology, M.d.l.Á.O.-R. and P.C.-M.; software, O.V.-P.; validation, O.V.-P., Z.V.-d.l.M. and J.C.-R.; formal analysis, O.V.-P.; investigation, Z.V.-d.l.M. and L.E.G.-M.; resources, J.C.-R.; data curation, A.V.-L.; writing—original draft preparation, M.d.l.Á.O.-R.; writing—review and editing, A.V.-L., K.N. and Z.V.-d.l.M.; visualization, P.C.-M.; supervision, A.V.-L. and M.R.T.-V.; project administration, M.R.T.-V.; funding acquisition, M.R.T.-V. All authors have read and agreed to the published version of the manuscript.

Funding: This research was funded by CONACYT-INIFAP-UDG, grant number SAGARPA-2010-147449.

Institutional Review Board Statement: Not applicable.

Informed Consent Statement: Not applicable.

Data Availability Statement: Not applicable.

Conflicts of Interest: The authors declare no conflict of interest.

References

1. Oliver, S.P.; Jayarao, B.M.; Almeida, R.A. Foodborne pathogens in milk and the dairy farm environment: Food safety and public health implications. *Foodborne Pathog. Dis.* **2005**, *2*, 115–129. [CrossRef]
2. Soria-Herrera, R.J.; Dominguez-Gonzalez, K.G.; Rumbo-Pino, R.; Piña-Lazaro, A.; Alvarez-Perez, J.J.; Rivera-Gutierrez, S.; Ponce-Saavedra, J.; Ortiz-Alvarado, R.; Gonzalez-Y-Merchand, J.A.; Yahuaca-Juarez, B.; et al. Occurrence of Nontuberculous Mycobacteria, *Salmonella, Listeria monocytogenes,* and *Staphylococcus aureus* in Artisanal Unpasteurized Cheeses in the State of Michoacan, Mexico. *J. Food Prot.* **2021**, *84*, 760–766. [CrossRef] [PubMed]
3. Cervantes-Escoto, F.; Villegas de Gante, A.; Cesin-Vargas, J.A.; Espinoza-Ortega, A. *Los Quesos Mexicanos Genuinos. Patrimonio Cultural Que Debe Rescatarse,* 1st ed.; Mundi Prensa Mexico: Mexico City, Mexico, 2008.
4. Torres-Vitela, M.R.; Mendoza-Bernardo, M.; Castro-Rosas, J.; Gomez-Aldapa, C.A.; Garay-Martinez, L.E.; Navarro-Hidalgo, V.; Villarruel-López, A. Incidence of *Salmonella, Listeria monocytogenes, Escherichia coli* o157:h7, and staphylococcal enterotoxin in two types of mexican fresh cheeses. *J. Food Prot.* **2012**, *75*, 79–84. [CrossRef]
5. González-Córdova, A.F.; Yescas, C.; Ortiz-Estrada, Á.M.; de los Ángeles De la Rosa-Alcaraz, M.; Hernández-Mendoza, A.; Vallejo-Cordoba, B. Invited review: Artisanal Mexican cheeses. *J. Dairy Sci.* **2016**, *99*, 3250–3262. [CrossRef] [PubMed]
6. Montel, M.C.; Buchin, S.; Mallet, A.; Delbes-Paus, C.; Vuitton, D.A.; Desmasures, N.; Berthier, F. Traditional cheeses: Rich and diverse microbiota with associated benefits. *Int. J. Food Microbiol.* **2014**, *177*, 136–154. [CrossRef] [PubMed]
7. Beresford, T.P.; Fitzsimons, N.A.; Brennan, N.L.; Cogan, T.M. Recent advances in cheese microbiology. *Int. Dairy J.* **2001**, *11*, 259–274. [CrossRef]
8. Callon, C.; Berdagué, J.L.; Dufour, E.; Montel, M.C. The effect of raw milk microbial flora on the sensory characteristics of Salers-type cheeses. *J. Dairy Sci.* **2005**, *88*, 3840–3850. [CrossRef]
9. Ndoye, B.; Rasolofo, E.A.; LaPointe, G.; Roy, D. A review of the molecular approaches to investigate the diversity and activity of cheese microbiota. *Dairy Sci. Technol.* **2011**, *91*, 495–524. [CrossRef]

10. Callon, C.; Saubusse, M.; Didienne, R.; Buchin, S.; Montel, M.C. Simplification of a complex microbial antilisterial consortium to evaluate the contribution of its flora in uncooked pressed cheese. *Int. J. Food Microbiol.* **2011**, *145*, 379–389. [CrossRef]

11. Mallet, A.; Guéguen, M.; Kauffmann, F.; Chesneau, C.; Sesboué, A.; Desmasures, N. Quantitative and qualitative microbial analysis of raw milk reveals substantial diversity influenced by herd management practices. *Int. Dairy J.* **2012**, *27*, 13–21. [CrossRef]

12. Chombo-Morales, P.; Kirchmayr, M.; Gschaedler, A.; Lugo-Cervantes, E.; Villanueva-Rodríguez, S. Effects of controlling ripening conditions on the dynamics of the native microbial population of Mexican artisanal Cotija cheese assessed by PCR-DGGE. *LWT-Food Sci. Technol.* **2016**, *65*, 1153–1161. [CrossRef]

13. Flores-Magallón, R.; Oliva-Hernández, A.A.; Narváez-Zapata, A.A. Characterization of microbial traits involved with the elaboration of the Cotija cheese. *Food Sci. Biotechnol.* **2011**, *20*, 997–1003. [CrossRef]

14. CDC. Centers for Disease Control and Prevention Outbreak of multidrug-resistant *Salmonella enterica* serotype Newport infections associated with consumption of unpasteurized Mexican-style aged cheese—Illinois, March 2006–April 2007. *MMWR Morb. Mortal. Wkly Rep.* **2008**, *57*, 432–435.

15. Toribio-Jiménez, J.; Moral, B.M.-D.; Echeverria, S.E.; Pineda, C.O.; Rodríguez-Barrera, M.Á.; Velázquez, M.E.; Román, A.R.; Castellanos, M.; Romero-Ramírez, Y. Biotype, antibiotype, genotype and toxin gene tsst-1 in *Staphylococcus aureus* isolated from Cotija cheese in the state of Guerrero, México. *Afr. J. Microbiol. Res.* **2014**, *8*, 2893–2897. [CrossRef]

16. van Asselt, E.D.; van der Fels-Klerx, H.J.; Marvin, H.J.P.; van Bokhorst-van de Veen, H.; Groot, M.N. Overview of Food Safety Hazards in the European Dairy Supply Chain. *Compr. Rev. Food Sci. Food Saf.* **2017**, *16*, 59–75. [CrossRef] [PubMed]

17. Guzman-Hernandez, R.; Contreras-Rodriguez, A.; Hernandez-Velez, R.; Perez-Martinez, I.; Lopez-Merino, A.; Zaidi, M.B.; Estrada-Garcia, T. Mexican unpasteurized fresh cheeses are contaminated with *Salmonella* spp., non-O157 Shiga toxin producing *Escherichia coli* and potential uropathogenic *E. coli* strains: A public health risk. *Int. J. Food Microbiol.* **2016**, *237*, 10–16. [CrossRef]

18. Escobar-Zepeda, A.; Sanchez-Flores, A.; Quirasco Baruch, M. Metagenomic analysis of a Mexican ripened cheese reveals a unique complex microbiota. *Food Microbiol.* **2016**, *57*, 116–127. [CrossRef] [PubMed]

19. Costard, S.; Espejo, L.; Groenendaal, H.; Zagmutt, F.J. Outbreak-Related Disease Burden Associated with Consumption of Unpasteurized Cow's Milk and Cheese, United States, 2009–2014-Volume 23, Number 6—June 2017-Emerging Infectious Disease journal-CDC. *Emerg. Infect. Dis.* **2017**, *23*, 957–964. [CrossRef] [PubMed]

20. Gould, L.H.; Mungai, E.; Barton-Behravesh, C. Outbreaks Attributed to Cheese: Differences between Outbreaks Caused by Unpasteurized and Pasteurized Dairy Products, United States, 1998–2011. *Foodborne Pathog. Dis.* **2014**, *11*, 545–551. [CrossRef]

21. Robinson, E.; Travanut, M.; Fabre, L.; Larréché, S.; Ramelli, L.; Pascal, L.; Guinard, A.; Vincent, N.; Calba, C.; Meurice, L.; et al. Outbreak of *Salmonella* newport associated with internationally distributed raw goats' milk cheese, france, 2018. *Epidemiol. Infect.* **2020**, *148*, E180. [CrossRef] [PubMed]

22. Ming, T.; Geng, L.; Feng, Y.; Lu, C.; Zhou, J.; Li, Y.; Zhang, D.; He, S.; Li, Y.; Cheong, L.; et al. ITRAQ-Based Quantitative Proteomic Profiling of *Staphylococcus aureus* under Different Osmotic Stress Conditions. *Front. Microbiol.* **2019**, *10*, 1–16. [CrossRef] [PubMed]

23. Gao, Z.; Daliri, E.B.M.; Wang, J.U.N.; Liu, D.; Chen, S.; Ye, X.; Ding, T. Inhibitory effect of lactic acid bacteria on foodborne pathogens: A review. *J. Food Prot.* **2019**, *82*, 441–453. [CrossRef]

24. Genigeorgis, C.; Carniciu, M.; Dutulescu, D.; Farver, T.B. Growth and Survival of *Listeria monocytogenes* in Market Cheeses Stored at 4 to 30 °C. *J. Food Prot.* **1991**, *54*, 662–668. [CrossRef] [PubMed]

25. García-Cano, I.; Serrano-Maldonado, C.E.; Olvera-García, M.; Delgado-Arciniega, E.; Peña-Montes, C.; Mendoza-Hernández, G.; Quirasco, M. Antibacterial activity produced by *Enterococcus* spp. isolated from an artisanal Mexican dairy product, Cotija cheese. *LWT-Food Sci. Technol.* **2014**, *59*, 26–34. [CrossRef]

26. Muhammad, Z.; Ramzan, R.; Abdelazez, A.; Amjad, A.; Afzaal, M.; Zhang, S.; Pan, S. Assessment of the antimicrobial potentiality and functionality of *Lactobacillus plantarum* strains isolated from the conventional inner mongolian fermented cheese against foodborne pathogens. *Pathogens* **2019**, *8*, 71. [CrossRef] [PubMed]

27. Rizzello, C.G.; Losito, I.; Gobbetti, M.; Carbonara, T.; De Bari, M.D.; Zambonin, P.G. Antibacterial activities of peptides from the water-soluble extracts of Italian cheese varieties. *J. Dairy Sci.* **2005**, *88*, 2348–2360. [CrossRef]

28. Fröhlich-Wyder, M.T.; Arias-Roth, E.; Jakob, E. Cheese yeasts. *Yeast* **2019**, *36*, 129–141. [CrossRef] [PubMed]

29. Roostita, T.; Suryaningsih, L.; Lengkey, H.A.W.; Pratama, A.; Utama, G.L. Isolation and identification of yeast in traditional cottage cheese with strawberry as coagulant. *Sci. Pap. Ser. D Anim. Sci.* **2017**, *LX*, 300–302.

30. Younis, G.; Awad, A.; Dawod, R.E.; Yousef, N.E. Antimicrobial activity of yeasts against some pathogenic bacteria. *Vet. World* **2017**, *10*, 979–983. [CrossRef]

31. Cardozo, M.C.; Fusco, Á.J.V.; Carrasco, M.S. Yeast microbiota in artisanal cheeses from Corrientes, Argentina. *Rev. Argent. Microbiol.* **2018**, *50*, 165–172. [CrossRef]

32. Yoon, Y.; Lee, S.; Choi, K.H. Microbial benefits and risks of raw milk cheese. *Food Control* **2016**, *63*, 201–215. [CrossRef]

Article

Lipid Remodeling in the Mitochondria upon Ageing during the Long-Lasting Cultivation of *Endomyces magnusii*

Elena P. Isakova [1,*], Natalya N. Gessler [1], Daria I. Dergacheva [1], Vera M. Tereshina [2], Yulia I. Deryabina [1] and Marek Kieliszek [3]

[1] Research Center of Biotechnology of the Russian Academy of Sciences, A.N. Bach Institute of Biochemistry, bld 33-2, Leninsky Prospect, 119071 Moscow, Russia; gessler51@mail.ru (N.N.G.); ddarya1993@gmail.com (D.I.D.); yul_der@mail.ru (Y.I.D.)

[2] Research Center of Biotechnology of the Russian Academy of Sciences, Winogradsky Institute of Microbiology, bld 33-2, Leninsky Prospect, 119071 Moscow, Russia; v.m.tereshina@inbox.ru

[3] Department of Food Biotechnology and Microbiology, Institute of Food Sciences, Warsaw University of Life Sciences—SGGW, Nowoursynowska 159c, 02-776 Warsaw, Poland; marek_kieliszek@sggw.edu.pl

* Correspondence: elen_iss@mail.ru; Tel.: +7-(495)-954-4008

Citation: Isakova, E.P.; Gessler, N.N.; Dergacheva, D.I.; Tereshina, V.M.; Deryabina, Y.I.; Kieliszek, M. Lipid Remodeling in the Mitochondria upon Ageing during the Long-Lasting Cultivation of *Endomyces magnusii*. *Appl. Sci.* **2021**, *11*, 4069. https://doi.org/10.3390/app11094069

Academic Editors: Francesca Silvagno and Burkhard Poeggeler

Received: 25 February 2021
Accepted: 28 April 2021
Published: 29 April 2021

Publisher's Note: MDPI stays neutral with regard to jurisdictional claims in published maps and institutional affiliations.

Abstract: In this study, we used *Endomyces magnusii* yeast with a complete respiratory chain and well-developed mitochondria system. This system is similar to the animal one which makes the yeast species an excellent model for studying ageing mechanisms. Mitochondria membranes play a vital role in the metabolic processes in a yeast cell. Mitochondria participate in the metabolism of several pivotal compounds including fatty acids (FAs) metabolism. The mitochondria respiratory activity, the membrane and storage lipids composition, and morphological changes in the culture during the long-lasting cultivation (for 168 h) were under investigation. High metabolic activity of *E. magnusii* might be related to the active function of mitochondria increasing in the 96- and 168-h growth phases. Cardiolipin (CL), phosphatidylethanolamine (PE), phosphatidylcholine (PC), and sterols (St) were dominant in the membrane lipids. The St and sphingolipids (SL) shares increased by a lot, whereas the CL and phosphatidylinositol (PI) + PE ones decreased in the membrane lipids. This was the main change in the membrane lipid composition during the cultivation. In contrast, the amount of PE and phosphatidylserine (PS) did not change. Index of Hydrogen Deficiency (IHD) of phospholipids (PL) FAs significantly declined due to a decrease in the linoleic acid share and an increase in the amount of palmitic and oleic acid. There were some storage lipids in the mitochondria where free fatty acids (FFAs) (73–99% of the total) dominated, reaching the highest level in the 96-h phase. Thus, we can conclude that upon long-lasting cultivation, for the yeast assimilating an "oxidative" substrate, the following factors are of great importance in keeping longevity: (1) a decrease in the IHD reduces double bonds and the peroxidation indices of various lipid classes; (2) the amount of long-chain FFAs declines. Moreover, the factor list providing a long lifespan should include some other physiological features in the yeast cell. The alternative oxidase activity induced in the early stationary growth phase and high mitochondria activity maintains intensive oxygen consumption. It determines the ATP production and physiological doses of reactive oxygen species (ROS), which could be regarded as a trend favoring the increased longevity.

Keywords: *Endomyces magnusii*; mitochondria; yeast; lipids; fatty acids; unsaturation degree

1. Introduction

Mitochondria participate in numerous cellular processes being critical for cell survival and death. Besides their best-known function of ATP generation, the mitochondria are entirely involved in cellular metabolism, partly maintaining calcium homeostasis and regulating cell adaption to various stresses. They are implicated in ROS signaling modulation, maintenance of oxidative homeostasis, and apoptosis regulation under stress conditions [1,2]. Indeed, mitochondria are involved in free radicals generation, with the respiratory chain as the leading ROS producer. Excessive ROS production is supposed to result from mitochondria dysfunction [3]. Mitochondria are indispensable organelles of any eukaryotic cell, determining the organism's fitness and physiology [4]. The damage of mitochondrial function is a universal sign of ageing in a eukaryotic cell [4,5]. Some mechanisms causing damage to mitochondria bioenergetics include the accumulation of mutations and deletions in mtDNA, oxidation of mitochondria proteins, destabilization of macromolecular super-complexes in the respiratory chain, some changes in the lipid composition of the membranes, and modulation in mitochondria dynamics due to imbalance in organelle fusion, fission, and mitophagy [5]. The major ROS sources are mitochondrial respiratory chain, non-enzymatic reactions involving oxygen, phagocytosis, prostaglandin synthesis, the P450 cytochrome system, and ionizing radiation (in mammals). Thus, an increase in ROS generation due to an increased metabolic rate is considered the lifespan's principal determiner.

Not only the abundant ROS generated in pathological mitochondria but also ROS produced in the normal ones can lead to oxidative damage of the cells causing the ageing process. The hypothesis known as the "free radical theory" of ageing (FRTA) [6] declares the major statements of ageing in any aerobes, which later result in the mitochondrial theory of ageing. The theory supposes that mitochondria are the primary ROS sources in a cell and consequently participate in ageing and lifespan regulation [7]. Nevertheless, some new data have led to a more thorough overview of mitochondria ROS function, which is now considered to be signaling molecules in numerous pivotal biological processes, including ageing and lifespan [8].

The molecular mechanisms of the response to oxidative stress and the role of ROS in ageing have been widely and keenly studied for a long time. Hence, using simple eukaryotic models, namely yeast organisms, can be helpful. The *Saccharomyces cerevisiae* yeast is widely used in research on ageing mechanisms, and it contributes to understanding basic cellular and molecular processes [9]. Budding yeast research largely contributed to comprehending the ageing process and age-related diseases [10]. Yeasts, being the simplest unicellular eukaryotes, have many features in common with mammalian and human cells that permit studying both chronological and replicative ageing [11]. Moreover, the response of the yeast cell to oxidative stress is similar to that of mammals, including the sites of ROS generation in the electron transport chain and the performance of the major antioxidant enzyme complex [12].

Until recently, it was unknown if the composition of the mitochondrial membrane lipids during ageing in yeast can result in mitochondrial dysfunction. However, recent studies showed that in the *S. cerevisiae* yeast, a metabolic pathway of ceramide and sphingolipid synthesis is an essential node of the signaling network to determine the replicative and chronological lifespan [13–15]. High content of triacylglycerols (TAGs) is another aspect of lipid metabolism defining the lifespan of chronologically senescent yeast cells [15,16]. The neutral lipids are synthesized in the endoplasmic reticulum (ER) and then accumulate into lipid bodies (Lb), which determines the longevity of senescent yeast cells regardless of any other signaling pathways [15]. Their lipid composition can also distinguish mitochondrial inner and outer membranes. The outer mitochondrial membrane is a smooth lipid-rich envelope with pore-forming proteins. In contrast, the highly folded inner mitochondrial membrane is protein-rich harboring mainly enzymes of the respiratory chain [15]. PI is present in the outer mitochondrial membrane at a considerable amount. By contrast, CL and PE are enriched in the inner mitochondrial membrane. The presence of high CL levels in the mitochondrial membranes suggests that this PL is essential for efficient oxidative phosphorylation. The unsaturation degree of membrane lipids was found to influence respiratory properties and cytochrome content of mitochondria from *S. cerevisiae*. It was shown that a decrease in the amount of unsaturated FAs also declined the activity of several mitochondrial enzymes.

Vladimir Titorenko's team has methodically studied the relationship between the mitochondria lipidome and the life span of the yeast. The authors declared that ageing changes of the mitochondria lipid composition in the yeast include. (1) The levels of PE, CL and monolysocardiolipin (MLCL) decrease in the mitochondrial membranes. (2) the levels of phosphatidic acid (PA), PS, PC and phosphatidylglycerol (PG) increase in the mitochondrial membranes [17,18]. The applied lithoholic bile acid was reported to penetrate into the mitochondria and accumulate in the inner mitochondria membrane. This leads to changes in the phospholipid composition of the mitochondria membranes [17–19]. The remodeling of mitochondrial phospholipids intensifies while a yeast culture ages and finally causes substantial changes in mitochondria membrane lipidome.

The composition of the mitochondria's respiratory chain in traditional yeast species, namely *S. cerevisiae* is quite invariable. In particular, their mitochondria have no complex I [20]. By contrast, the *Endomyces magnusii* yeast being an obligate aerobe possesses a complete respiratory system similar in its mitochondrial system to the animal one. The *E. magnusii* cannot grow in the presence of the inhibitors of the mitochondria transcription (ethidium bromide) and mitochondria translation (erythromycin). Under these conditions, the growth should undergo only at the expense of substrate and glycolytic phosphorylation. In our experiments, the glycerol-containing medium-low concentrations of ethidium bromide (5–15 µg/mL) led to the complete inhibition of the growth. We assayed the number of generations (Figure S3B). We could see similar results in the experiments with erythromycin, which in low concentration (5 mg/mL) blocked the growth of the *E. magnusii* (Figure S3A).

In the yeast with fermentative metabolism types such as *S. cerevisiae*, the membrane and mitochondrial apparatus are ill-developed. There are only scanty large mitochondria with small cristae. Their mitochondria have much fewer cristae and trend to irregularity in their shape, structure, and packing. However, the yeast of strongly-pronounced aerobic metabolism, namely *E. magnusii,* possesses well-developed membrane apparatus with abundant complex mitochondria with numerous cristae [21]. According to the above-mentioned statement, we consider that the *E. magnusii* species is a unique model close to the animal in the mitochondrial system that makes it a convenient model for studying some mechanisms, including ageing.

Previously, we demonstrated high viability capacity in the *E. magnusii* yeast upon chronological ageing [22]. For the 20-day cultivation of the strain in glycerol-containing (1%) medium, there were no signs of growth culture inhibition. The analysis of cell survival during culture ageing showed that after 168 h of growth (7 days), it decreased by not more than 10–15% compared to the logarithmic growth stage [22]. The yeast population in the stationary growth stage consists of long-lived quiescent cells, which can enter the cycle of cell fission cycle, and non-quiescent cells, which are short-lived and incapable of dividing, similar to the phenotype of ageing mammalian cells [13,23].

Considering the hypothesis of yeast population heterogeneity upon its shift into the stationary growth phase, our data, and the key role of the mitochondria in the ageing regulation, we wanted to systematically study the alterations in the lipid amount and composition, assayed the mitochondria metabolic activity, and the changes in cell morphology of *E. magnusii* mitochondria during long-term cultivation.

2. Results

2.1. Long-Lasting Growth of the E. magnusii Yeast

Figure 1 shows the growth curve of *E. magnusii* yeast upon assimilating 1% glycerol. The dynamics of cell growth displayed classical logarithmic dependence with a well-pronounced lag phase (from 0 to 6 h of cultivation), a phase of logarithmic (exponential) growth (from 6 to 20 h of cultivation), a diauxic shift phase defining the transition of the population into the stationary growth phase (from 22 to 30 h of cultivation), early stationary growth phase (from 30 to 48 h of cultivation), and late (48 h) and stationary phase (from 48 to 200 h of cultivation) (Figure 1).

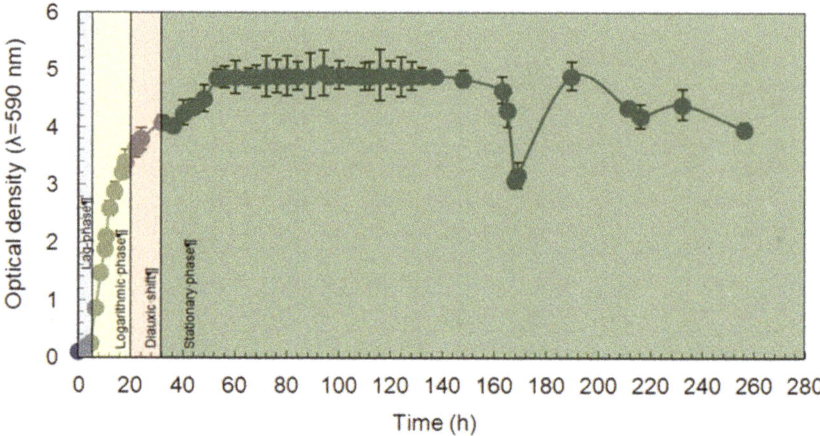

Figure 1. The growth curve of *E. magnusii* yeast in glycerol-containing medium grown for 260 h. Absorbance was assayed every two hours in cell suspension at the wavelength of 590 nm. The curve is representative of four biological experiments. (The figure was partially used in the article [Microorganisms 2020, 8, 91; doi:10.3390/microorganisms8010091]).

Microscopic analysis of the morphology of *E. magnusii* cells in various growth stages showed that it varies greatly. In the logarithmic growth stage, the cells are round and oval, sometimes fusiform or pear-shaped, either with well-visible buds (in the process of division) or budded daughter cells (Figure 2A), whereas in the stationary stage (after 30 h of cultivation) the number of daughter cells decreased to 5% of the total number (Figure 2B). The population of the exponential phase had about 4% actively dividing cells with buds at the poles, about 10% cells with detached immature buds, and about 20% young daughter cells, which are smaller and regularly spherical (Figure 2A). In the stationary growth phase, the number of actively budding and young cells nearly halved, there was some heterogeneity in the cell population, some cells changed their shape, and some single pseudomycelial and mycelial forms appeared (Figure 2B). It should be noted that during long-lasting cultivation, the nuclei of *E. magnusii* cells changed their shape (increased by 1.5 times). In the stationary phase, the nuclei increased in their volume and became more round (Figure 2C,H). The morphometric features of this growth stage indicated the intensive growth and fission of *E. magnusii* cells (Table S1 in the Supplementary Section). At the same time, the population was highly heterogeneous without any signs of pathological changes or inclusions (Figure 2D). Moreover, the cells at this stage got a distinct thickened cell wall, large vacuoles, and small dense cytoplasmic inclusions, which are likely to be autophagic vacuoles or bodies (Figure 2H). It should be emphasised that the energy status of the cells in the logarithmic and stationary growth phases, confirmed by potentiometric staining, was maintained at a high level (Figure 2E,F).

Summarising all the obtained results including the changes in morphology, along with the functional heterogeneity of the *E. magnusii* culture, shown before [21], we studied the lipid amount and composition in the mitochondria at different growth stages upon ageing.

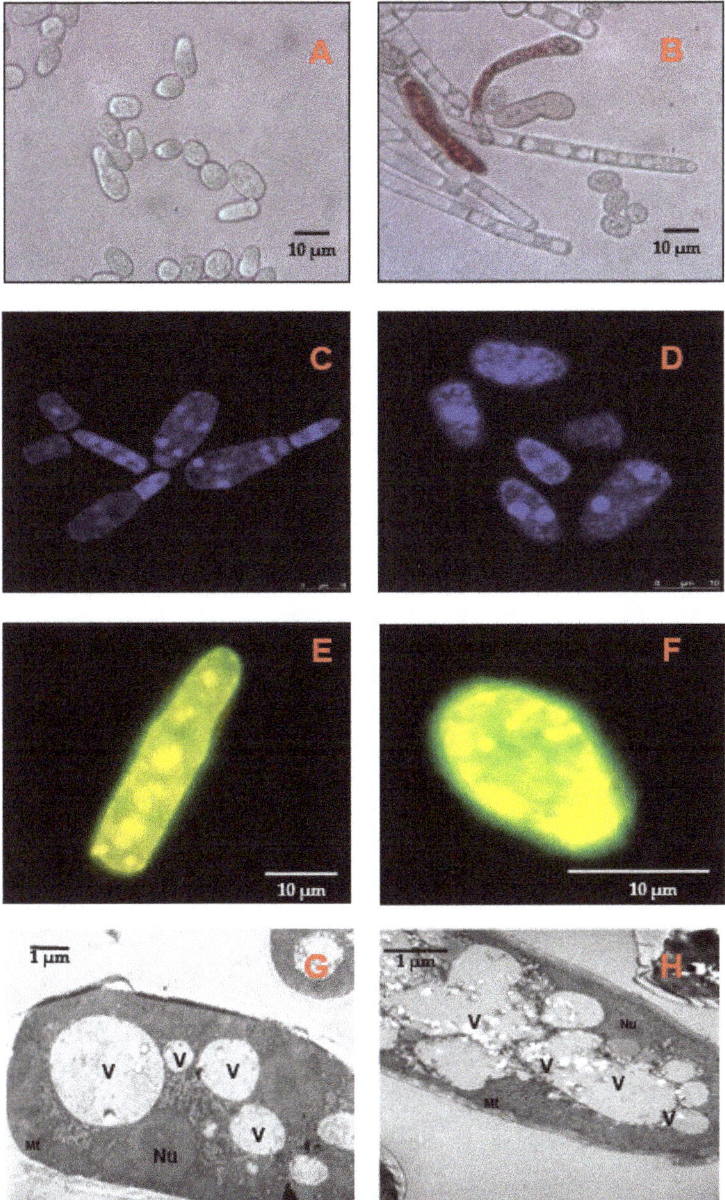

Figure 2. Micro images of the *E. magnusii* cells in the logarithmic (**A,C,E,G**) and late stationary (**B,D,F,H**) growth stages. (**A,B**)—The cells were stained with neutral red; (**C,D**). Fluorescent micro-images of the cells, labelled with 0.3 µM DAPI for DNA. (**E,F**) the potential-dependent stain of the mitochondria in *E. magnusii* cells by 5 mM JC-1. The regions of high mitochondrial polarization are bright yellow due to the concentrated dye. The cells were examined after 30 min. The incubation medium contained 0.01 M phosphate buffer saline (PBS) and 1% glycerol, pH 7.4. To examine the Rh123-stained preparations, filters 02, 15 (Zeiss) were used (magnification 100×). Photos were taken using an AxioCam MRc camera. (**G,H**)-ultrastructure of *E. magnusii* cells. Nu—nucleus; V—vacuole; Mt–mitochondria; CW–cell wall.

2.2. Respiratory Activity of E. magnusii Yeast Mitochondria

We designed a unique method for mitochondria isolation modified especially for the yeast. The test for fraction contamination included electron microscopy of the organelles and the polarographic assay of the mitochondria function: (1) the electron microscopy images of the *E. magnusii* yeast (Figure S1 in the Supplementary Section) showed that the mitochondria fraction was pure and free from any inclusions and other organelles. Moreover, the mitochondria in the photos had a regular structure with well-developed cristae and intact outer and inner membranes; (2) the mitochondria could oxidize the respiratory substrates (pyruvate+malate) (Figure S2A in the Supplementary Section) and generate a high membrane potential without any co-factors application (NAD(P)H, cytochrome c) (Figure S2B in the Supplementary Section); (3) the assay of the mitochondria activity showed no endogenous respiration (without any substrate application), indicating neither other membranes nor enzymes (Figure S2A (a) in the Supplementary Section); (4) the mitochondria demonstrated high respiratory rates and phosphorylation, ADP/O ratios close to the theoretically expected maxima (Table S1 in the Supplementary Section) upon oxidation of NAD-dependent substrates. It says about good mitochondria integrity after the isolation. We could detect the activity of adenilate kinase, which is in the inter-membrane mitochondria space and synthesizes ADP from AMP and ATP (Figure S2C in the Supplementary Section). The enzyme cannot work if the membrane is damaged. However, its activity was observed even in the hypotonic incubation medium. We are also sure that the mitochondria fraction was pure because we modified the method of isolation of the mitochondria, especially, to bind and remove the lipid fraction from the suspension. In the mitochondrial isolation and washing medium, we use 1% bovine serum albumin (FAs free fraction V), which binds FAs in the medium. Moreover, the high degree of mitochondria coupling indicates no lipid contamination, as any lipid compounds lead to powerful uncoupling of respiration and phosphorylation [24]. We get live mitochondria with respiratory controls close to the maximum that is impossible if the fraction is contaminated (Table 1, Table S2 in the Supplementary Section).

Table 1. Respiratory activity and the inhibitory analysis of the *E. magnusii* yeasts mitochondria in different growth phases upon ageing.

Time, h	Respiration Rate, ng-Atom Consumed O per 1 mg of Protein				
	$V4_1$ *	V3 **	$V4_2$ *	+KCN *** Inhibition (% of Control)	+KCN + SHAM **** Inhibition (% of Control)
17	19.6 ± 0.63 [f]	58.1 ± 0.08 [b]	16.95 ± 1.23 [f]	100	100
24	27.9 ± 1.68 [f]	164.8 ± 1.82 [c]	65.28 ± 4.67 [e]	87.26	100
48	70.19 ± 1.68 [e]	205.19 ± 1.82 [b]	70.28 ± 4.67 [e]	88.79	100
96	140.0 ± 14.14 [cd]	311.18 ± 18.59 [a]	139.26 ± 15.7 [cd]	72.68	100
168	60.85 ± 4.6 [e]	127.14 ± 18.13 [d]	72.68 ± 14.18 [e]	50.72	100

* $V4_1$, $V4_2$-the respiration rate in the second state (initial respiration) before and after the phosphorylation cycle. ** V3-the respiration rate in state 3 (the phosphorylation after the ADP addition). *** KCN was added in the concentration of 4 mM. **** SHAM was added in the concentration of 2 mM. The incubation medium for the experiments contained 0.6 M mannitol, 1 mM Tris-phosphate (pH 7.4); 1 mM EDTA, 20 mM pyruvate, 5 mM malate, and mitochondria corresponding to 0.4 mg protein/mL. [a–f] Means with the same letter did not differ significantly. Values are mean ± S.E.M from five independent experiments and three analytical replicates.

Changes in respiratory activity is the main indicator of the cellular energy status. The mitochondria of most plants, fungi, and yeasts possess an alternative pathway of electron transport that is induced if the main (cytochrome) pathway is inhibited by KCN, azide, or antimycin A [25,26]. The electron flux is switched off at the reduced ubiquinone site and is specifically inhibited by hydroxamic acid derivatives. The dynamics of the oxygen uptake rate by *E. magnusii* mitochondria showed that it varied significantly during the growth and ageing of the culture (Table 1).

The respiration rate in the state 4 (initial respiration) was minimal at the logarithmic growth stage, slightly increased at the 24-h stage, increased threefold in the 48-h phase, and reached its peak by 96 h of growth. The respiration rate of mitochondria in the third state (phosphorylation after the ADP addition) was also altered, similar to that in the second one. However, the coupling degree of the mitochondria insignificantly decreased at the 24-h growth stage compared to that in the log-stage.

The proportion of the alternative electron transfer pathway showed that the mitochondria respiration at the mid-logarithmic stage was extremely sensitive to the inhibitor KCN. However, in the 24- and 48-h phases, we observed the induction of AO upon inhibiting the main (cytochrome) pathway. The level of the KCN-resistant respiration was 12–13% being blocked by the specific AO inhibitor salicylhydroxamic acid (SHAM). At the 96-h and 168-h phases of cultivation the share of cyanide resistance increased up to about 30 and 50%, respectively.

According to the model proposed by *Bahr and Bonner* [27], the alternative pathway becomes active only if the main pathway of electron transport cannot work (in case of inhibition or electron saturation), i.e., the activity of the KCN-resistant pathway is regulated by the main cytochrome chain. It can be assayed by its participation in the total cell respiration. The AO level indicates the respiration inhibited by the KCN + SHAM system, and its contribution into the total cell respiration is defined as the respiration share inhibited by SHAM only without KCN [27]. The attempts to assess the proportion of the alternative pathway in vivo corroborated no hypothesis of constitutive induction of AO. SHAM did not induce the respiratory rate inhibition in any experiments significantly (Table 1). Based on the data, we could suppose that during the growth and ageing of *E. magnusii*, the alternative pathway induction triggers; however, it is not likely to be related to the decrease in the cytochrome pathway activity.

To comprehend the mechanism of high energy activity of yeast mitochondria for 20 days, we tried to determine the alterations in the main lipid components of the *E. magnusii* mitochondria during long-lasting cultivation.

2.3. Storage Lipids Profile of E. magnusii Yeast in the Different Growth Phases

Figure 3 shows the total content and composition of neutral lipids in *E. magnusii* mitochondria. The major lipid fractions were presented by FFAs and diacylglycerols (DAGs). The FFAs level was high in all the mitochondria samples and reached 73–99% of the total lipids (Figure 3A). The maximum DAGs amount was found in the logarithmic growth phase. Of note that a certain share of them (10–15% of the total lipids) was also revealed at the later growth stages. The minority comprised TAGs and sterol esters (ESt). TAGs were detected only in the early growth stages (17 h of growth) while ESt appeared in the late stationary stages and their amount nearly doubled by 168 h of cultivation compared to that in the 96-h phase.

The amount of total storage lipids remained relatively stable during the growth culture up to 96 h when it increased nearly twofold with a concurrent decrease by 168 h of cultivation (Figure 3B).

In the next step of the study, we assayed the membrane lipids profile in the *E. magnusii* mitochondria.

2.4. Membrane Lipids Profile of E. magnusii Mitochondria during Long-Lasting Cultivation

The amount and composition of the mitochondrial membrane lipids also changed significantly during growth and ageing of the culture. The membrane lipid amount reached its peak of 118.120 mg/g mitochondrial protein in the early growth stationary stage (48 h of cultivation) (Figure 4B), whereas the minimum content of 59.086 mg/g mitochondrial protein was in the deep stationary phase (96 h). The dynamics of the mitochondria membrane lipids showed a gradual increase by 48 h of cultivation with a nearly twofold drop in the deep stationary phase. The main fractions were sterols St (10–40%), PE (10–15%), PC (13–22%), CL (47%), and lysophosphatidylethanolamines + PI (up to 13%) (Figure 4A).

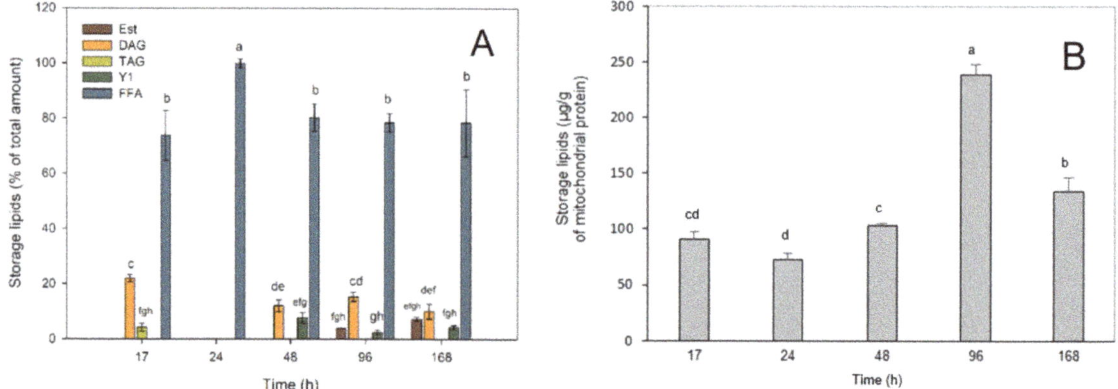

Figure 3. The storage lipid composition in *E. magnusii* yeast in the different growth phases during long-lasting cultivation. (**A**)—the share of each storage lipid fraction,%; TAGs—triacylglycerols; DAGs—diacylglycerols; FFAs—free fatty acids; ESt—sterol esters; Y1—unknown fraction (**B**)—the share of each storage lipid fraction (mg/g). [a–h] Means with the same letter did not differ significantly. Values are mean ± S.E.M from three independent experiments and three analytical replicates.

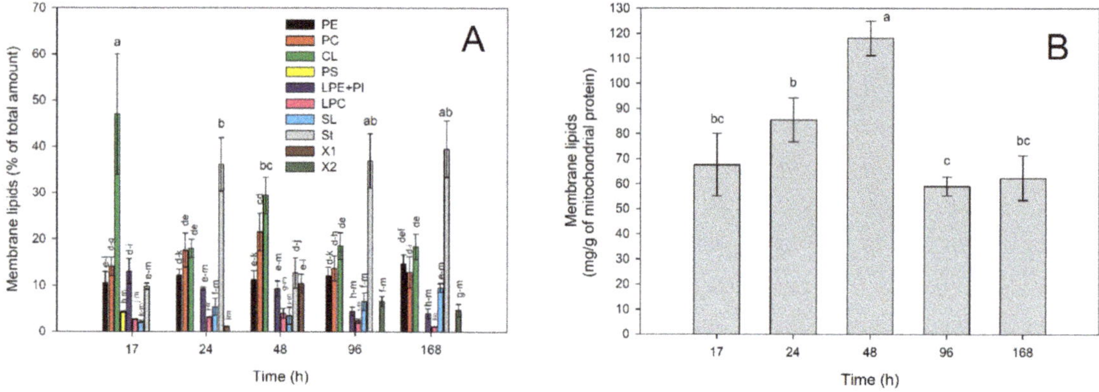

Figure 4. Membrane lipid composition of the mitochondria from *E. magnusii* cells raised in the different growth phases upon ageing during the seven-day cultivation. (**A**)—the share of each membrane lipid fraction; PE—phosphatidylethanolamines; PC—phosphatidylcholines; CL—cardiolipins; PS—phosphatidylserine; LPE + PI—lysophosphatidylethanolamine + phosphatidylinositols; LPC—lysophosphatidylcholines; SL—sphingolipids; St—sterols; X1, X2—unknown; (**B**)—the total membrane lipid content. [a–m] Means with the same letter did not differ significantly.

The PE fraction remained constant at the level of about 10% throughout the experiment with a concurrent increase up to 15% in the late stationary phase (168 h of cultivation). The PC fraction gradually increased up to 22% in the 48-h growth stage and nearly halved in the 168-h growth stage while the CL content reached its peak in the logarithmic growth stage (47%), decreased by more than two times in the 24-h stage (up to 18%) with a slight increase in the 48-h growth phase (up to 29.4%), and stabilized in the late stationary phase at 18%.

The LPE + PI fraction decreased during the cultivation from 13% to 4%. The St level changed in waves in different growth phases, reaching its maximum of 36–39% in the 24-h, 96-h, and 168-h growth phases (Figure 4A). The minor components of the lipid spectrum with the share of less than 10%, comprised lysophosphatidylcholines (LPC) and sphingolipids (SL). It is noteworthy that two unidentified lipid components were revealed: X1 was in the 24- and 48-h growth phases, and X2 was in 96- and 168-h growth stages.

The overall degree of acyl residue unsaturation, Index of Hydrogen Deficiency (IHD) in phospholipids determines the fluidity of the membrane lipid bilayer, which in turn may affect the adaptation and survival of the yeast under stress conditions, including ageing. Thus, we next assayed the FFAs composition and amount of the membrane lipids.

2.5. FAs of the Main PLs in the Different Growth Phases

To determine the degree of unsaturation, four main PLs fractions (PE, PC, CL, and PA), the share of which in the membrane lipids comprised more than 6%, were chromatographically isolated, and their FAs composition was analyzed. Figures 5 and 6 show the changes in the FAs composition and IHD of the mitochondrial membrane lipids in *E. magnusii*. The overall IHD in the membrane lipids was below 1.0 during the whole experiment. At the same time, the degree of unsaturation reached its maximum in the 48-h growth stage and its minimum at the 24-h growth stage (Figure 5B). Moreover, in the late stationary phase, the IHD was a bit higher than 0.6.

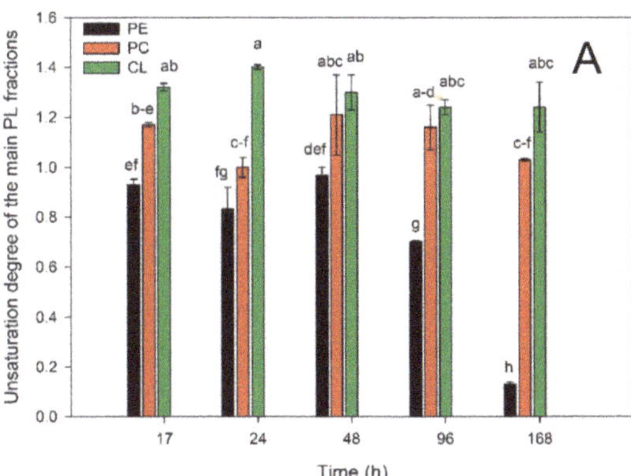

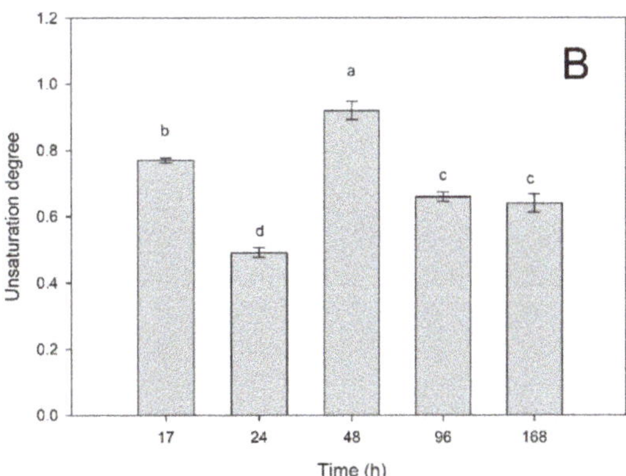

Figure 5. The unsaturation degree of the *E.magnusii* mitochondria from the cells raised in different growth phases (**B**). The unsaturation degree in the main PLs fractions of the *E. magnusii* mitochondria from the cells raised in different growth phases (**A**); PE— phosphatidylethanolamines; PC—phosphatidylcholines; CL—cardiolipins. [a–h] Means with the same letter did not differ significantly. Values are mean ± S.E.M from three independent experiments and three analytical induction replicates.

During long-lasting cultivation, the IHD changed due to significant alterations in the unsaturation of some fractions of the membrane lipids. Thus, IHD of PE remained at the same level in all the growth stages except for the 168-h stage, where it decreased sevenfold (Figure 5A). However, it was extremely stable in the PC and CL fractions, which dominate in the PLs of the mitochondrial membranes.

Dominating FAs in the mitochondrial PLs were palmitic (C16:0), oleic (C18:1), and linoleic (C18:2) acids (Figure 6A). In general, during the seven-day cultivation, we could observe the following consistent pattern: (1) palmitic acid (16:0) increased while the level of palmitoleic acid (C16:1) decreased; (2) oleic acid (C18:1) gradually increased with a sharp drop in the 168-h growth stage; and (3) the level of linoleic acid significantly decreased during cultivation being the highest in the early logarithmic stage (Figure 6A).

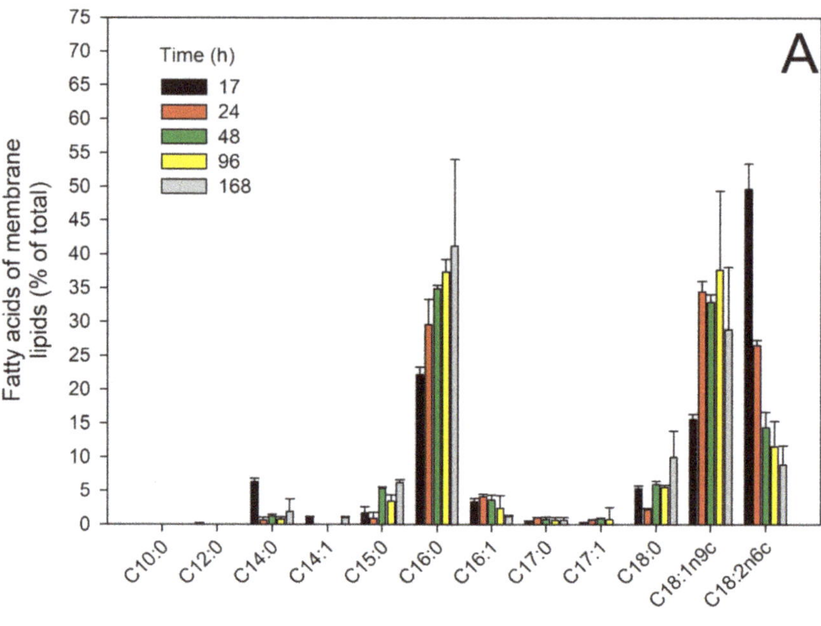

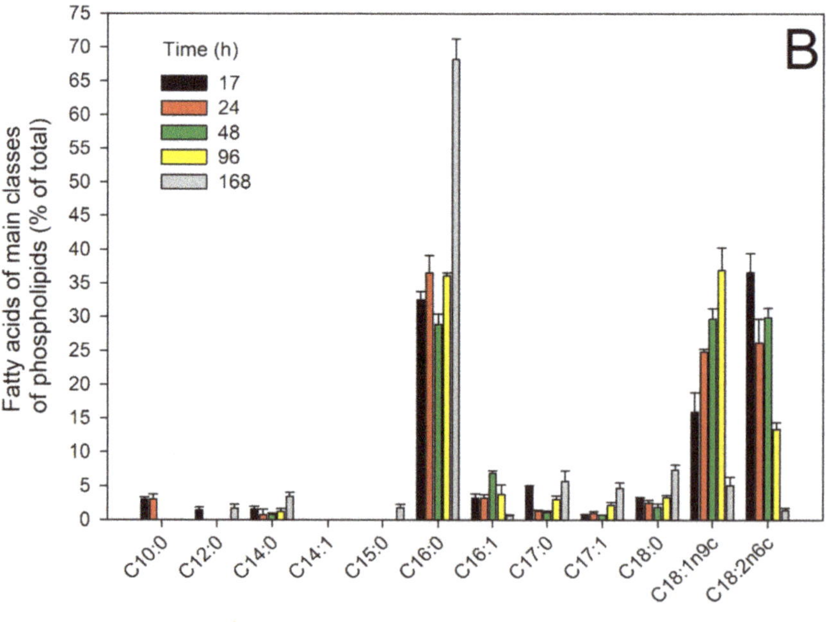

Figure 6. *Cont.*

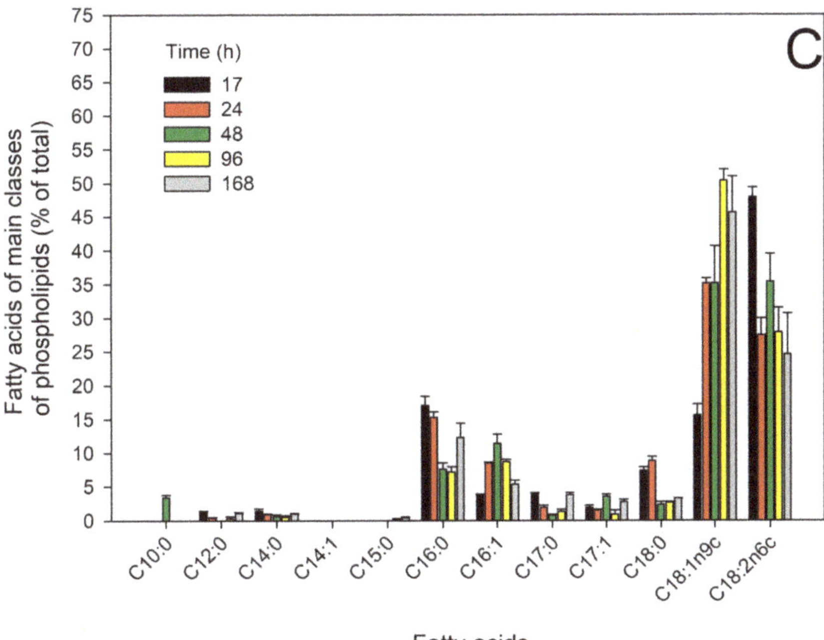

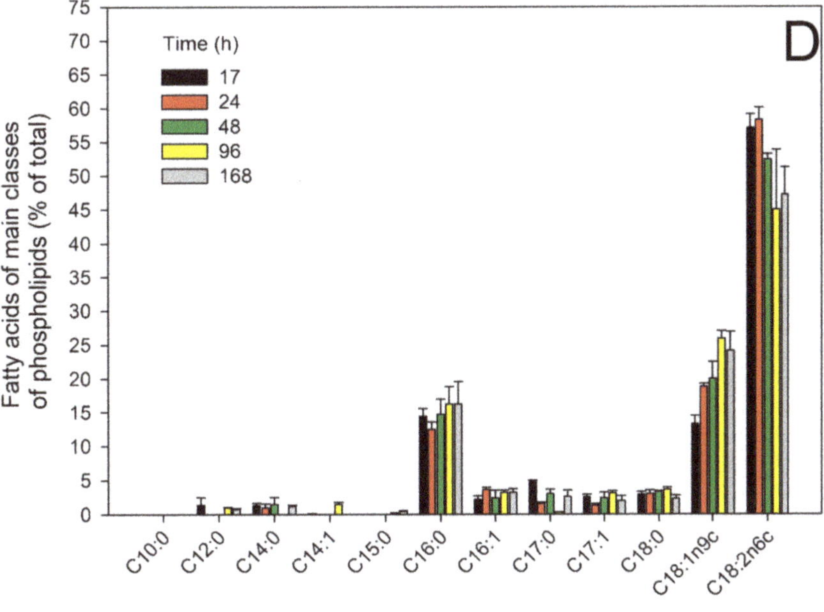

Figure 6. *Cont.*

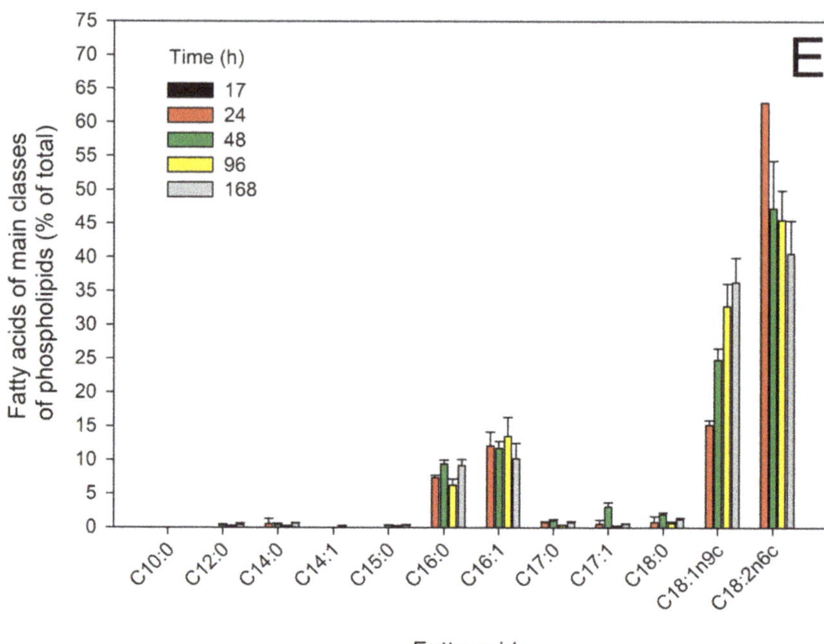

Figure 6. Fatty acid composition of the main membrane PLs in *E. magnusii* mitochondria in the different growth phases (**A**); PE—phosphatidylethanolamines (**B**); PC—phosphatidylcholines (**C**); CL—cardiolipins (**D**); PA—phosphatidic acids fraction (**E**). Values are mean ± S.E.M from three independent experiments and three analytical replicates.

The linoleic acid level in the PE fraction compared to that in the logarithmic stage decreased by 30%, 20%, and 63% in the 24-, 48-, and 96-h growth stages, respectively. Furthermore, by 168 h of cultivation, its level decreased by 26 times. On the contrary, the oleic acid level increased by 1.5, 1.96, and 2.3 times in the 24-, 48-, and 96-h growth stages, respectively, followed by more than a threefold decline in the 168-h growth stage (Figure 6B–E). The palmitic acid level in the PE fraction compared to that in the logarithmic stage doubled in the 168-h stage, but the palmitic acid level in the CL fraction remained constant of about 14–16% throughout the experiment. As for the oleic acid in the CL fraction, its level increased nearly twofold as the culture was ageing and reached 24–25% of the total fatty acid amount in the late stationary phase (Figure 6B–E).

At the same time, the linoleic acid level in the CL fraction decreased during the growth and ageing by about 10% (Figure 6B–E). As for PC, the palmitic acid level decreased by about two times in the 48-h stage and remained the same until the 96-h growth stage compared to that in the logarithmic stage. The amount of the oleic acid in the PC fraction increased by 2.3, 3.2, and 2.9 times in the 24-, 48-, and 96-h growth stages, respectively. Meanwhile, the linoleic acid level in the PC nearly halved during the growth and ageing of the culture. It is noteworthy that these FAs levels in the phosphatidic acids (PA) fraction also varied significantly. For example, the linoleic acid level decreased by more than 20% while the amount of the oleic acid doubled during the cultivation.

In the stage of the shift to the stationary growth phase, rather significant changes in the heptadecenoic acid (C17:1) level in the PE fraction were observed. Its amount increased by 6–8 times in the 96-h and 168-h growth stages compared to that in the logarithmic phase. Moreover, the main membrane PLs comprised negligibly small amounts of short-chain FAs, namely lauric acid (C12:0), myristoleic acid (C14:1), and pentadecenoic acid (C15:0).

3. Discussion

E. magnusii is a classical lower eukaryote of aerobic metabolism. It has a well-developed mitochondria system with a complete respiratory chain containing the invariable first coupling point [28]. It is capable of assimilating lots of substrates, including those substrates that determine the oxidative type of metabolism. The physiological features make significant advantages for the strain compared to the traditional *S. cerevisiae* yeast, used as a convenient tool to simulate processes in the unicellular eukaryotes model.

Earlier, we performed a systematic assay of the redox status of the *E. magnusii* yeast cultivated using two different types of substrates: a so-called "respiratory" substrate of glycerol and "fermentation" type substrate of glucose [22]. We revealed a significant difference in the survival of *E. magnusii* culture grown in glucose-containing medium, which showed a gradual decrease from 85% in the early stationary growth phase to 55% in the 168-h phase. The culture grown using glycerol showed a consistently high level of viability [22].

The data obtained suggest that the survival of the yeast culture during long-term cultivation and their high metabolic activity may be related to the active function of mitochondria. Their biogenesis is observed upon aerobic growth on non-fermentable substrates [22]. The yeast is capable of aerobic growth while using glycerol as a carbon source. The transition from "fermentative" to "respiratory" growth requires significant transcriptional rearrangements [29,30], which triggers the genes encoding the respiratory chain components, the elements of oxidative, osmotic, and general anti-stress responses, the enzymes of protein synthesis, glycerol assimilation and so on [20,31]. The mechanisms of yeast adaptation to various types of metabolism have been thoroughly studied, and it has been shown that 176 proteins of the *S. cerevisiae* mitochondria change while shifting from the fermentation type of metabolism (glucose, galactose) to the respiratory one (lactate) [32].

Analysis of the dynamics of the respiratory activity in *E. magnusii* mitochondria using the "oxidative" glycerol showed that it was maintained at a high level during the experiment (Table 1). It should be noted that the coupling parameters a bit decreased, however, the overall mitochondria functions kept up at a high level. The mitochondria AO is activated in the 96- and 168-h growth phases. The results agreed well with our published data on the dynamics of the ROS generation with the maximum in the 96-h growth stage [22]. Thus, the high ROS level in our studies could provoke the induction of the alternative cyanide-resistant pathway, which, in turn, is an element of the antioxidant defense system. This was shown in some papers [33–35]. The cyanide-resistant pathway of electron transport is not known to be related to energy storage and leads to a decrease in the ROS generation in the respiratory system, reducing the impact of the stress on the cell. Thus, this pathway induction is one of the ways for the cell to avoid excessive ROS generation under oxidative stress [36,37]. A large amount of storage lipids in the mitochondria, namely FFAs and DAGs, are considered as the second aspect of the high adaptation of *E. magnusii* upon long-lasting cultivation (Figure 3A,B). Abundant Lb in the yeast cells in the logarithmic (5–6 per a cell) and early stationary growth phases (6–7 per a cell), forming a structural complex with the mitochondria and nucleus (Figure 7A,B) testify to this.

The protective role of storage lipids, particularly TAGs and unsaturated FAs, is well known. TAGs can inhibit the yeast chronological ageing due to their accumulation in the Lb, which allows most of the unsaturated FAs to be deposited in TAGs by esterification [16,38]. In addition, the esterification of unsaturated FAs into TAGs slows down yeast chronological ageing by eliciting the liponecrotic cells. Some neutral lipids serve as a source of energy and precursors for membrane lipids synthesis [39]. The "model of secretory vesicles" based on the development of Lb from the secretory vesicles filled with TAGs is one of the possible models for forming Lb in a yeast cell. Lbs are usually about 300–400 nm in diameter and covered with a PL monolayer containing few proteins [39], mainly the enzymes of lipid metabolism [40]. Of note, the ratio of TAGs to ESt in Lbs from *S. cerevisiae* is usually 1:1 [41], and their TAGs core is surrounded by several ESt envelopes [42].

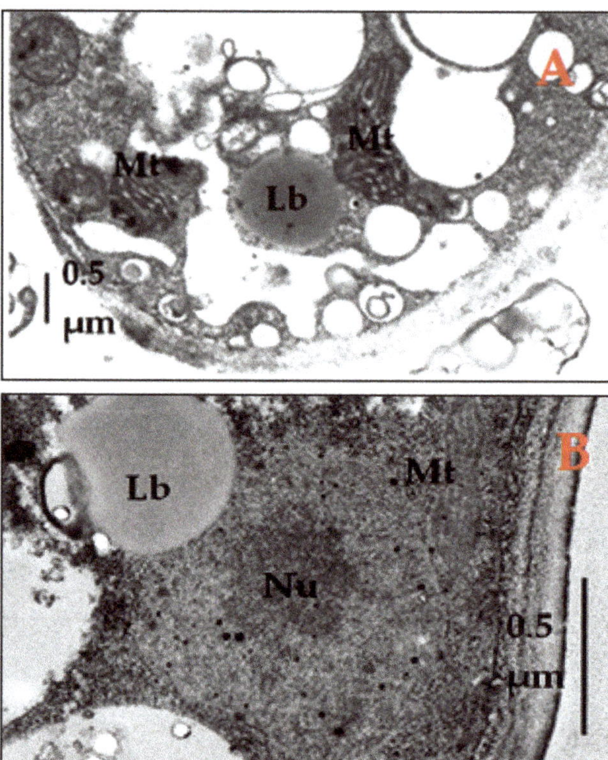

Figure 7. Image of *E. magnusii* yeast cells in the logarithmic (**A**) and early stationary growth phases (**B**). Nu—nucleus; Mt—mitochondria; Lb—lipid bodies.

In our studies, we found TAGs as a minority (about 4% of the total lipids) only in the mitochondria of the logarithmic growth stage (Figure 2A). During cultivation, they were replaced by FFAs with the highest level in the 24-h stage. DAGs kept up at about the same level (about 10–15%) throughout the experiment. It is probable that such an alteration in the storage lipids composition in the mitochondria is due to the depletion of the storage lipid resource during growth. In a yeast cell, there are three main pathways for synthesis of DAGs: (1) PA removal by phosphatidate phosphatase (*PAH1*); (2) degradation of phospholipids by phospholipases (*LRO1*, *DGA1*, *ARE1/2*); and (3) deacetylation of TAGs (*TGL2/3/4/5*, *AYR1*) [39]. We suppose that upon intensive growth, the third pathway of DAGs formation seems preferable, taking into account the disappearance of TAGs in the stationary growth phase.

At the same time, the high FFAs level in the *E. magnusii* mitochondria could be related to the presence of a minority of ESt (Figure 2A). The FFAs synthesis in yeast occurs via three main pathways: (1) de novo synthesis, (2) the complex and storage lipid degradation, and (3) external uptake [39]. Taking our data altogether, including the growing conditions of *E. magnusii*, we posit that in our experiments, the FFAs synthesis occurs de novo in the cytosol and mitochondria with a concurrent elongation and desaturation in the ER. The initial stage of the FFA synthesis is triggered by the acetyl-CoA carboxylase (the cytosolic enzyme is encoded by *ACC1*, the mitochondrial one is by *HFA1*). In this reaction, acetyl-CoA is carboxylated and forms malonyl-CoA serving as a two-carbon building block for the next FFAs synthesis reactions [39].

CL performs a lot of cell functions, being associated with all the major proteins of the mitochondrial respiratory system, and thereby increases the efficacy of the electron flux and the ADP/ATP exchange [40]. This PL corrects the catalytic activity and stability of the interactive proteins [41], is crucial for the biogenesis of mitochondria proteins [42], promotes mitochondrial fission/fusion [38], and participates in the cristae structure and morphology formation [43]. According to some researchers' data [43,44], the CL amount in the mitochondrial fraction of *S. cerevisiae* reaches 7.2% of the total membrane PL one, while for the mitochondria from *E. magnusii*, its level was much higher. In the logarithmic growth stage, the CL fraction made up about 50%, and in the stationary ones, it was not less than 18–20% (Figure 3A). It is probable that such a permanently high level of the CL fraction provides the high mitochondrial activity. In 2012, *Rostovtseva and Bezrukov* [45] showed that CL-rich areas of the outer mitochondrial membrane exhibited a higher activity of the mitochondrial VDAC porin involved in eliminating the ROS from the mitochondria. The CL fatty acids composition impacts VDAC activity greatly. This fact can be considered as the third most important point of the lipid composition adaptation in the yeast mitochondrial upon long-term cultivation.

Nevertheless, it is noteworthy that the PLs composition of the mitochondria in *E. magnusii* differs significantly from that in *Saccharomycetes*, where the dominating PLs fractions were PC, PE, and PI. Their shares made up 40.4% (PC), 26.7% (PE), and 14.6% (PI) [46,47]. In the *E. magnusii* mitochondria, these fractions are not dominating and reach no more than 10–20% of the total PLs amount (Figure 3A). However, among the major components, besides the CL fraction, the St share made up 40% in some samples. St are essential for supporting membrane integrity and eukaryotic cell viability. Taking into account our results on the "LB + MITO + Nucleus" complex in *E. magnusii* cells (Figure 7), we could speculate that the St fraction plays a crucial role in the functional interaction of the organelles upon ageing (Figure 2F–D).

Finally, the composition of FFAs in PLs is another important point of the lipid profile adaptation in the yeast mitochondria to the ageing processes. The main findings while assaying the FFAs composition of the mitochondria membrane lipids upon ageing were: (1) a smooth increase in the level of the palmitic acid (C16:0) against a fall in the palmitoleic acid (C16:1); (2) a gradual increase in the oleic acid amount (C18:1) with a sharp drop in the 168-h growth phase; and (3) a significant increase in the stearic acid level (C18:0) with a simultaneous decrease in the linoleic acid level (C18:2) throughout the whole experiment (Figure 5A).

Earlier, *Kieliszek* et al. [48] and his team showed a significant increase in the margarine (C17:0) and heptadecenoic (C17:1) acids level in the *Candida utilis* yeast grown either using 5% glycerol as a carbon source or enriched with selenium (20 mg/L) by two times and by 25%, respectively. Additionally, abundant margarine (C17:0; 12.19%) and heptadecenoic (C17:1; 9.31%) FAs were found in the *Yarrowia lipolytica* yeast when it was cultivated in batches using glycerol as a carbon source. Our findings showed that in the extremophilic *Y. lipolytica* cells under alkaline stress, the CL fraction had similar changes, namely the amount of saturated acyl residues in it increased [49]. An increase in the saturated FAs amount in the mitochondria of the culture under stress is likely to provide the integrity and rigidity of the membranes according to the homeoviscous adaptation hypothesis, which suggests these changes in the membrane lipid profile should facilitate its necessary fluidity [50].

4. Materials and Methods

4.1. Yeast Strain and Culture Conditions

E. magnusii yeast VKM Y261 strain was grown in batches of 100 mL in glycerol- (1%) containing media of the following composition (g/l): $MgSO_4$-0.5, $(NH_4)_2SO_4$-0.3, KH_2PO_4-8.6, NaCl-0.1, $CaCl_2$-0.05, yeast extract–2.0, L-histidine–2.75 mg, L-methionine–2.75 mg, and L-tryptophan–2.75 mg at 28 °C as described previously [51]. Absorbance was assessed in cell suspension at the wavelength of 590 nm (A_{590}) using a Specol-11 spectrophotometer (Carl Zeiss, Oberkochen, Germany). Cells were harvested at different stages of growth:

logarithmic (A_{590} = 2.6–2.7), early stationary (24 h of growth, A_{590} = 4.0–4.1), late stationary (48 h of growth, A_{590} = 4.5–4.6), deep stationary 1 (96 h of growth, A_{590} = 4.4–4.7), deep stationary 2 (168 h of growth, A_{590} = 4.4–4.7).

4.2. Potential-Dependent Staining

Potential-dependent staining of mitochondria in the *E. magnusii* cells raised in the different growth phase by JC-1. Cells were incubated with 0.5 μM JC-1 and examined in 0, 15, 20, and 30 min. Incubation medium contained 0.01 M PBS, pH 7.4; 1% glycerol or glucose, respectively. Regions of high mitochondrial polarization are indicated by red fluorescence due to the concentrated dye. To examine the JC-1-stained preparations, filters 02, 15 (Zeiss, Oberkochen, Germany) were used (magnification ×100). The photos were taken using an AxioCam MRC camera (Microvisioneer, Esslingen am Neckar, Germany)

4.3. Staining with Neutral Red

Yeast cells were suspended in PBS, and a 200 μL sample of the cell suspension was mixed with 100 μL neutral red (0.1 mg/mL stock solution, dissolved in a 2% dihydrate sodium citrate solution) and incubated for 5 min at room temperature. Viability was examined under a light microscope using Gorjaev's chamber (×400) from at least 1.000 cells in one biological replicate. Viable cells were colorless, and dead ones were red.

4.4. Transmission Electron Microscopy (TEM)

TEM analysis of untreated *E. magnusii* yeast cells was performed as described previously [15]. Briefly, the yeast cells were raised in the logarithmic or stationary (24 h) growth phase, precipitated, fixed with 2.5% glutaraldehyde in 0.1 M sodium phosphate buffer (pH 7.2) for 2 h, and then post-fixed in 1% OsO_4 for an hour at room temperature. After dehydration, the samples were embedded in Epon 812. Ultrathin sections were prepared with an LKB-8800 ultratome using diamond knives. Thereafter, the sections were stained with uranyl acetate for 60 min and post-stained as described previously, and examined with a Jeol (JEM-100B) and Hitachi U-12 electron microscopes (Hitachi, Tokyo, Japan).

4.5. Isolation of Mitochondria

Mitochondria were isolated using the method described in [52] with minor modifications. The mitochondria thus obtained met all known criteria of physiological intactness, as inferred from high respiratory rates and ADP/O ratios close to their theoretically expected maxima. Mitochondria were fully active for at least 4 h after being isolated when kept on ice. Briefly, cells were harvested at different growth phases, washed twice with ice-cold water, resuspended (0.1 g wet cells/mL) in pre-spheroplast buffer (50 mM Tris-HCl buffer; pH 8.6, 4 mM dithiothreitol), incubated at room temperature for 10–15 min, then diluted with ice-cold water, pelleted at 3000× *g* for 10 min, washed twice to remove excess dithiothreitol, and incubated at 28 °C using gentle stirring for 15–20 min in spheroplast buffer (10 mM HEPES-buffer, pH 7.2, 1.1 M sorbitol) with Novozym 20 T from *Trichoderma harzianum* (Sigma-Aldrich, St. Louis, MO, USA) (2.5 mg/g cells) and lytic enzymes from snail gut juice (50 mg/g cells). Spheroplast formation was monitored by measuring the osmotic fragility in distilled water. The spheroplasts were rapidly cooled, pelleted by centrifugation at 3.000× *g* for 10 min, washed gently twice in post-spheroplast buffer (1.2 M sorbitol, 5 mM $MgSO_4$, pH adjusted to 7.2), resuspended (0.1 g wet cells/mL) in grinding buffer (10 mM Tris-HCl, pH 7.2, containing 0.4 M mannitol, 1 mM EDTA, 0.05 mM EGTA, and 4 mg/mL BSA), and disrupted in an all-glass *Dounce* homogenizer (Kontes, Vineland, NJ, USA) with a low clearance pestle. The suspension was diluted with isolation buffer (10 mM Tris-HCl, pH 7.2, 0.6 M mannitol, 0.05 mM EDTA, 0.05 mM EGTA, and 4 mg/mL BSA) and centrifuged at 2000× *g* for 10 min. The supernatant was centrifuged once more at 7000× *g* for 20 min. The resulting pellet was washed in 10 mM Tris-HCl; pH 7.2, containing 0.6 M mannitol and 4 mg/mL BSA, resuspended in a smaller volume of the same buffer, and used immediately.

Mitochondrial protein was assayed by the method of Bradford with bovine serum albumin as the standard.

4.6. Respiration Assessment

Oxygen consumption by the yeast cells was assessed in vitro at 25 °C using electrodes covered by fluoroplastic film at a constant potential of 660 mV. Analysis of respiratory activity was performed using a multichannel microelectrode polarograph with data-analysis software Record-4 (Institute of Cell Biophysics of the Russian Academy of Sciences, Puschino, Russia). Oxygen consumption in mitochondrial suspensions was monitored polarographically with a Clark-type electrode in a medium containing 0.6 M mannitol, 1 mM Tris phosphate (pH 7.4), 1 mM EDTA, 20 mM pyruvate, 5 mM malate, and mitochondria corresponding to 0.4 mg protein/mL. All shown data traces are representative of four to six replicates.

4.7. Preparation and Analysis of Lipids

The lipids were extracted from the mitochondria by the Nichols method described in [53], which involved extraction with isopropanol and the isopropanol–chloroform mixture (1:1 and 1:2) at 70 °C, evaporation in a rotary evaporator, and extraction of the residue with chloroform-methanol (1:1) supplemented with 5% sodium chloride solution and water to remove water-soluble substances. After separating the mixture with a vortex, we dried the chloroform layer by passing it through water-free sodium sulphate, evaporated, and desiccated with a vacuum pump. The resulting pellet dissolved in a small amount of chloroform-methanol (2:1) was stored at −21 °C. The composition of storage lipids was assayed using an ascending thin layer chromatography on glass plates with silica gel 60 (Merck KGaA, Darmstadt, Germany). To separate storage lipids, the hexane: sulphuric ether: acetic acid (85:15:1) system [54] was used. To separate PLs and SLs SI60 Silica thin layer chromatography plates were activated and developed in two dimensions, first with chloroform/methanol/water (65/25/4, by volume) and second with chloroform/acetone/methanol/acetic acid/water 50/20/10/10/5, by volume) [55]. The lipids (100–200 µg) were applied to a plate. Lipid quantities were determined using the following standards: PC (Sigma, Saint-Luis, MO, USA) for PLs, a glycoceramide mixture (Larodan, Solna, Sweden) for SLs, and ergosterol (Sigma, Saint-Luis, MO, USA) for St. Samples of glycoceramides (5 and 10 µg) and PC (10 and 20 µg) were applied on the plates in the second direction. To develop the stains, the chromatograms were sprayed with 5% sulfuric acid in ethanol, followed by heating up to 180 °C. Quantitative densitometric analysis of the lipids was performed using the Dens software package (Lenkhrom, Sankt Peterburg, Russia) in the linear approximation regime using the calibration curves constructed with the standard solutions. Lipid data are presented in µg/g mitochondrial protein. PLs were identified using individual markers and qualitative tests for amino groups (with ninhydrin), choline-containing phospholipids (with the Dragendorff reagent), and glycolipids (with α-naphthol). Neutral lipids were identified with individual markers for MAGs, DAGs, and TAGs, St (ergosterol), FFAs, and hydrocarbons (Sigma, Saint-Luis, MO, USA). SLs were detected in the glycolipid fraction by the saponification method [56].

To assess the FAs composition of PLs, separate PLs were isolated using TLC with two plates, eluted with chloroform/methanol (1/1, *v/v*) for a night. Then, the supernatant was decanted, evaporated, 1 mL toluene and 2 mL of 2.5% H_2SO_4 dissolved in methanol and kept for two hours at +70 °C. FAs methyl ethers were extracted with hexane, dried, and analyzed by a Kristall 5000.1 gas chromatograph (Chromatek, Moscow, Russia) using an Optima-240 (60 m × 0.25 mm) capillary column (Macherey-Nagel GmbH and Co., Dueren, Germany). The temperature program was set from +130 to +240 °C. Eluting FAs were identified using the Supelco 37 Component FAME Mix (a mixture of FA methyl esters) (Supelco, Bellefonte, PA, USA). The degree of unsaturation was calculated by the equation: $\Delta/\text{mole} = 1.0 \times (\% \text{ monoene})/100 + 2.0 \times (\% \text{ dienes})/100 + 3.0 \times (\% \text{ trienes})/100$ [57].

4.8. Statistical analysis

The experiments were performed in biological triplicates with a standard error of less than 5%. The influence of pH and temperatures on soluble carbohydrates and lipids was estimated using one-way ANOVA with R (R Core Team 2016). The significance of differences between the mean values in each group was tested by Tukey's test. Values were considered significant at $p < 0.05$.

5. Conclusions

The recent review by *Medkour* et al. [15], using the correlation profiling of lipidomes in various tissues of long- and short-lived mammalian species, stated the following tendencies of so-called "lipidomic signature" for increasing longevity and delayed ageing in mammals and humans: (1) A decrease in the degree of FAs unsaturation, which reduces both double bonds and the peroxidation indices of various lipid classes. (2) Declined concentrations of long-chain FFAs. (3) An increased ratio of monounsaturated (MUFA) to polyunsaturated (PUFA) FAs. (4) Decreased levels of some SLs, some LPCs, and PCs, as well as highly polyunsaturated TAGs and DAGs (5). An increased amount of some sphingomyelins and cholesterol esters, as well as TAGs and DAGs with low IHD. The authors concluded that identifying the key trends in the lipidomic signature is an essential first step towards the detection of lipid biomarkers for healthy ageing and extended life span. However, whether any of the trends mentioned above have a cause-and-effect relationship in slowing the ageing process and increase in life expectancy should be determined. Summarizing the obtained results, we can conclude that, with respect to yeast mitochondria, the first and second statements, and partially the fifth one, are doubtlessly true for the long-lasting cultivation and the initial stages of ageing in the lower eukaryotes using an "oxidative" type glycerol. Moreover, the list of the factors favoring a long lifespan should include some other physiological parameters of yeast cells. The mitochondrial AO activity induced in the early stationary growth phase and high mitochondrial activity maintaining intensive mitochondrial respiration, which in turn determines the ATP production and physiological doses of ROS that should be added to the list of the trends providing increased longevity.

Supplementary Materials: The following are available online at https://www.mdpi.com/article/10.3390/app11094069/s1. Table S1: The dinamics of different cell types, Figure S1: Micro images of transmission electron microscopy of the *E. magnusii* mitochondria fraction, Figure S2: A - Amperometric recording of oxygen consumption by the E. magnusii mitochondria respiring on pyruvate + malate. Numbers adjacent to traces are respiration rates in ng-atoms of O/min/mg of mitochondrial protein. The incubation medium contained 0.6 M mannitol, 0.2 mM Tris-phosphate, pH 7.2; 20 mM pyruvate + 5 mM malate as respiratory substrates, and mitochondria corresponding to 0.5 mg mitochondrial protein, added at MITO. The frame in the Figure (a) shows the curve of oxygen consumption without substrate application (endogenous respiration). B - Recording of $\Delta\Psi$ generated by the *E. magnusii* mitochondria respiring on a 20 mM pyruvate + 5 mM malate. The incubation medium contained 0.4 M mannitol, 0.1 M KCl, 20 mM Tris-acetate, 0.4 mg of mitochondria protein, pH 7.4. C – the demonstration of the adenilate kinase activity in the intact mitochondria (the incubation medium composition is as in A), Table S2: The phosphorylating activities of the mitochondria using different substrates, Figure S3: Effect of erythromycin (A) and ethidium bromide (B) on the *E. magnusii* cell growth.

Author Contributions: Conceptualization: E.P.I.; methodology: E.P.I., N.N.G., and D.I.D.; validation: V.M.T.; formal analysis: M.K.; resources: E.P.I. and V.M.T.; writing—original draft preparation: E.P.I.; writing—review and editing: M.K.; supervision: Y.I.D.; project administration: E.P.I. and Y.I.D. All authors have read and agreed to the published version of the manuscript.

Funding: This research received no external funding.

Institutional Review Board Statement: Not applicable.

Informed Consent Statement: Not applicable.

Data Availability Statement: Not applicable.

Conflicts of Interest: The authors declare no conflict of interest.

References

1. Spinelli, J.B.; Haigis, M.C. The multifaceted contributions of mitochondria to cellular metabolism. *Nat. Cell Biol.* **2018**, *20*, 745–754. [CrossRef]
2. Rogov, A.G.; Goleva, T.N.; Epremyan, K.K.; Kireev, I.I.; Zvyagilskaya, R.A. Propagation of Mitochondria-Derived Reactive Oxygen Species within the *Dipodascus magnusii* Cells. *Antioxidants* **2021**, *10*, 120. [CrossRef]
3. Nunnari, J.; Suomalainen, A. Mitochondria: In sickness and in health. *Cell* **2012**, *148*, 1145–1159. [CrossRef] [PubMed]
4. Annesley, S.J.; Fisher, P.R. Mitochondria in Health and Disease. *Cells* **2019**, *8*, 680. [CrossRef] [PubMed]
5. López-Otín, C.; Blasco, M.A.; Partridge, L.; Serrano, M.; Kroemer, G. The hallmarks of aging. *Cell* **2013**, *153*, 1194–1217. [CrossRef] [PubMed]
6. Harman, D. Aging: Overview. *Ann. N. Y. Acad. Sci.* **2001**, *928*, 1–21. [CrossRef] [PubMed]
7. Jang, J.Y.; Blum, A.; Liu, J.; Finkel, T. The role of mitochondria in aging. *J. Clin. Investig.* **2018**, *128*, 3662–3670. [CrossRef] [PubMed]
8. Akbari, M.; Kirkwood, T.B.L.; Bohr, V.A. Mitochondria in the signaling pathways that control longevity and health span. *Ageing Res. Rev.* **2019**, *54*, 100940. [CrossRef]
9. Eleutherio, E.; Brasil, A.A.; França, M.B.; de Almeida, D.S.G.; Rona, G.B.; Magalhães, R.S.S. Oxidative stress and aging: Learning from yeast lessons. *Fungal Biol.* **2018**, *122*, 514–525. [CrossRef]
10. Kaeberlein, M. Lessons on longevity from budding yeast. *Nature* **2010**, *464*, 513–519. [CrossRef] [PubMed]
11. Oliveira, A.V.; Vilaça, R.; Santos, C.N.; Costa, V.; Menezes, R. Exploring the power of yeast to model aging and age-related neurodegenerative disorders. *Biogerontology* **2017**, *18*, 3–34. [CrossRef]
12. Herrero, E.; Ros, J.; Bel, G.; Cabiscol, E. Redox control and oxidative stress in yeast cells. *Biochim. Biophys. Acta* **2008**, *1780*, 1217–1235. [CrossRef] [PubMed]
13. Huang, X.; Withers, B.R.; Dickson, R.C. Sphingolipids and lifespan regulation. *Biochim. Biophys. Acta* **2014**, *1841*, 657–664. [CrossRef] [PubMed]
14. Jazwinski, S.M. Mitochondria to nucleus signaling and the role of ceramide in its integration into the suite of cell quality control processes during aging. *Ageing Res. Rev.* **2015**, *23*, 67–74. [CrossRef] [PubMed]
15. Medkour, Y.; Dakik, P.; McAuley, M.; Mohammad, K.; Mitrofanova, D.; Titorenko, V.I. Mechanisms underlying the essential role of mitochondrial membrane lipids in yeast chronological aging. *Oxid. Med. Cell. Longev.* **2017**, *2017*, 2916985. [CrossRef] [PubMed]
16. Handee, W.; Li, X.; Hall, K.W.; Deng, X.; Li, P.; Benning, C.; Williams, B.L.; Kuo, M.-H. An energy-independent pro-longevity function of triacylglycerol in yeast. *PLoS Genet.* **2016**, *12*, e1005878. [CrossRef]
17. Beach, A.; Richard, V.R.; Leonov, A.; Burstein, M.T.; Bourque, S.D.; Koupaki, O.; Juneau, M.; Feldman, R.; Iouk, T.; Titorenko, V.I. Mitochondrial membrane lipidome defines yeast longevity. *Aging* **2013**, *5*, 551–574. [CrossRef] [PubMed]
18. Burstein, M.T.; Titorenko, V.I. A mitochondrially targeted compound delays aging in yeast through a mechanism linking mitochondrial membrane lipid metabolism to mito-chondrial redox biology. *Redox Biol.* **2014**, *2*, 305–307. [CrossRef] [PubMed]
19. Leonov, A.; Arlia-Ciommo, A.; Bourque, S.D.; Koupaki, O.; Kyryakov, P.; Dakik, P.; McAuley, M.; Medkour, Y.; Mohammad, K.; Di Maulo, T.; et al. Specific changes in mitochondrial lipidome alter mitochondrial proteome and increase the geroprotective efficiency of lithocholic acid in chronologically aging yeast. *Oncotarget* **2017**, *8*, 30672–30691. [CrossRef]
20. Roberts, G.G.; Hudson, A.P. Rsf1p is required for an efficient metabolic shift from fermentative to glycerol-based respiratory growth in *S. cerevisiae. Yeast* **2009**, *26*, 95–110. [CrossRef]
21. Deryabina, Y.; Isakova, E.; Sekova, V.; Antipov, A.; Saris, N.E. Inhibition of free radical scavenging enzymes affects mito-chondrial membrane permeability transition during growth and aging of yeast cells. *J. Bioenerg. Biomembr.* **2014**, *46*, 479–492. [CrossRef] [PubMed]
22. Isakova, E.P.; Matushkina, I.N.; Popova, T.N.; Dergacheva, D.I.; Gessler, N.N.; Klein, O.I.; Semenikhina, A.V.; Deryabina, Y.I.; La Porta, N.; Saris, N.L. Metabolic remodeling during long-lasting cultivation of the *Endomyces magnusii* yeast on oxidative and fermentative substrates. *Microorganisms* **2020**, *8*, 91. [CrossRef] [PubMed]
23. Werner-Washburne, M.; Roy, S.; Davidson, G.S. Aging and the survival of quiescent and non-quiescent cells in yeast stationary-phase cultures. In *Aging Research in Yeast*; Subcellular Biochemistry; Breitenbach, M., Jazwinski, S., Laun, P., Eds.; Springer: Dordrecht, The Netherlands, 2011; Volume 57. [CrossRef]
24. Polcic, P.; Sabová, L.; Kolarov, J. Fatty acids induced uncoupling of *Saccharomyces cerevisiae* mitochondria requires an intact ADP/ATP carrier. *FEBS Lett.* **1997**, *412*, 207–210. [CrossRef]
25. Medentsev, A.G.; Arinbasarova, A.Y.; Akimenko, V.K. Regulation and physiological role of cyanide-resistant oxidases in fungi and plants. *Biochemistry* **1999**, *64*, 1230–1243.
26. Rogov, A.G.; Zvyagilskaya, R.A. Physiological role of alternative oxidase (from yeasts to plants). *Biochemistry* **2015**, *80*, 400–407. [CrossRef] [PubMed]
27. Bahr, J.T.; Bonner, W.D., Jr. Cyanide-insensitive respiration. II. Control of the alternate pathway. *J. Biol. Chem.* **1973**, *248*, 3446–3450. [CrossRef]
28. Zelenshchikova, V.A.; Burbaev, D.S.; Zviagil'skaia, R.A. Reverse electron transfer in mitochondria of the yeast *Endomyces magnusii* grown on glycerol. *Biokhimiia* **1983**, *48*, 186–192.
29. DeRisi, J.L.; Iyer, V.R.; Brown, P.O. Exploring the metabolic and genetic control of gene expression on a genomic scale. *Science* **1997**, *278*, 680–686. [CrossRef] [PubMed]

30. Roberts, G.G.; Hudson, A.P. Transcriptome profiling of *Saccharomyces cerevisiae* during a transition from fermentative to glycerol-based respiratory growth reveals extensive metabolic and structural remodeling. *Mol. Genet. Genom.* **2006**, *276*, 170–186. [CrossRef]

31. Ohlmeier, S.; Kastaniotis, A.J.; Hiltunen, J.K.; Bergmann, U. The yeast mitochondrial proteome, a study of fermentative and respiratory growth. *J. Biol. Chem.* **2004**, *279*, 3956–3979. [CrossRef]

32. Renvoisé, M.; Bonhomme, L.; Davanture, M.; Valot, B.; Zivy, M.; Lemaire, C. Quantitative variations of the mitochondrial proteome and phosphoproteome during fermentative and respiratory growth in *Saccharomyces cerevisiae*. *J. Proteom.* **2014**, *106*, 140–150. [CrossRef]

33. Veiga, A.; Arrabaça, J.D.; Loureiro-Dias, M.C. Cyanide-resistant respiration, a very frequent metabolic pathway in yeasts. *FEMS Yeast Res.* **2003**, *3*, 239–245. [CrossRef]

34. Vanlerberghe, G.C. Alternative oxidase: A mitochondrial respiratory pathway to maintain metabolic and signaling homeostasis during abiotic and biotic stress in plants. *Int. J. Mol. Sci.* **2013**, *14*, 6805–6847. [CrossRef] [PubMed]

35. Saha, B.; Borovskii, G.; Panda, S.K. Alternative oxidase and plant stress tolerance. *Plant Signal Behav.* **2016**, *11*, e1256530. [CrossRef]

36. Van Aken, O.; Giraud, E.; Clifton, R.; Whela, J. Alternative oxidase: A target and regulator of stress responses. *Physiol. Plant* **2009**, *137*, 354–361. [CrossRef]

37. Lushchak, V.I. Adaptive response to oxidative stress: Bacteria, fungi, plants, and animals. *Comp. Biochem. Physiol. C Toxicol. Pharmacol.* **2011**, *153*, 175–190. [CrossRef] [PubMed]

38. Li, X.; Handee, W.; Kuo, M.H. The slim, the fat, and the obese: Guess who lives the longest? *Curr. Genet.* **2017**, *63*, 43–49. [CrossRef]

39. Klug, L.; Daum, G. Yeast lipid metabolism at a glance. *FEMS Yeast Res.* **2014**, *14*, 369–388. [CrossRef] [PubMed]

40. Baile, M.G.; Lu, Y.W.; Claypool, S.M. The topology and regulation of cardiolipin biosynthesis and remodeling in yeast. *Chem. Phys. Lipids* **2014**, *179*, 25–31. [CrossRef]

41. Wenz, T.; Hielscher, R.; Hellwig, P.; Schagger, H.; Richers, S.; Hunte, C. Role of phospholipids in respiratory cytochrome bc (1) complex catalysis and supercomplex formation. *Biochim. Biophys. Acta* **2009**, *1787*, 609–616. [CrossRef]

42. Joshi, A.S.; Zhou, J.; Gohil, V.M.; Chen, S.; Greenberg, M.L. Cellular functions of cardiolipin in yeast. *Biochim. Biophys. Acta* **2009**, *1793*, 212–218. [CrossRef]

43. Joshi, A.S.; Thompson, M.N.; Fei, N.; Huttemann, M.; Greenberg, M.L. Cardiolipin and mitochondrial phosphatidylethanolamine have overlapping functions in mitochondrial fusion in *Saccharomyces cerevisiae*. *J. Biol. Chem.* **2012**, *287*, 17589–17597. [CrossRef]

44. Mileykovskaya, E.; Dowhan, W. Cardiolipin membrane domains in prokaryotes and eukaryotes. *Biochim. Biophys. Acta* **2009**, *1788*, 2084–2091. [CrossRef]

45. Rostovtseva, T.K.; Gurnev, P.A.; Chen, M.; Bezrukov, S.M. Membrane lipid composition regulates tubulin interaction with mitochondrial voltage-dependent anion channel. *J. Biol. Chem.* **2012**, *287*, 29589–29598. [CrossRef] [PubMed]

46. Zinser, E.; Sperka-Gottlieb, C.D.; Fasch, E.V.; Kohlwein, S.D.; Paltauf, F.; Daum, G. Phospholipid synthesis and lipid composition of subcellular membranes in the unicellular eukaryote *Saccharomyces cerevisiae*. *J. Bacteriol.* **1991**, *173*, 2026–2034. [CrossRef] [PubMed]

47. Horvath, S.E.; Wagner, A.; Steyrer, E.; Daum, G. Metabolic link between phosphatidylethanolamine and triacylglycerol metabolism in the yeast *Saccharomyces cerevisiae*. *Biochim. Biophys. Acta* **2011**, *1811*, 1030–1037. [CrossRef] [PubMed]

48. Kieliszek, M.; Błażejak, S.; Bzducha-Wróbel, A.; Kot, A.M. Effect of selenium on lipid and amino acid metabolism in yeast cells. *Biol. Trace Elem. Res.* **2019**, *187*, 316–327. [CrossRef]

49. Sekova, V.Y.; Dergacheva, D.I.; Isakova, E.P.; Gessler, N.N.; Tereshina, V.M.; Deryabina, Y.I. Soluble sugar and lipid readjustments in the *Yarrowia lipolytica* yeast at various temperatures and pH. *Metabolites* **2019**, *9*, 307. [CrossRef] [PubMed]

50. Glatz, A.; Pilbat, A.M.; Németh, G.L.; Vince-Kontár, K.; Jósvay, K.; Hunya, Á.; Udvardy, A.; Gombos, I.; Péter, M.; Balogh, G.; et al. Involvement of small heat shock proteins, trehalose, and lipids in the thermal stress management in *Schizosaccharomyces pombe*. *Cell Stress Chaperones* **2016**, *21*, 327–333. [CrossRef] [PubMed]

51. Deryabina, Y.I.; Zvyagilskaya, R.A. The Ca^{2+}-Transport system of Yeast (*Endomyces magnusii*) mitochondria: Independent pathways for Ca2+ uptake and release. *Biochemistry* **2000**, *65*, 1352–1356. [CrossRef] [PubMed]

52. Bazhenova, E.N.; Saris, N.E.; Zvyagilskaya, R.A. Stimulation of the yeast mitochondrial calcium uniporter by hypotonicity and by ruthenium red. *Biochim. Biophys. Acta* **1998**, *1371*, 96–100. [CrossRef]

53. Nichols, B.W. Separation of the lipids of photosynthetic tissues: Improvements in analysis by thin-layer chromatography. *Biochim. Biophys. Acta* **1963**, *70*, 417–422. [CrossRef]

54. Kates, M. Techniques of lipidology: Isolation, analysis and identifcation of lipids. In *Laboratory Techniques in Biochemistry and Molecular Biology*; Work, T.S., Work, E., Eds.; North-Holland: Amsterdam, The Netherlands, 1972; pp. 267–610.

55. Benning, C.; Huang, Z.H.; Gage, D.A. Accumulation of a novel glycolipid and a betaine lipid in cells of *Rhodobacter sphaeroides* grown under phosphate limitation. *Arch. Biochem. Biophys.* **1995**, *317*, 103–111. [CrossRef] [PubMed]

56. Kates, M. *Techniques of Lipidology: Isolation, Analysis, and Identification of Lipids*, 2nd ed.; Elsevier: Amsterdam, The Netherlands, 1988; Volume 464. [CrossRef]

57. Weete, J.D. Introduction to fungal lipids. In *Fungal Lipid Biochemistry*; Kritchevsky, D., Ed.; Springer: Berlin/Heidelberg, Germany, 1974.

Review

Nutritional and Health Potential of Probiotics: A Review

Muhammad Modassar Ali Nawaz Ranjha [1,*], **Bakhtawar Shafique** [1], **Maria Batool** [2], **Przemysław Łukasz Kowalczewski** [3,*], **Qayyum Shehzad** [4,5], **Muhammad Usman** [5], **Muhammad Faisal Manzoor** [6], **Syeda Mahvish Zahra** [1,7], **Shazia Yaqub** [1] and **Rana Muhammad Aadil** [8,*]

[1] Institute of Food Science and Nutrition, University of Sargodha, Sargodha 40100, Pakistan; BakhtawarShafique111@gmail.com (B.S.); mahvish.zahra@aiou.edu.pk (S.M.Z.); shaziaft743@gmail.com (S.Y.)

[2] Institute of Diet and Nutritional Sciences, University of Lahore, Lahore 54000, Pakistan; mbatool215@gmail.com

[3] Department of Food Technology of Plant Origin, Poznań University of Life Sciences, 31 Wojska Polskiego St., 60-624 Poznań, Poland

[4] National Engineering Laboratory for Agri-Product Quality Traceability, Beijing Technology and Business University, Beijing 100048, China; qayyum.shehzad.5@gmail.com

[5] Beijing Advanced Innovation Center for Food Nutrition and Human Health, School of Food and Health, Beijing Technology and Business University, Beijing 100048, China; ch.usman1733@gmail.com

[6] School of Food and Biological Engineering, Jiangsu University, Zhenjiang 212013, China; faisaluos26@gmail.com

[7] Department of Environmental Design, Health and Nutritional Sciences, Allama Iqbal Open University, Islamabad 44000, Pakistan

[8] National Institute of Food Science and Technology, University of Agriculture Faisalabad, Faisalabad 38000, Pakistan

* Correspondence: ModassarRanjha@gmail.com (M.M.A.N.R.); przemyslaw.kowalczewski@up.poznan.pl (P.Ł.K.); muhammad.aadil@uaf.edu.pk (R.M.A.)

Citation: Ranjha, M.M.A.N.; Shafique, B.; Batool, M.; Kowalczewski, P.Ł.; Shehzad, Q.; Usman, M.; Manzoor, M.F.; Zahra, S.M.; Yaqub, S.; Aadil, R.M. Nutritional and Health Potential of Probiotics: A Review. *Appl. Sci.* **2021**, *11*, 11204. https://doi.org/10.3390/app112311204

Academic Editors: Patrizia Messi and Anabela Raymundo

Received: 26 September 2021
Accepted: 23 November 2021
Published: 25 November 2021

Publisher's Note: MDPI stays neutral with regard to jurisdictional claims in published maps and institutional affiliations.

Abstract: Several products consist of probiotics that are available in markets, and their potential uses are growing day by day, mainly because some strains of probiotics promote the health of gut microbiota, especially *Furmicutes* and *Bacteroidetes*, and may prevent certain gastrointestinal tract (GIT) problems. Some common diseases are inversely linked with the consumption of probiotics, i.e., obesity, type 2 diabetes, autism, osteoporosis, and some immunological disorders, for which the disease progression gets delayed. In addition to disease mitigating properties, these microbes also improve oral, nutritional, and intestinal health, followed by a robust defensive mechanism against particular gut pathogens, specifically by antimicrobial substances and peptides producing probiotics (AMPs). All these positive attributes of probiotics depend upon the type of microbial strains dispensed. Lactic acid bacteria (LAB) and *Bifidobacteria* are the most common microbes used, but many other microbes are available, and their use depends upon origin and health-promoting properties. This review article focuses on the most common probiotics, their health benefits, and the alleviating mechanisms against chronic kidney diseases (CKD), type 1 diabetes (T1D), type 2 diabetes (T2D), gestational diabetes mellitus (GDM), and obesity.

Keywords: probiotics; health benefits; alleviating mechanism; CKD; T1D; T2D; obesity

1. Introduction

Good health is a fundamental need of human beings [1]. Humans' priority was to rely on natural resources for good health outcomes [2–5]. In the 20th century, it was observed that healthy children fed mothers' milk had *Bifidobacteria* in their gut microbiota [6]. It established a positive association between health and gut microbiota, and numerous studies have been conducted to investigate the proper mechanism of this association, and some found that some bacteria have a positive correlation with health [7]. The term probiotics were first used in 1965 by Stillwell and Lilley to describe these beneficial bacteria [8]. At the beginning of the 20th century, a Nobel laureate in Paris named Élie Metchnikoff, a professor by profession noticed the health-enhancing properties of fermented dairy products. He then reported that the presence of lactic acid bacteria in fermented dairy products helps keep the defensive system activated, resulting in higher longevity of its consumers [9].

Some crucial characteristics of probiotics were introduced in the 1980s, and these include: (a) strains to have a beneficial impact, (b) non-toxic, non-allergic, and non-pathogenic, (c) available in large quantity as viable cells, (d) suitable for the environment of the gut, and (e) storable as well as stable [10]. Although the most commonly used probiotics are bacteria, especially lactic acid, molds and yeasts can also be probiotics [11]. In 2001, the Food and Agriculture Organization and the World Health Organization council defined probiotics, which was later refined in 2014 by Hill et al. as "live micro-organisms that, when administered in adequate amounts, confer a health benefit on the host," which can be understood to mean that probiotic strains must be (i) sufficiently characterized; (ii) safe for the intended use; (iii) supported by at least one positive human clinical trial conducted according to generally accepted scientific standards or as per recommendations and provisions of local/national authorities when applicable, and (iv) alive in the product at an efficacious dose throughout shelf life [12]. Examples include living bacteria containing *Lactobacillus*, *Bifidobacterium*, and some others [13].

From the start, probiotics have been considered beneficial to the health of gut microbiota, studies have also confirmed their positive associations with many other chronic diseases [14]. Although *lactobacillus* is the most commonly used probiotic, many other beneficial microbes are present, and their usage depends upon their origin and health-friendly properties [15]. All bacterial strains have specific properties; some are useful in treating obesity, some treat diabetes mellitus (DM), some help with CKD, and some deal with osteoporosis. In addition to these highly prevalent diseases, several studies have confirmed the beneficial role of probiotics in the case of autism, irritable bowel syndrome (IBS), and wound healing [16,17].

Probiotics have also been investigated to promote oral health and strengthen the immunological system [18–20]. Additionally, they are an essential contributor in the field of agriculture as well as in food processing [21].

In this article, we review and discuss some essential probiotics that can be useful for human health, how they affect an individual's nutritional status, and how probiotics perform their action to prevent disorders. In addition, the health-promoting role of probiotics and action mechanisms in the case of several of the most prevalent diseases are reviewed.

1.1. Nutritional Impacts of Probiotics

Lactic acid bacteria (LAB) most commonly ferment fruit, vegetables, and cereals. It impacts taste, and, when fermenting rice bran, sprouts of bean sprouts, and buckwheat, it also produces bioactive components that benefit inflammatory, compromised immunity, glycemic imbalance, and fatigue conditions [22].

1.2. Nutritional Impacts of Probiotics in Inflammation

Currently, people are highly concerned about their health. There is an increasing focus on disease prevention compared to cures. The consideration of probiotics and their health benefits began about a century ago. It was discovered that the people of Bulgaria and Russia lived longer than other populations because they use sour milk that contains beneficial bacteria [23]. Probiotics play an important role in preventing and improving diseases like allergies, liver disorders, intestinal ailments, and metabolic syndromes leading to diabetes, cardiovascular diseases, and obesity [24]. *Escherichia coli* and LAB are extensively used to treat inflammatory bowel disease, colon cancer, and constipation because lactic acid bacteria directly deliver cytokines to the target sites within the host [21,25,26]. Investigators indicated a significant inhibiting role of non-pathogen apoptosis induction within carcinoma cells, which helps protect against colon cancer (HGC-27) and human colonic cancer cells (Caco-2, DLD-1, HT-29), especially with the action of *Escherichia coli* K-12 strains, *Lactobacillus rhamnosus*, and *Bifidobacterium latis* [27].

1.3. Nutritional Impacts of Probiotics in Dental Carries

LAB probiotics in cheese seem to reduce the number of mutant streptococci in saliva and therefore have a beneficial impact on dental carries; the LAB in cheese also prevents the demineralization of enamel and dental plaque [28]. However, the decrease in streptococci number is independent of the strain used, and the effect is not the same in all studies. Therefore, no exclusive report has been made on dental carries [29].

The health of an individual depends upon the status of a healthy microbiota. Pathogenic bacteria cause inflammation during acute infection, whereas symbiotic bacteria regulate the immune responses to such inflammations and protect the host from different diseases [30].

1.4. Nutritional Impacts of Probiotics in Obesity, Diabetes and Associated Issues

Gut bacteria play a vital role in the pathogenesis of bile duct diseases. Probiotics are also used for the treatment of liver diseases [31,32]. The result of a study on a boy suffering from primary sclerosing cholangitis treated with the combination of drugs and probiotics showed improvement in symptoms and laboratory tests [33]. According to the result of a study on a group suffering from type 2 diabetes mellitus, eating yogurt that contains probiotics (*Lactobacillus acidophilus* La5 and *Bifidobacterium lactis* Bb12) improved fasting glucose level and antioxidant ability, and it was therefore concluded that probiotic yogurt has a positive impact on patients with type 2 diabetes [34]. *Bifidobacteria* play a role in the development of obesity. Mothers who are overweight give birth to neonates with low levels of *Bifidobacterium*. A low number of *Bifidobacterium* at birth is associated with overweight issues later in childhood. The *Bifidobacterium* level in obese adults is lower as compared to that in lean people [35]. The gut microbiota provides opportunities for the fermentation of non-digestible compounds like fibers that support the microbes that produce short-chain fatty acids and gases [36]. The short-chain fructo-oligosaccharides increase intestinal magnesium absorption in postmenopausal women in whom magnesium deficiency leads to osteoporosis [37].

Probiotics also prevent and treat gastrointestinal inflammatory conditions like inflammatory bowel diseases, irritable bowel syndrome [38], allergies, and respiratory disorders [39]. When probiotics were administered in pregnant women, no increase or decrease in the risk of preterm birth and other pregnancy-related adverse effects was seen [40]. *E. coli*, *Staphylococcus aureus*, *Bifidobacterium*, and *Firmicutes* have been reported by researchers to have an impact on obesity, thus can be used as obesity inhibitors [41]. Type 2 diabetes (T2D) is one of the most prevalent diseases globally that can be cured by using probiotics as an adjuvant. It was proved that probiotics exhibit anti-diabetic and anti-inflammatory properties and therefore play a vital role in diabetes prevention and treatment. Probiotics have also been proven beneficial in autistic children, especially for GIT disturbance caused by this disorder. Bone mineral density has also been proven to be affected by these gut microbes, as these microbes lead to enhanced absorption of various nutrients and minerals. These microbes improve human health and improve the health of plants and crops, increasing the overall yield of the land [42].

Traditional probiotics have marginal ameliorative effects on diseases, so the next generation probiotics (NGP) now serve as new preventive and therapeutic tools. These NGP include *Prevotella copri* for controlling insulin resistance, *Akkermansia muciniphila*, and *Bacteroides thetaiotaomicron* that are used to reverse obesity and insulin resistance, and *Bacteroides fragilis* is used to reduce inflammation and facilitates anticancer effects [43].

2. Probiotics and Their Benefits

The use of probiotics in the form of live bacteria for health promotion in animals and human beings is emerging daily. Today, a vast array of fermented food items and beverages is available, accounting for approximately one-third of worldwide human diets [44]. The level of probiotics in foods range from 2 to 20 g/day depending on the component and desired effect and can be added to different food products, including cereals, biscuits, bread, sauces, yogurts, and drinks [45]. Curd is considered the most preferred source of probiotics, as it is globally consumed [46]. Interest and development of functional foods consisting of both probiotics and prebiotics have increased due to increased awareness of their health-promoting properties. They positively affect gut health and decrease the risk of diseases, which is why they are used as therapy [47].

All probiotic products have different nutritional and therapeutic characteristics, due to various conditions, such as the genetic make-up of the strain, amount of the probiotics used in the product, the purpose it is used for, and its shelf life. The selection of probiotic strain depends on its production, impact, and health benefits in the host [48]. A probiotic food must have 10^6 CFU/g of probiotic micro-organism to achieve a health benefit. The dosage recommended for human consumption is 10^7–10^9 CFU/mg/day. It is known that the effect of probiotic food consumption depends on the specific strain used in that product [49]. To get the beneficial strains of probiotics, the following genera are of great importance: *Lactobacillus, Escherichia coli, Bifidobacterium, Enterococcus, Saccharomyces, Pediococcus, Streptococcus*, and *Leuconostoc*. The most common micro-organisms used as probiotics are lactic acid bacteria (LAB) and *Bifidobacteria* [50].

Strains of lactobacillus species that are commonly found in saliva samples include *L. paracasei, L. plantarum*, and *L. rhamnosus*. The *bifidobacterial* species are the anaerobes that are also found in the oral cavity, and both species are found in breast milk and are generally regarded as safe [51]. Several elements, such as nutrition, age, environmental difficulties, incompatibilities, illnesses, and treatment routes, strongly impact gut microbiota growth, maintenance, and functionality [52]. When selecting the strain it must be kept in mind that it should be originated from target and natural microflora as it is vital for its survival in an acidic environment during its travel to the intestine [53,54].

Probiotics have a positive effect on the immune system of the host and it was proven that they influence the healthy bacteria present in the gut or intestine [55]. Probiotics improve the immunity system by modifying the humoral and cellular immune response [56]. Figure 1 summarizes the benefits of different probiotics on health.

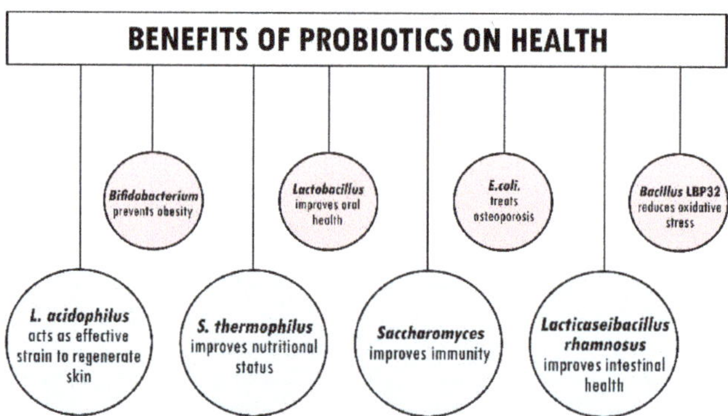

Figure 1. Benefits of probiotics on health [57].

The strains that are being selected also depend on the biosafety level, which means they should not be toxic or pathogenic [58]. They should be tested for safety parameters, including antibiotic susceptibility, antibiotic resistance gene, and hemolytic activity. Antimicrobial production is an important feature of probiotics against pathogens, but non-optimal antimicrobial activity will disrupt the healthy microbiota in the intestines and pose harmful effects [59]. Bile present in higher concentrations resulted in the lower growth of strains [60]. Probiotics can alter the pH of the surrounding environment, and hence they can compete with the pathogens present there.

In the same way, probiotics adhere to the adhesion sites on the mucosa, which decreases the chances of adhesion of pathogens and decreases the cases of probiotics being washed out [61]. Different tests are performed to identify the strain of bacteria, such as biochemical and molecular tests, then further techniques are used to differentiate between the strains of the same species, such as polymerase chain reaction (PCR) and gene sequencing tests [62,63]. Other tests are performed, such as hemolytic tests to determine whether the organisms are destroying red blood cells or not and platelet aggregation tests, which are crucial factors for pathogen activity [64]. Recommended storage temperature for probiotic foods is 4–5 °C, and the product must be used according to the information noted on the label, which should be clear [65].

2.1. Significant Effects of Probiotics on Oral Health Status

Probiotics primarily contribute to gastrointestinal health, but their diversity has been increasing and now they greatly contribute to oral health care [66]. Oral disease treatment is one of the most expensive health care treatments; that is why new approaches are being used to treat these infections, such as the use of probiotics [67]; in other words, oral lactic acid bacteria have been characterized after isolation and purification to use for oral health, especially to cure periodontal diseases, caries, and halitosis (bad breath) [19].

These ailments are some of the most widespread diseases throughout the world, but the treatment of these infections require potent antibiotics that cause severe side effects in the gastrointestinal tract. The use of probiotics as a relatively less harmful treatment has been advised to avoid such side effects [68].

Additionally, the buccal cavity is an excellent bacteriological medium, where micro-organisms can grow very easily if certain factors are out of balance due to poor oral health, poor diet, or immunodeficiency. These parameters alter the pH of the mouth and create a favorable environment for micro-biota growth. Micro-organisms are deposited on the oral cavity like biofilm and create a bacterial hold and aggressively colonize the buccal cavity [69].

Probiotic strains of *Bifidobacterium*, *Lactobacillus*, and *Streptococcus* are primarily used to prevent or cure oral infections. Plaque or dental biofilms in the buccal cavity leads to poor oral health, but lactic acid bacteria are probiotics that interact with that biofilm/plaque and destroy the causative agents with LAB's antimicrobial activity [70].

Indirectly, probiotics enhance immunity and strengthen the immune system by the interaction of lactic acid bacteria with immuno-competent cells, which leads to modification in the production of cytokines and ensuing effects on the overall immune system [71]. Lactobacilli used as a probiotic in mild gingival inflammation causes a great decline of Interleukin-8 discharge in the gingival crevicular fluid [72]. Probiotics have been used in the treatment of periodontitis and gingivitis: they modify the buccal cavity homeostasis, with possible numerous health benefits such as combatting dysbiosis and moderating inflammatory pathways to minimize the inflammation of periodontitis [73,74].

Probiotics are also being used in the treatment of halitosis; for this purpose, bacteriocins producing *Streptococcus salivarius* K12 strain were developed. This strain treats the mouth odor by actively helping de-colonize bacteria like *Solobacterium moorei*, *Parvimonas micra*, and *Eubacterium sulci*, which produce volatile sulfur compounds(VSCs) on the tongue [75,76].

A new study demonstrated the effects of widely commercially used chewing gum containing probiotic bacteria (xylitol), which greatly impacted plaque and gingival scores by reducing *Streptococcus mutants* counts in plaque and saliva [77]. Probiotics played an essential role in the clinical manifestation of dental disorders and periodontal diseases and perhaps impacted halitosis as well. Probiotics provide long and short-term therapeutic effects. Advance studies about particular strains and oral micro-biome transplants could also increase probiotics' role or efficacy in oral health, and probiotics consumed as part of the diet may also enhance oral health and strong oral immunity or hygiene [78,79].

2.2. Improvement of Intestinal Health through Probiotics

Probiotics are known as important micro-organisms that are used to improve intestinal health, including beneficial micro-organisms known to retard the production of bacterial enzymes that cause colon cancers [80]. Colonic microflora is very important for human health [81]. These bacteria promote the normal functioning in the intestine and maintain the host's health [82]. Consumption of pre and probiotics alters gut microbiota by facilitating or inhibiting microbe growth [83].

These bacteria first colonize the intestinal track and then help establish the immune system with balanced cell responses [84]. During pregnancy, prebiotics and probiotics are supplemented in pregnant females to protect the fetus from many autoimmune diseases and syndromes (AIDS). These supplements immunize the child and protect against AIDS [85]. Recent investigations have shown that human intestinal microbiota comprises over 1000 microbes [86]. In one study performed on obese and diabetic patients, scientists concluded that their gut microbiota differs from non-diabetic patients and this modification occurred as a result of dietary changes of diabetics versus non-diabetics [87]. Several studies have shown that the microbiota influences energy homeostasis and controls body weight [88].

Drug and probiotic interference are well known; studies have revealed the probiotics and warfarin interaction. Intestinal bacteria are well known for the production of vitamin K, but absorption of antibiotics disturb the gut flora and result in the deficiency of vitamin K [89]. According to the study, many external factors like medication, radiotherapy, stress, and infections disturb the gut microflora growth and their ratios [90]. Other essential functions played by probiotics include improvement in flu/influenza; protection from dental caries; prevention of tonsillitis, respiratory infections, and urogenital health problems; and help in wound healing and throat infection through immunomodulatory action. Currently, probiotics are being sold in markets as dietary supplements proven to be so-called standard drug therapy [91].

2.3. Role of Probiotics in Development of Immunity

Probiotics and prebiotics exert beneficial effects on humans' health, including immune modulating capacity, modulation of cellular metabolism, epithelial barrier functions, or proliferation [92]. The growth of the gut microbiome is a dynamic process, and early colonization of *Bacteroides* and *Bifidobacterium* species might play a crucial role in immune regulation. Factors that can affect the early colonization of gut microbes in neo-nates include the mother's diet, antibiotic treatments, method of delivery, and surrounding environment [93]. On the surface of immune cells, Toll-like receptors (TLR) behave differently depending on the immune system's response, which allows distinguishing between pathogenic and native gut microbiota. Probiotics and their effector molecules affect the gut barrier by different methods such as modulation of mucus production, reduction in bacterial adhesion, and induction of IgA [94].

3. Association of Probiotics in Prevention of Diseases

Probiotics are very helpful in preventing chronic diseases by mediating their effects. They show positive effects on gut health and help in skin-related problems such as burns, scars, infections, wounds. They increase the skin's innate immunity and help to regenerate healthy skin [95]. The action of Saccharomyces cerevisiae dressing improved burn skin healing significantly as demonstrated by [96], whereas a hypothetical model of intestinal microbiota influencing wound healing through gut-brain-skin axes explains damaged tissue repair [97]. The distinction in gut microbiota constitution and decreased levels of *Bifidobacteria* and *lactobacillus* in infants' guts leads to the onset of allergic symptoms. Both the reports of specific probiotic strains and their results indicate that they can be used in the prevention of eczema. Dysbiosis, also known as dysbacteriosis, is a term used for imbalance in microbes inside the body, such as impaired microbiota, which is often related to inflammatory bowel disease, colonic cancer, metabolic syndrome, and allergic reactions. Improving the gut microbiota balance by different nutritional concepts or by ingesting specific micro-organisms led to significant improvement in health and decreased the risk of diseases or changing treatment mode [98].

The development of irritable bowel syndrome is associated with deviation in intestinal homeostasis, whose outcome is the uncontrolled immune response to gut microbiota by intestinal immune cells and epithelial cells, which results in complications including ulcers and fibrosis. A prebiotic is a valuable food substance that can be used to facilitate the growth of beneficial bacteria that modifies intestinal microbiota. Both probiotics and prebiotics are helpful for irritable bowel syndrome [97].

Probiotics have been involved in the healing process of intestinal ulcers and infected cutaneous wounds. Skin microbiota acts as a defensive barrier and can regulate the skin's inflammatory response to minor epidermal injury by decreasing and promoting cytokine production to maintain healthy skin [99]. The process through which probiotics show positive effects includes directly killing the pathogen, increasing the epithelial barrier, competitive displacement of pathogenic bacteria, and induction of fibroblasts [100].

Probiotics are also very beneficial for burn patients; they can reduce the bacterial load on the ulcer area [101]. Skin injuries cause disturbance in microbiota levels and increase the prevalence of bacteria that exert adverse effects on wounds. Additionally, having a wound causes stress, which results in alterations of neuro-endocrinal responses and impairs wound healing [102]. Chronic wounds are those that are difficult to heal and can exert a burden not only on the patient but also on the health care system, for example, diabetic foot ulcers (DFU), venous leg ulcers (VLU), and decubitus ulcers (DU). In chronic wounds, polymicrobial biofilms that promote pathogenic microbial growth and interfere in the process of wound healing are abundant and play an important role in the development of impaired wounds [103].

Probiotics play an important role in the treatment of autism by affecting the microbiome, which is present in the gut and responsible for imbalanced neuro-developmental conditions, like autism. If gene *Shank3* is disturbed by gut microbes, it affects a person's behavior and can lead to autism [104]. There are numerous medicines are available for treatment, but they have side effects; to avoid these side effects, probiotics are used as an alternative therapy. Probiotics alter the gene that is responsible for neurodevelopment and maintains the gut environment to treat this disease [105]. This bacterial disorder promotes different pathophysiological gastrointestinal syndromes like bowel syndrome, obesity, diarrhea, and food allergies [106].

Dietary biotic aid is used to maintain the GIT flora and relieve pain, vomiting, nausea, and bowel syndrome. The most potent probiotic strains used to treat GIT disorder are *L. rhamnosus GG, L. reuteri 17938, VSL #3,* and *Bifidobacteria* species [107]. With probiotic intake, stool PCR tests of autistic children indicated increased colony count of Bifidobacteria and lactobacilli with a major bodyweight loss and great progress in the severity of autism and gastrointestinal disorders [108]. In pregnancy, Interleukin-6, 17a cytokines prompt autism spectrum disorder. Probiotics play an important role in inhibiting the production of cytokines and preventing the autism spectrum disorder induced by maternal immune activation [109].

Osteoporosis is a disease that affects the skeletal system, manifesting as low bone mass density, deterioration of the skeletal system, and greater bone brittleness and sensitivity to crack. Most cracks are in the distal forearm, femur, and back. These cracks especially occur in postmenopausal women [110]. The cause of bone loss is due to the low level of the estrogen hormone because estrogen performs a significant role in developing and sustaining bones [111].

In America, many people face the problem of osteoporosis, and most people are at high risk of low bone mass. Osteoporosis can occur in males and females at any period of life, but it predominantly occurs in older females [112]. Probiotics act as a therapeutic agent that helps treat many bone diseases such as osteoporosis and rheumatoid arthritis [113]. There are many mechanisms of probiotics that affect bones; the most latent influence of probiotics on bone happens through the integration of vitamins. In the metabolism of calcium, vitamin D, C, K, and folate are linked and are necessary for bone development [114].

L. reuteri is a probiotic that plays an essential role in alleviating osteoporosis and improving bone density [115]. Osteocalcin serum (OC), osteoblasts, i.e., bone composition parameters, low serum C-terminal telopeptide (CTX) and osteoclasts, i.e., bone resorption parameters are increased by *B. longum* supplementation. *B. longum* changed the structure of the femur [116]. When bone loss in postmenopausal osteoporosis occurs due to the estrogen level being inadequate, intestinal and bacterial antigens traverse the arbitrated intestinal epithelium wall and inaugurate the immune response. Probiotics boost the intestinal epithelial wall, balance the deviant host immune responses, support intestinal calcium absorption, restore intestinal microbial diversity to prevent bone resorption, and help in the latest production of the estrogen-like substance [117]. Action and effects of probiotics on numerous diseases that have been reported by in vivo analysis are shown in Table 1.

Table 1. Effect and action mechanisms of probiotics to prevent diseases reported in various studies.

Probiotics	Type of Probiotics	Subjects of Administration	Duration of Intervention	Diseases	Potential Effects	Mechanism of Therapeutic Action	References
Lactobacillus	*L. acidophilus*	Humans	4 weeks	T2D	Improved the homoeostasis of glucose	Preservation and reduction of insulin sensitivity	[118]
	Limosilactobacillus reuteri	Ovx Mice	4 weeks	Menopausal bone loss	Loss in vertebral and femur bone was prevented, improvement in the density of bone	T-cells induced signals of osteoclast suppression by causing the reduction in osteoclastogenesis	[119]
	Lacticaseibacillus casei	Neonates	12 months	Enteric Colonization	Fungal diseases such as enteric colonization is preventable with the consumption of probiotic	Modification in the ecology of fungi carries out the application of mechanisms potentially by LCC in the gut. Increased mucosal responses of IgA include a significant fungi exclusion and reduction in colonization ability as well	[120]
	L. crispatus	Women	4 weeks	Recurrent UTI	Prevention of urinary tract infection such as vaginal flora infection and pregnancy issues due to urinary tract infection could be resolved	Probiotic was given as a vaginal suppository to indicate the reduction of infection in the urinary tract and enhanced the IgA	[121]
	L. gasseri	Rats	12 weeks	Obesity	Obesity and other weight gain problems could be resolved by oral uptake	Probiotic extracted from the milk of human breast effectively imposed influence on adipose tissues either by destroying cells or reducing their number	[122]
	Lactiplantibacillus plantarum	Mice	4 weeks	Spontaneous Colitis	Development of colitis due to the deficiency of IL usually under SPF conditions was prevented	Decrease the establishment of inflammatory colony inflammation due to mucosal IL	[123]
	Lacticaseibacillus rhamnosus	Epithelial cells	NR	Rotavirus	The development of an efficient immune system prevented rotavirus	Inoculation of IPEC-J2 cells with probiotics reduce the risk of rotavirus due to the anti-inflammatory properties	[124]

Table 1. *Cont.*

Probiotics	Type of Probiotics	Subjects of Administration	Duration of Intervention	Diseases	Potential Effects	Mechanism of Therapeutic Action	References
Bifidobacterium	*B. lactis*	Mice	12 weeks	Obesity	Reduced fat mass and weight gain	Significantly reduces adherence of mucosal bacteria in caecum and ileum	[125]
	B. bifidum	Mice	4 weeks	Inflammatory bowel disease	Assisted in controlling unusual responses of immunity in the tissues of intestine	Reduces lymphocyte infiltration and ameliorated the goblet cells reduction	[126]
	B. adolescentis	Cells	NR	Melanogenesis	Acted as antioxidant and inhibitory properties of melanoma made the *B. adolescentis*, a novel whitening agent for skin	Inhibition of tyrosinase action would lead to the reduction in melanogenesis, such as the melanoma process of cells	[127]
	B. infantis	Mice	7 days	Inflammatory bowel disease	The permeability of intestine could be reduced to treat IBD	The reduction in the infiltration of neutrophils and the inflammatory colon is reported	[128]
	B. breve	Mice	NR	Alzheimer's disease	Exhibit beneficial effects on peripheral tissues and the central nervous system and manages the neurodegenerative disorders	Non-viable bacterium metabolite or its components partially treat the cognitive decline while the specific probiotic suppresses the expressions and inflammation in the hippocampus	[129]
	B. longum	Mice	10 days	Gut derived sepsis	Treat opportunistic infection and significantly treat immunocompromised patients	Less *P. aeruginosa* viable count in the jejunum is reported and significantly repressed the *P. aeruginosa* adherence to the cell's monolayers	[130]
Other Species	*Escherichia coli*	Humans	12 weeks	Irritable bowel syndrome	Demonstrates beneficial effects to reduce the action of the syndrome of irritable bowel	Shows its action particularly in patients with enteric microflora having alteration, for instance, after antibiotics' administration or gastroenterocolitis	[131]
	Streptococcus thermophilus	Infants	5 months	Diarrhea	The formula that is obtained by fermentation can cause a reduction in the severe diarrhea	A combination of *Streptococcus thermophilus* and Bifidobacterium breve are fermented and interact with the immune system of the intestine to prevent acute diarrhea	[132]
	Bacillus subtilis	Rabbits	7 weeks	Immunodeficiency	Improvement in the mechanism of defense and immunity	Increases the weight of spleen and thymus prominently, provide innate immunity and induction of immunity on RK-13 cells	[133]
	Lactococcus lactis	Mice	5 days	Colitis	Acute and chronic colitis could be effectively managed to prevent damage to epithelial tissues	*L. lactis* which releases TFF are involved in the Intragastric administration at the colonic mucosa, in distinction to the administration of TFF which is purified, demonstrated to heal and prevent acute colitis induced by DSS	[134]

NR = not reported, EcN = Escherichia coli Nissle, Ovx = ovariectomized, RK-13 cells = rabbit kidney epithelial cell line, TFF = trefoil factors, DSS = dextran sodium sulfate, LCC = Lacticaseibacillus casei, IgA = Immunoglobulin A, UTI = urinary tract infection, IL = Interleukin, SPF = specific pathogen-free, IPEC-J2 = non-transformed porcine jejunum epithelial cell line, IBD = inflammatory bowel disease, AD = Alzheimer's disease.

3.1. Action Mechanism of Probiotics in Reduction of Obesity

Obesity is due to the union of microflora in the rectum area due to excessive absorption. Several new studies have confirmed that human intestinal micro-organisms play an important role in obesity through energy production and absorption of nutrients, and it has been shown that the duodenal microbiota is different in obese compared to lean persons. The upsurge of some gut microbial taxa such as *Escherichia coli*, *Staphylococcus aureus*, and other general bacterial gut fungi with *Bifidobacterium* has been shown to impact obesity [135].

Firmicutes are used directly as healing adjuvants, sold as named probiotics, prebiotics, or, usually, functional foods. In the United States, these products are characterized by the Food and Drug Administration and are generally recognized as safe. It is assumed that the gut of a fetus throughout the intrauterine period is deprived of any bacterial groups, i.e., it is thoroughly germ-free; however, during parturition, the transfer of micro-organisms from the mother to the fetus gut seems to occur and basic microbiota were found in it [136].

The mechanism credited with managing an upsurge in body fat was the capability of microbiota in extracting energy from food and controlling the host's energy balance. Deprivation of dietary polysaccharides and fiber by *Firmicutes* and *Bacteroides* in the gut resulted in the manufacture of SCFAs, such as acetate, propionate, and butyrate. Propionate is a significant energy source for the host from a mixture of glucose and lipids in the liver. A survey of 61 primary studies indicated differences in the microbial formation between overweight and non-obese subjects and the potential mechanisms involved [137].

Alterations are made homeostasis during the utility of dietary consumption and storehouse of lipids due to the balance of the abdominal microbiota. The microbiota is higher on metabolic pathways connected to obesity, including intonation with probiotics and prebiotics. This rise in *Bifidobacteria*, generally followed by weight gain and the increase in parameters linked to obesity [138].

Probiotics play a role in stopping the spread of bacteria and maintaining the gut environment to restore its normal flora. The ecology of the gut can be disturbed by numerous antibiotics plus some dietary products. Mostly probiotic strains come from gram-positive bacteria to treat gut disorders. To reduce tumor necrosis factor in the host, *L. reuteri* 6475 is important, and it also plays a vital role in promoting bone health and limiting bone absorption. Probiotic *Limosilactobacillus reuteri* has anti-inflammatory properties, particularly anti-TNFα properties [139].

Results demonstrated that these species reduce (LDL) low-density lipoprotein and (TC) total cholesterol, resulting in decreased coronary heart disease. A low level of leptin decreases obesity. LAB supplements also decrease LDL-C, aspartate aminotransferase (AST), alanine transaminase (ALT), high density lipoprotein (HDL), glucose, lipase, and triglycerides when a high fecal count is reported. LAB supplements also reduce intestinal microflora (tryptophanase, β-glucuronidase, and β-glucosidase). *Lactobacillus acidophilus* probiotics help ovariectomized subjects, intensify the microstructure of cortical and trabecular bone, expand the mineral density and diversity in bones, and prevent obesity [140].

Obesity levels has increased from 1980 to 2014 [141]. Obesity has an association with the high availability of energy and ambient temperature. An imbalance between intake and expenditure of energy is one of the significant causes of obesity. Gut microbiota influences the metabolism of the whole body by disturbing energy balance. They cause inflammation and barrier function in the gut, assimilating regulatory signals of central and peripheral food intake, thus enhancing body weight. Several physiologic functions are linked with probiotics that contribute to gut microbiota health and affect appetite, composition, food intake, metabolic functions, and body weight through gut bacterial community modulation and gastrointestinal pathways [142]. Figure 2 explains the anti-obesity mechanism of probiotic *bifidobacterium* spp., which is supplemented in the form of LAB supplement including *B. pseudocatenulatum* SPM 1204, *B. longum* SPM 1205, and *B. longum* SPM 1207.

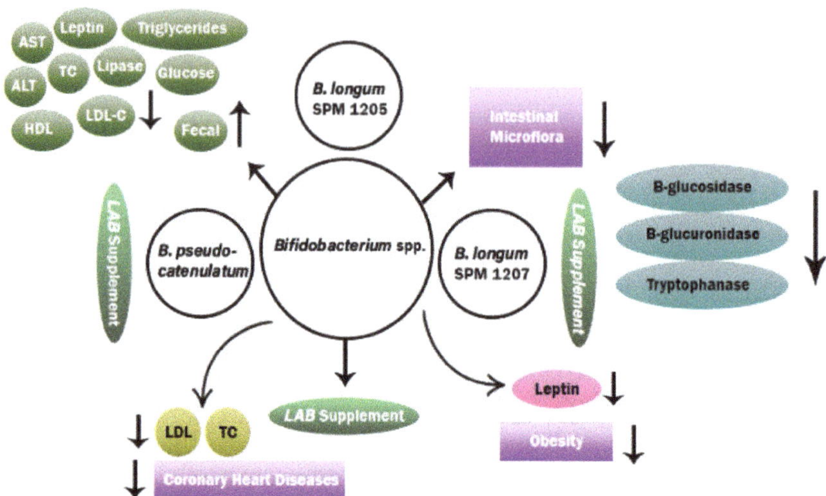

Figure 2. Probiotics (*Bifidobacterium* spp.) against obesity (adapted from [143]).

3.2. Action Mechanism of Probiotics in Minimizing T1D, T2D, GDM

A current major health problem is diabetes mellitus [144]. Failure in insulin secretion and insulin action leads to a state of hyperglycemia, which is a metabolic disease that ultimately results in diabetes mellitus [145]. Between 5 and 10% of people have type 1 diabetes, also known as term insulin-dependent diabetes or juvenile-onset diabetes, due to the cellular-arbitrated autoimmune disorder in these β-cells of the pancreas [146].

Diabetes mellitus type 2, also known as insulin resistance, is when the patient does not utilize the insulin correctly. As a result, the pancreas generates additional insulin to compensate, but it cannot impel the production of sufficient insulin to sustain the appropriate level of blood glucose. Gut-microbiota contours in diabetic patients are not regular. A spatial association of human metagenome investigation revealed an essential relationship between metabolic pathways in type 2 diabetes subjects, specific gut microbes, and bacterial genes the level of *Lactobacillus* spp. that were very different than that of the non-diabetic subject [147].

Glycated hemoglobin (HbA1c) values, fasting glucose level, and concentration of *Lactobacillus* species are positively correlated. *Clostridium* species concentration in the gut is positively correlated with high-density lipoprotein (HDL) cholesterol level and negatively related to fasting glucose, HbA1c, insulin concentration, C-peptide, and plasma triglyceride levels [148]. Gram-positive bacteria and coagulase-negative staphylococci were prevalent in diabetic subjects in higher proportions, primarily in diabetics with retinopathy [149].

In the development of T2D, many cytokines are included. In the intestinal immune system, there are fundamental signals that are involved in suppressing the physiological state of action in the gut and affect the probiotics in glucose metabolism [150]. There is enough research to recommend that focalization of diabetes plays an important role in oxidative damage and anti-oxidative ability. In diabetic rats, inhibition of lipid peroxidation and increases in the antioxidant content of glutathione, superoxide dismutase, catalase, and glutathione peroxidase due to probiotics decreases oxidative damage. Probiotics also can prevent insulin resistance due to their anti-diabetic properties, which, in turn, increases the liver's natural T killer cells [151].

Probiotics also can improve inflammation by accentuating TNF-α expression, insulin stability, and overcoming NF-kB tying activity. Probiotics increase the bioavailability of gliclazide, which improves glucose metabolism and also helps repress the intestinal assimilation of glucose and modulate the action of the autonomic nervous system. If the colonial microbial diversity increases, it helps with upholding the probity of the gastrointestinal lining, improving glucose homeostasis, reducing inflammation, sustaining insulin production, and accumulating nutrients from the diet [152].

Adverse effects are associated with gestational diabetes mellitus (GDM) for both mother and infant during pregnancy. It is difficult, but GDM can be prevented by adopting a healthy lifestyle. The bacteria composite, e.g., the gut microbiome is available in intestines that alter the pathways for glucose and lipid metabolism, and in some host-inflammatory settings, gut microbiome alteration was shown to affect responses. Probiotics are one of the ways to modify the gut microbiome. They cause alterations in the gut microbiome but also modify the pregnancy metabolic environment [153].

Lactobacillus has an immunomodulatory effect on the host immune system. Subdued expression of osteoclastogenic factors (IL-6, IL-17, TNF-α, and RANKL) and enhanced expression of anti-osteoclastogenic factors are due to the *Lactobacillus acidophilus*. *Lactobacillus acidophilus* also has therapeutic, osteo-protective action in bone well-being (via tweaking Treg-Th17 cell balance) in postmenopausal osteoporosis as well as preventing diabetes mellitus [154]. Figure 3 illustrates the mechanism action of probiotics (*Bacteroides dorei, Bifidobacterium*, and *Lactobacillus rhamnoses*) against type-1-diabetes (T1D), type-2-diabetes (T2D), and gestational diabetes mellitus (GDM).

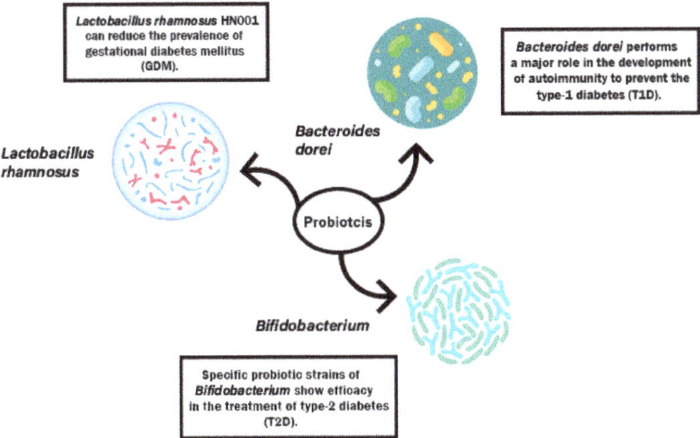

Figure 3. Mechanism action of probiotics (*Bacteroides dorei, bifidobacterium* and *Lactobacillus rhamnoses*) against type-1-diabetes, type-2-diabetes, and gestational diabetes mellitus adapted from [125,154–156].

3.3. Mechanism of Action of Probiotics against CKD

Probiotics play a vital role in reducing the risk and prevention of chronic diseases. There is an increasing concern in probiotics and prebiotics addition in the cases of chronic kidney disease (CKD) [157]. CKD increases at fluctuating rates, conditional on etiology and state of uremia. CKD is associated with consistent fatal disease build-up, and finally requires expensive treatments including peritoneal dialysis, kidney transplantation, and hemodialysis, which may prolong life [158].

The risk factor in the development of renal stones is known as hyperoxaluria, which can cause major damage to kidney function. The formation of oxalates primarily occurs. To ameliorate this condition, lactobacilli are given in the form of supplements that may decrease the formation of stones and reduce the chances of urolithiasis. *Lactobacillus casei* HY2743 and *L. casei* HY7201 may prevent the formation of oxalate [159].

Probiotics help the gut to maintain the luminal pH, produce antimicrobial peptides, influence function blockage by greater production of mucus and intonation of the host immune system, and transform the gut microflora. Gut microbiota plays an important role in the regulation of bone mass. The gut microbiota's effect on bone mass is propitiated through impacts on the immune system, whereas a change controls osteoclasto-genesis. In normal conditions bone-forming osteoblasts and bone-resorbing osteoclasts are constantly remodeled; when an imbalance occurs in this process, osteoporosis occurs [160]. Figure 4 represents the mechanism action of *Lactobacillus* spp. against oxalate stones.

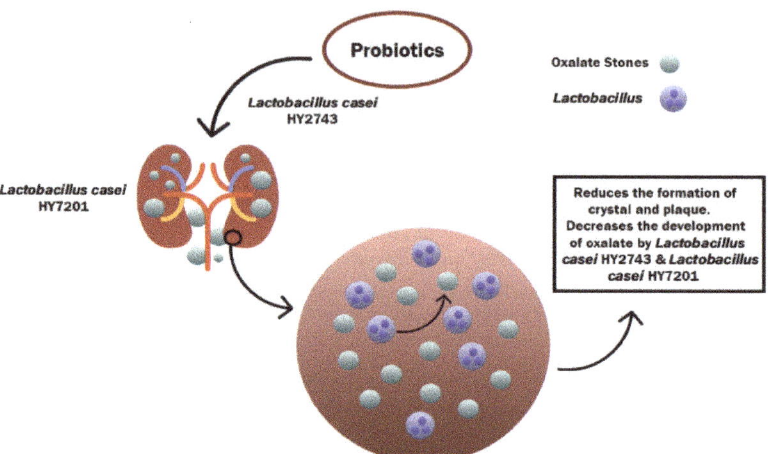

Figure 4. Prevention of kidney stones (oxalate) by probiotic (*Lactobacillus* spp.). Adapted from [161].

4. Probiotics for Animal Health

Sturdy pathogens such as Campylobacter and Salmonella can have a detrimental effect on the wellness and health of animals, as well as their development and reproductive abilities. These harmful bacteria can be transported via the food chain and into the human digestive organs, where they can cause serious human illnesses [162]. Numerous investigations on farm animals and normal or stressed humans have been conducted, which have proven the impact of various feed additives, and/or putative probiotics on good bacteria and decreased pathogen load in chickens, pigs, and ruminating animals [163].

5. Safety of Probiotics

A bacterial strain's safety, its source, antibiotic-resistant characteristics, and absolute lack of pathogenicity associated with virulent cultures all contribute to the safety profile's primary foundation; the rest is performance [164]. Probiotic performance promotes a variety of pathways, including adhesion to epithelial cells, decrease in gastrointestinal permeability, and immunoregulatory impacts [165]. Probiotics are not metabolized, have no potential for transference to animal-derived foods, and so do not result in the creation of residues. Due to the absence of their explicit and/or indirect transit from the gut into the animal body, they do not affect metabolic activities and therefore have no adverse effect [166].

6. Conclusions

Awareness regarding beneficial bacteria in the human gut originated in the 20th century, and by that time probiotics were considered an important player in intestinal and oral health. These gut microbes are associated with higher longevity. This associated effect between probiotics and long life is the basis of present research trends which are exploring the health-promoting properties and action mechanisms of gut microbes in cases of several prevalent diseases. Probiotics have anti-inflammatory properties because they can interfere with the working of cytokines and chemokines. Due to this vital property, they have been investigated in various diseases, especially in the case of chronic inflammatory diseases and GIT disorders. These gut microbes have a direct link with the overall nutritional status of a person as they help in the digestion and absorption of various nutrients, and therefore help regulate cholesterol levels and pancreatic functions. Due to their interference with cytokines and inflammation stimulation factors, they also tend to affect the immune system. Probiotics also tend to affect the gut barrier and thus also play a defensive role in human health.

7. Future Recommendations

More molecular studies are needed to ensure the safe use of some strains of bacteria, as insufficient data are available on the safety aspects of the probiotic strains. Not all microbes are beneficial and some of them can result in severe outcomes. A thorough study of the strain type used, its impact on health, and its interactions should be studied. Metabolomics exploration of human microbiome functionality with the involvement of bioinformatics may be a new horizon of research, without strong evidence the use of microbes as probiotics should not be encouraged.

Author Contributions: All authors have contributed to the final manuscript by division of responsibilities, i.e., literature search, compilation, write-up, final version development, English language editing, figures and table formation, formatting and reference styling, and proofreading. All authors have read and agreed to the published version of the manuscript.

Funding: This research received no external funding.

Institutional Review Board Statement: Not applicable.

Informed Consent Statement: Not applicable.

Data Availability Statement: Not applicable.

Conflicts of Interest: The authors declare no conflict of interest.

References

1. Guidance on Community Mental Health Services: Promoting Person-Centred and Rights-Based Approaches. Available online: https://www.who.int/publications/i/item/9789240025707 (accessed on 30 October 2021).
2. Ranjha, M.M.A.N.; Amjad, S.; Ashraf, S.; Khawar, L.; Safdar, M.N.; Jabbar, S.; Nadeem, M.; Mahmood, S.; Murtaza, M.A. Extraction of polyphenols from apple and pomegranate peels employing different extraction techniques for the development of functional date bars. *Int. J. Fruit Sci.* **2020**, *20*, S1201–S1221. [CrossRef]
3. Ranjha, M.M.A.N.; Irfan, S.; Lorenzo, J.M.; Shafique, B.; Kanwal, R.; Pateiro, M.; Arshad, R.N.; Wang, L.; Nayik, G.A.; Roobab, U.; et al. Sonication, a potential technique for extraction of phytoconstituents: A systematic review. *Process* **2021**, *9*, 1406. [CrossRef]
4. Nadeem, H.R.; Akhtar, S.; Ismail, T.; Sestili, P.; Lorenzo, J.M.; Ranjha, M.M.A.N.; Jooste, L.; Hano, C.; Aadil, R.M. Heterocyclic aromatic amines in meat: Formation, isolation, risk assessment, and inhibitory effect of plant extracts. *Foods* **2021**, *10*, 1466. [CrossRef] [PubMed]
5. Ranjha, M.M.A.N.; Irfan, S.; Nadeem, M.; Mahmood, S. A comprehensive review on nutritional value, medicinal uses, and processing of banana. *Food Rev. Int.* **2020**, 1–27. [CrossRef]
6. Salvatori, G. Effect of breast and formula feeding on gut microbiota shaping in newborns. *Front. Cell. Infect. Microbiol.* **2012**, *2*, 1–4. [CrossRef]
7. Grover, S.; Rashmi, H.M.; Srivastava, A.K.; Batish, V.K. Probiotics for human health—New innovations and emerging trends. *Gut Pathog.* **2012**, *4*, 15. [CrossRef]
8. Gupta, V.; Garg, R. Probiotics. *Indian J. Med. Microbiol.* **2009**, *27*, 202–210. [CrossRef]
9. Stambler, I. A History of Life-Extensionism in the Twentieth Century. Ph.D. Thesis, Bar Ilan University, Ramat Gan, Israel, 2017.
10. Amara, A.A.; Shibl, A. Role of probiotics in health improvement, infection control and disease treatment and management. *SAUDI Pharm. J.* **2013**, *23*, 107–114. [CrossRef]
11. Suskovic, J.; Kos, B.; Beganovic, J.; Pavunc, A.L. Antimicrobial activity—The most important property of probiotic and starter lactic acid bacteria. *Food Technol. Biotechnol.* **2010**, *9862*, 296–307.
12. Hill, C.; Guarner, F.; Reid, G.; Gibson, G.R.; Merenstein, D.J.; Pot, B.; Morelli, L.; Canani, R.B.; Flint, H.J.; Salminen, S.; et al. Expert consensus document: The International Scientific Association for Probiotics and Prebiotics consensus statement on the scope and appropriate use of the. *Nat. Rev. Gastroenterol. Hepatol.* **2014**, *11*, 506–514. [CrossRef]
13. Mei, G.; Carey, C.M.; Tosh, S.; Kostrzynska, M. Utilization of different types of dietary fibres by potential probiotics. *Can. J. Microbiol.* **2011**, *57*, 857–865. [CrossRef] [PubMed]
14. Aggarwal, J.; Swami, G.; Kumar, M. Probiotics and their effects on metabolic diseases: An update. *J. Clin. Diagn. Res.* **2012**, *7*, 173–177. [CrossRef] [PubMed]

15. Mu, Q.; Tavella, V.J.; Luo, X.M. Role of *Lactobacillus reuteri* in human health and diseases. *Front. Microbiol.* **2018**, *9*, 1–17. [CrossRef] [PubMed]
16. Khani, S.; Hosseini, H.M.; Taheri, M.; Nourani, M.R.; Fooladi, A.A.I. Probiotics as an alternative strategy for prevention and treatment of human diseases: A review. *Inflamm. Allergy-Drug Targets* **2012**, *11*, 79–89. [CrossRef] [PubMed]
17. Veerappan, G.R.; Betteridge, J.; Young, P.E. Probiotics for the treatment of inflammatory bowel disease. *Curr. Gastroenterol. Rep.* **2012**, *14*, 324–333. [CrossRef]
18. Roobab, U.; Aadil, R.M.; Zeng, X. Ultrasound. In *Advances in Noninvasive Food Analysis*; CRC Press: Boca Raton, FL, USA, 2019. [CrossRef]
19. Haukioja, A. Probiotics and oral health. *Eur. J. Dent.* **2010**, *4*, 348–355. [CrossRef]
20. Pagnini, C.; Saeed, R.; Bamias, G.; Arseneau, K.O.; Pizarro, T.T. Probiotics promote gut health through stimulation of epithelial innate immunity. *Proc. Natl. Acad. Sci. USA* **2010**, *107*, 1–6. [CrossRef]
21. Neffe-Skocińska, K.; Rzepkowska, A.; Szydłowska, A.; Kołożyn-Krajewska, D. Trends and possibilities of the use of probiotics in food production. In *Alternative and Replacement Foods*; Academic Press: Amsterdam, The Netherlands, 2018; ISBN 9780128114469.
22. Kim, K.M.; Yu, K.W.; Kang, D.H.; Suh, H.J. Anti-stress and anti-fatigue effect of fermented rice bran. *Phyther. Res.* **2002**, *16*, 700–702. [CrossRef]
23. Sales-Campos, H.; Soares, S.C.; Freire, C.J. An introduction of the role of probiotics in human infections and autoimmune diseases. *Crit. Rev. Microbiol.* **2019**, *45*, 413–432. [CrossRef]
24. Eslami, M.; Yousefi, B.; Kokhaei, P.; Hemati, M.; Nejad, Z.R.; Arabkari, V.; Namdar, A. Importance of probiotics in the prevention and treatment of colorectal cancer. *J. Cell. Physiol.* **2019**, *234*, 17127–17143. [CrossRef]
25. Behnsen, J.; Deriu, E.; Sassone-Corsi, M.; Raffatellu, M. Probiotics: Properties, examples, and specific applications. *Cold Spring Harb. Perspect. Med.* **2013**, *3*, a010074. [CrossRef]
26. Śliżewska, K.; Markowiak-Kopeć, P.; Śliżewska, W. The role of probiotics in cancer prevention. *Cancers* **2020**, *13*, 20. [CrossRef] [PubMed]
27. Altonsy, M.O.; Andrews, S.C.; Tuohy, K.M. Differential induction of apoptosis in human colonic carcinoma cells (Caco-2) by Atopobium, and commensal, probiotic and enteropathogenic bacteria: Mediation by the mitochondrial pathway. *Int. J. Food Microbiol.* **2010**, *137*, 190–203. [CrossRef] [PubMed]
28. Twetman, S.; Keller, M.K. Advances in dental research probiotics for caries prevention and control. *Adv. Dent. Res.* **2012**, *24*, 98–102. [CrossRef]
29. Mahasneh, S.A.; Mahasneh, A.M. Probiotics: A promising role in dental health. *Dent. J.* **2017**, *5*, 26. [CrossRef] [PubMed]
30. Louis, P.; Duncan, S.H. The role of the gut microbiota in nutrition and health. *Nat. Rev. Gastroenterol. Hepatol.* **2012**, *9*, 577–589. [CrossRef]
31. Velayudham, A.; Dolganiuc, A.; Ellis, M.; Petrasek, J.; Kodys, K.; Mandrekar, P.; Szabo, G. VSL#3 probiotic treatment attenuates fibrosis without changes in steatohepatitis in a diet-induced NASH model in mice. *Hepatology* **2009**, *49*, 989. [CrossRef]
32. Eslamparast, T.; Eghtesad, S.; Hekmatdoost, A.; Poustchi, H. Probiotics and nonalcoholic fatty liver disease. *Middle East J. Dig. Dis.* **2013**, *5*, 129. [PubMed]
33. Schrumpf, E.; Kummen, M.; Valestrand, L.; Greiner, T.U.; Holm, K.; Arulampalam, V.; Reims, H.M.; Baines, J.; Bäckhed, F.; Karlsen, T.H.; et al. The gut microbiota contributes to a mouse model of spontaneous bile duct inflammation. *J. Hepatol.* **2016**, *66*, 382–389. [CrossRef]
34. Ejtahed, H.S.; Niafar, M.; Mofid, V. Effect of probiotic yogurt containing *Lactobacillus acidophilus* and *Bifidobacterium lactis* on lipid profile in individuals with type 2 diabetes mellitus. *J. Dairy Sci.* **2011**, *94*, 3288–3294. [CrossRef]
35. Da Silva, C.C.; Monteil, M.A.; Davis, E.M. Overweight and Obesity in children are associated with an abundance of firmicutes and reduction of bifidobacterium in their gastrointestinal microbiota. *Child. Obes.* **2020**, *16*, 204–210. [CrossRef]
36. Wang, H.; Ren, P.; Mang, L.; Shen, N.; Chen, J.; Zhang, Y. In vitro fermentation of novel microwave-synthesized non-digestible oligosaccharides and their impact on the composition and metabolites of human gut microbiota. *J. Funct. Foods* **2019**, *55*, 156–166. [CrossRef]
37. Slevin, M.M.; Allsopp, P.J.; Magee, P.J.; Bonham, M.P.; Naughton, V.R.; Strain, J.J.; Duffy, M.E.; Wallace, J.M.; Sorley, E.M.M. Supplementation with calcium and short-chain fructo-oligosaccharides affects markers of bone turnover but not bone mineral density in postmenopausal women. *J. Nutr.* **2014**, *144*, 297–304. [CrossRef] [PubMed]
38. Verna, E.C.; Lucak, S. Use of probiotics in gastrointestinal disorders: What to recommend? *Therap. Adv. Gastroenterol.* **2010**, *3*, 307. [CrossRef] [PubMed]
39. Tortora, A.; Gabrielli, M. The role of intestinal microbiota and the immune system. *Eur. Rev. Med. Pharmacol. Sci.* **2013**, *17*, 323–333.
40. Wolff, B.J.; Price, T.K.; Joyce, C.J.; Wolfe, A.J.; Mueller, E.R. Oral probiotics and the female urinary microbiome: A double-blinded randomized placebo-controlled trial. *Int. Urol. Nephrol.* **2019**, *51*, 2149–2159. [CrossRef] [PubMed]
41. Abenavoli, L.; Scarpellini, E.; Colica, C.; Boccuto, L.; Salehi, B.; Sharifi-rad, J.; Aiello, V.; Romano, B.; De Lorenzo, A.; Izzo, A.A. Gut microbiota and obesity: A Role for Probiotics. *Nutrients* **2019**, *11*, 2690. [CrossRef]
42. Marques, A.M. Preclinical relevance of probiotics in type 2 diabetes: A systematic review. *Int. J. Exp. Pathol.* **2020**, *101*, 68–79. [CrossRef]

43. O'Toole, P.W.; Marchesi, J.R.; Hill, C. Next-generation probiotics: The spectrum from probiotics to live biotherapeutics. *Nat. Microbiol.* **2017**, *2*, 17057. [CrossRef]

44. Borresen, E.C.; Henderson, A.J.; Kumar, A.; Weir, T.L.; Ryan, E.P. Fermented foods: Patented approaches and formulations for nutritional supplementation and health promotion. *Recent Pat. Food Nutr. Agric.* **2012**, *4*, 134–140. [CrossRef]

45. Dekumpitiya, N.; Gamlakshe, D.; Indrika, S.; Jayaratne, D.L. Identification of the microbial consortium in Sri Lankan buffalo milk curd and their growth in the presence of prebiotics. *J. Food Sci. Technol.* **2016**, *9*, 20–30. [CrossRef]

46. Granato, D.; Branco, G.F.; Nazzaro, F.; Faria, A.F.; Cruz, A.G. Functional foods and nondairy probiotic food development: Trends, concepts, and products. *Compre. Rev. Food Sci. Food Saf.* **2010**, *9*, 292–302. [CrossRef]

47. Sullivan, J.N.O.; Rea, M.C.; Sullivan, J.N.O.; Rea, M.C.; Hill, C.; Ross, R.P. Protecting the outside: Biological tools to manipulate the skin microbiota. *Microbiol. Ecol.* **2020**, *96*. [CrossRef]

48. Ghishan, F.K.; Kiela, P.R. From probiotics to therapeutics: Another step forward? *J. Clin. Invest.* **2011**, *121*, 2149–2152. [CrossRef]

49. Islam, S.U. Clinical uses of probiotics. *Medicine* **2016**, *95*, e2658. [CrossRef] [PubMed]

50. Harneet Singh, H.S. Probiotics—An emerging concept. *Int. J. Sci. Res. Publ.* **2014**, *4*, 1–3.

51. Salvetti, E.; Torriani, S.; Felis, G.E. The genus *Lactobacillus*: A taxonomic update related papers. *Probiotics Antimicrob. Proteins* **2012**, *4*, 217–226. [CrossRef]

52. Zommiti, M.; Feuilloley, M.G.J.; Connil, N. Update of probiotics in human world: A nonstop source of benefactions till the end of time. *Micro-organisms* **2020**, *8*, 1907. [CrossRef]

53. Shokryazdan, P.; Jahromi, M.F.; Liang, J.B.; Wan, Y.; Shokryazdan, P.; Faseleh, M.; Boo, J.; Wan, Y. Probiotics: From isolation to application. *J. Am. Coll. Nutr.* **2017**, *36*, 666–676. [CrossRef]

54. Shewale, R.N.; Sawale, P.D.; Khedkar, C.D.; Singh, A. Selection criteria for probiotics: A review Department of Dairy Microbiology College of Dairy Technology, Pusad, India. *Int. J. Probiotics Prebiotics* **2014**, *9*, 17–22.

55. Jiang, T.; Li, H.-S.; Han, G.G.; Singh, B.; Kang, S.-K. Oral delivery of probiotics in poultry using pH-sensitive tablets. *J. Microbiol. Biotechnol.* **2017**, *27*, 739–746. [CrossRef] [PubMed]

56. Dhama, K.; Verma, V.; Sawant, P.M.; Tiwari, R.; Vaid, R.K.; Chauhan, R.S. Applications of probiotics in poultry: Enhancing immunity and beneficial effects on production performances and health—A Review. *J. Immunol. Immunopathol.* **2011**, *13*, 1–19.

57. Kechagia, M.; Basoulis, D.; Konstantopoulou, S.; Dimitriadi, D.; Gyftopoulou, K.; Skarmoutsou, N.; Fakiri, E.M. Health benefits of probiotics: A review. *Int. Sch. Res. Not.* **2013**, *2013*. [CrossRef] [PubMed]

58. Venugopalan, V.; Shriner, K.A.; Wong-Beringer, A. Regulatory oversight and safety of probiotic use. *Emerg. Infect. Dis.* **2010**, *16*, 1661–1665. [CrossRef] [PubMed]

59. Santacroce, L.; Charitos, I.A.; Bottalico, L. A successful history: Probiotics and their potential as antimicrobials. *Expert Rev. Anti. Infect. Ther.* **2019**, *17*, 635–645. [CrossRef]

60. Sivamaruthi, B.S.; Fern, L.A.; Siti, D.; Rashidah, N.; Hj, P.; Chaiyasut, C. The influence of probiotics on bile acids in diseases and aging. *Biomed. Pharmacother.* **2020**, *128*, 110310. [CrossRef]

61. Chaucheyras-Durand, F.; Fonty, G. Establishment of cellulolytic bacteria and development of fermentative activities in the rumen of gnotobiotically-reared lambs receiving the microbial additive Saccharomyces cerevisiae CNCM I-1077. *Reprod. Nutr. Dev.* **2001**, *41*, 57–68. [CrossRef]

62. Galanis, A.; Kourkoutas, Y.; Tassou, C.C.; Chorianopoulos, N. Detection and identification of probiotic *Lactobacillus plantarum* strains by multiplex PCR using RAPD-derived primers. *Int. J. Mol. Sci.* **2015**, *16*, 25141–25153. [CrossRef]

63. Sattler, V.A.; Mohnl, M.; Klose, V. Development of a strain-specific real-time PCR Assay for enumeration of a probiotic *Lactobacillus reuteri* in chicken feed and intestine. *PLoS ONE* **2014**, *9*, e90208. [CrossRef]

64. Collins, J.; Van Pijkeren, J.-P.; Svensson, L.; Claesson, M.; Sturme, M.; Li, Y.; Cooney, J.; Van Sinderen, D.; Walker, A.; Parkhill, J.; et al. Fibrinogen-binding and platelet-aggregation activities of a *Lactobacillus salivarius* septicaemia isolate are mediated by a novel fibrinogen-binding protein. *Mol. Microbiol.* **2012**, *85*, 862–877. [CrossRef]

65. Fenster, K.; Freeburg, B.; Hollard, C.; Wong, C.; Laursen, R.R.; Ouwehand, A.C. The production and delivery of probiotics: A review of a practical approach. *Micro-organisms* **2019**, *7*, 83. [CrossRef]

66. Allaker, R.P.; Stephen, A.S. Use of probiotics and oral health. *Curr. Oral Health Rep.* **2017**, *4*, 309–318. [CrossRef]

67. Hung, A. Probiotics and Oral health: A Sytematic Review. *Int. J. Sci. Eng. Res.* **2011**, *2*.

68. Lin, T.; Lin, C.; Pan, T.; Lin, T. The implication of probiotics in the prevention of dental caries. *Appl. Microbiol. Biotechnol.* **2017**, *102*, 577–586. [CrossRef] [PubMed]

69. Ramburrun, P.; Pringle, N.A.; Dube, A.; Adam, R.Z.; Souza, S.D.; Aucamp, M. Recent advances in the development of antimicrobial and antifouling biocompatible materials for dental applications. *Materials* **2021**, *14*, 3167. [CrossRef] [PubMed]

70. Saraf, K.; Shashikanth, M.C.; Priya, T.; Sultana, N.; Chaitanya, N.C.S.K. Probiotics—Do they have a role in medicine and dentistry? *J. Assoc. Phys. India* **2010**, *58*, 488–490.

71. Hashemi, A.; Villa, C.R.; Comelli, E.M. Probiotics in early life: A preventative and treatment approach. *Food Funct.* **2016**, *7*, 1752–1768. [CrossRef]

72. Lefevre, M.; Racedo, S.M.; Denayrolles, M.; Ripert, G.; Lobach, A.R.; Simon, R.; Pélerin, F.; Jüsten, P.; Urdaci, C. Safety assessment of *Bacillus subtilis* CU1 for use as a probiotic in humans. *Regul. Toxicol. Pharmacol.* **2016**, *83*, 54–65. [CrossRef]

73. Vera, R.; Visitacion, D.; Guzman, G.; Acids, D.; Nutrition, M. Probiotics prevent dysbiosis and the raise in blood pressure in genetic hypertension: Role of short-chain fatty acids. *Mol. Nutr. Food Res.* **2020**, *64*, e1900616. [CrossRef]

74. Ducatelle, R.; Eeckhaut, V.; Haesebrouck, F.; Van Immerseel, F. A review on prebiotics and probiotics for the control of dysbiosis: Present status and future perspectives. *Anim. Int. J. Anim. Biosci.* **2015**, *9*, 43–48. [CrossRef]

75. Bustamante, M.; Oomah, B.D.; Mosi-Roa, Y.; Rubilar, M. Probiotics as an adjunct therapy for the treatment of halitosis, dental caries and periodontitis. *Probiotics Antimicrob. Proteins* **2019**, *12*, 325–334. [CrossRef]

76. Shringeri, P.I.; Fareed, N.; Battur, H.; Khanagar, S. Role of probiotics in the treatment and prevention of oral malodor/halitosis: A systematic review. *J. Indian Assoc. Public Health Dent.* **2019**, *17*, 90. [CrossRef]

77. Nagamine, Y.; Hasibul, K.; Ogawa, T.; Tadab, A.; Kamitoric, K.; Hossainc, A.; Yamaguchid, F.; Tokudae, M.; Kuwaharab, T.; Miyakea, M. D-tagatose effectively reduces the number of streptococcus mutans and oral bacteria in healthy adult subjects: A chewing gum pilot study and randomized clinical trial. *Acta Med. Okayama* **2020**, *74*, 307–317. [PubMed]

78. Faujdar, S.S.; Mehrishi, P.; Sharma, A. Role of probiotics in human health and disease: An update role of probiotics in human health and disease: An update. *Int. J. Curr. Microbiol. Appl. Sci.* **2016**, *5*, 328–344. [CrossRef]

79. Reddy, R.S.; Swapna, L.A.; Ramesh, T.; Singh, T.R.; Vijayalaxmi, N.; Lavanya, R. Bacteria in oral health—Probiotics and prebiotics: A review. *Int. J. Biol. Med. Res.* **2011**, *2*, 1226–1233.

80. Sharma, V.; Sharma, N.; Sheikh, I.; Kumar, V.; Sehrawat, N.; Yadav, M. Probiotics and prebiotics having broad spectrum anticancer therapeutic potential: Recent trends and future perspectives. *Curr. Pharmacol. Rep.* **2021**, *7*, 67–79. [CrossRef]

81. Manuel, J.; Rodrı, M.A.; Macı, M.G.M.F.A.; Microflora, D.Á.E.Á. Soy isoflavones and their relationship with microflora: Beneficial effects on human health in equol producers. *Phtochem. Rev.* **2013**, *12*, 979–1000. [CrossRef]

82. Makki, K.; Deehan, E.C.; Walter, J.; Bäckhed, F. Review the impact of dietary fiber on gut microbiota in host health and disease. *Cell Host Microbe* **2018**, *23*, 705–715. [CrossRef]

83. Team, R.; Guarner, F.; Khan, A.G.; Garisch, J.; Africa, S.; Eliakim, R.; Gangl, A.; Thomson, A.; France, J.K.; Lemair, T.; et al. World gastroenterology organisation global guidelines probiotics and prebiotics October 2011. *J. Clin. Gastroenterol.* **2012**, *46*, 119–129.

84. Zheng, D.; Liwinski, T.; Elinav, E. Interaction between microbiota and immunity in health and disease. *Cell Res.* **2020**, *30*, 492–506. [CrossRef]

85. Salonen, A.; Philippou, E. Impact of diet on human intestinal microbiota and health. *Annu. Rev. Food Sci. Technol.* **2014**, *5*, 6–24. [CrossRef] [PubMed]

86. Daniali, M.; Nikfar, S.; Abdollahi, M. Expert review of endocrinology & metabolism a brief overview on the use of probiotics to treat overweight and obese patients. *Expert Rev. Endocrinol. Metab.* **2020**, *15*, 1–4. [CrossRef] [PubMed]

87. Michael, D.R.; Jack, A.A.; Masetti, G.; Davies, T.S.; Loxley, K.E.; Kerry-Smith, J.; Plummer, J.F.; Marchesi, J.R.; Mullish, B.H.; McDonald, J.; et al. A randomised controlled study shows supplementation of overweight and obese adults with lactobacilli and bifidobacteria reduces bodyweight and improves. *Sci. Rep.* **2020**, *10*, 1–12. [CrossRef] [PubMed]

88. Tan, H.; Toole, P.W.O. Impact of diet on the human intestinal microbiota. *Curr. Opin. Food Sci.* **2014**, *2*, 71–77. [CrossRef]

89. Scaldaferri, F.; Gerardi, V.; Lopetuso, L.R.; Del Zompo, F.; Mangiola, F.; Boškoski, I.; Bruno, G.; Petito, V.; Laterza, L.; Cammarota, G.; et al. Gut microbial flora, prebiotics, and probiotics in IBD: Their current usage and utility. *Bio Med Res. Int.* **2013**, *2013*. [CrossRef]

90. Lee, E.; Song, E.; Nam, Y.; Lee, S. Probiotics in human health and disease: From nutribiotics to pharmabiotics. *J. Microbiol.* **2018**, *56*, 773–782. [CrossRef] [PubMed]

91. Ohashi, Y.; Ushida, K. Health-beneficial effects of probiotics: Its mode of action. *Anim. Sci. J.* **2009**, *80*, 361–371. [CrossRef]

92. Martin, R.; Makino, H.; Yavuz, A.C.; Ben-Amor, K.; Roelofs, M.; Ishikawa, E.; Kubota, H.; Swinkels, S.; Sakai, T.; Oishi, K.; et al. Early-life events, including mode of delivery and type of feeding, siblings and gender, shape the developing gut microbiota. *PLoS ONE* **2016**, *11*, e0158498. [CrossRef]

93. Lin, L.; Zhang, J. Role of intestinal microbiota and metabolites on gut homeostasis and human diseases. *BMC Immunol.* **2017**, *18*, 2. [CrossRef]

94. Ozyurt, V.H.; Otles, S. Properties of probiotics and encapsulated probiotics in food. *Acta Sci. Pol. Technol. Aliment.* **2014**, *13*, 413–424. [CrossRef]

95. Delgado, S.; Gueimonde, M.; Margolles, A. Probiotics, gut microbiota, and their influence on host health and disease. *Mol. Nutr. Food Res.* **2017**, *61*, 1600240. [CrossRef]

96. Oryan, A.; Jalili, M.; Kamali, A.; Nikahval, B. The concurrent use of probiotic micro-organism and collagen hydrogel/scaffold enhances burn wound healing: An in vivo evaluation. *Burns* **2018**, *44*, 1775–1786. [CrossRef] [PubMed]

97. Lukic, J.; Chen, V.; Strahinic, I.; Begovic, J.; Lev-tov, H.; Davis, C.; Tomic-canic, M.; Pastar, I. Probiotics or pro-healers the role of beneficial bacteria in tissue repair. *Wound Repair Regen.* **2017**, *25*, 912–922. [CrossRef]

98. Salonen, A.; De Vos, W.M.; Palva, A. Gastrointestinal microbiota in irritable bowel syndrome: Present state and review gastrointestinal microbiota in irritable bowel syndrome: Present state and perspectives. *Microbiology* **2010**, *156*, 3205–3215. [CrossRef] [PubMed]

99. Iacono, A.; Mattace, G.; Berni, R.; Calignano, A.; Meli, R. Probiotics as an emerging therapeutic strategy to treat NAFLD: Focus on molecular and biochemical mechanisms. *J. Nutr. Biochem.* **2011**, *22*, 699–711. [CrossRef]

100. Nole, K.L.B.; Mph, E.Y.; Keri, J.E. Probiotics and prebiotics in dermatology. *J. Am. Dermatol.* **2014**, *22*, 699–711. [CrossRef]

101. Holmes, C.J.; Plichta, J.; Gamelli, R.L.; Radek, K.A. Dynamic role of host stress responses in modulating the cutaneous microbiome: Implications for wound healing and infection. *Wound Health Soc.* **2014**, *4*, 24–37. [CrossRef]

102. Tsiouri, M.G. Human microflora, probiotics and wound healing. *Biochem. Pharmacol.* **2017**, *19*, 33–38. [CrossRef]

103. Moratalla, A.Z.; De Lagrán, M.M.; Dierssen, M. Neurodevelopmental disorders: 2021 update. *Neuropathology* **2021**, *6*. [CrossRef]

104. Navarro, F.; Liu, Y.; Rhoads, J.M.; Navarro, F.; Liu, Y.; Rhoads, J.M. Can probiotics benefit children with autism spectrum disorders? *World J. Gasrtoenterol.* **2016**, *22*, 10093–10102. [CrossRef]

105. Daulatzai, M.A. Chronic functional bowel syndrome enhances gut-brain axis dysfunction, neuroinflammation, cognitive impairment, and vulnerability to dementia. *Neurochem. Res.* **2014**, *39*, 624–644. [CrossRef] [PubMed]

106. Roman, A.P.; Abalo, R.; Marco, E.M.; Cardona, D. Probiotics in digestive, emotional and pain-related disorders. *Behav. Pharmacol.* **2018**, *29*, 103–119. [CrossRef] [PubMed]

107. Larroya-García, A.; Navas-Carrillo, D.; Orenes-Piñero, E. Impact of gut microbiota on neurological diseases: Diet composition and novel treatments. *Critical* **2019**, *59*, 3102–3116. [CrossRef]

108. Hori, T.; Matsuda, K. Probiotics: A dietary factor to modulate the gut microbiome, host immune system, and gut-brain interaction. *Micro-organisms* **2020**, *8*, 1401. [CrossRef] [PubMed]

109. Liu, W.; Li, M.; Yi, L. Identifying children with autism spectrum disorder based on their face processing abnormality: A machine learning framework. *Autrism Res.* **2016**, *9*, 888–898. [CrossRef]

110. Collins, F.L.; Rios-Arce, N.D.; Schepper, J.D.; Parameswaran, N.; McCabe, L.R. The potential of probiotics as a therapy for osteoporosis. *Microbiol. Spectr.* **2017**, *5*, 213–233. [CrossRef]

111. Svensson, H. *Finding Ways Forward with Pain as a Fellow Traveler Older Women' S Experience of Living with Osteoporotic Vertebral Compression Fractures and Back Pain*; Institute of Health and Care Sciences Sahlgrenska Academy at the University of Gothenburg: Gothenburg, Sweden, 2018; ISBN 9789162904647.

112. Han, B. *Therapy of Social Medicine*; Springer: Berlin/Heidelberg, Germany, 2016; ISBN 9789812877475.

113. Traber, M.G.; Atkinson, J. Vitamin E, antioxidant and nothing more. *Free Radic. Biol. Med.* **2007**, *43*, 4–15. [CrossRef]

114. Li, P.; Sundh, D.; Lorentzon, M. Metabolic alterations in older women with low bone mineral density supplemented with *Lactobacillus reuteri*. *JBMR Plus* **2021**, *5*, 1–14. [CrossRef] [PubMed]

115. Yu, J.; Wong, H.S.; Cheung, W.-H. The role of gut microbiota in bone homeostasis a systematic review of preclinical animal studies. *Bone Jt. Res.* **2021**, *10*, 51–59. [CrossRef]

116. Ohlsson, C.; Engdahl, C.; Fåk, F.; Andersson, A.; Windahl, S.H.; Farman, H.H.; Movérare-Skrtic, S.; Islander, U.; Sjögren, K. Probiotics protect mice from ovariectomy-induced cortical bone loss. *PLoS ONE* **2014**, *9*, e92368. [CrossRef]

117. Anderson, J.L.; Miles, C.; Tierney, A.C. Effect of probiotics on respiratory, gastrointestinal and nutritional outcomes in patients with cystic fi brosis: A systematic review. *J. Cyst. Fibros.* **2016**, *16*, 186–197. [CrossRef] [PubMed]

118. Andreasen, A.S.; Larsen, N.; Pedersen-Skovsgaard, T.; Berg, R.; Møller, K.; Svendsen, K.D.; Jakobsen, M.; Pedersen, B.K. Effects of *Lactobacillus acidophilus* NCFM on insulin sensitivity and the systemic inflammatory response in human subjects. *Br. J. Nutr.* **2010**, *104*, 1831–1838. [CrossRef] [PubMed]

119. Britton, R.A.; Irwin, R.; Quach, D.; Schaefer, L.; Zhang, J.; Lee, T.; Parameswaran, N. ProbioticL. reuteriTreatment prevents bone loss in a menopausal ovariectomized mouse model. *J. Cell. Physiol.* **2014**, *229*, 1822–1830. [CrossRef] [PubMed]

120. Romeo, M.G.; Romeo, D.M.; Trovato, L.; Oliveri, S.; Palermo, F.; Cota, F.; Betta, P. Role of probiotics in the prevention of the enteric colonization by Candida in preterm newborns: Incidence of late-onset sepsis and neurological outcome. *J. Perinatol.* **2011**, *31*, 63–69. [CrossRef]

121. Grin, P.M.; Kowalewska, P.M.; Alhazzani, W.; Fox-robichaud, A.E. Lactobacillus for preventing recurrent urinary tract infections in women: Meta-analysis. *Can. J. Urol.* **2013**, *20*, 6607–6614.

122. Kang, J.; Yun, S.; Park, H. Effects of *Lactobacillus gasseri* BNR17 on body weight and adipose tissue mass in diet-induced overweight rats. *J. Microbiol.* **2010**, *48*, 712–714. [CrossRef]

123. Keubler, L.M.; Buettner, M.; Häger, C.; Bleich, A. A multihit model: Colitis lessons from the interleukin-10–deficient mouse. *Inflamm. Bowel Dis.* **2015**, *21*, 1967–1975. [CrossRef]

124. Liu, F. Porcine Small Intestinal Epithelial Cell Line (IPEC-J2) of rotavirus infection as a new model for the study of innate immune responses to rotaviruses and probiotics. *Viral Immunol.* **2010**, *23*, 135–149. [CrossRef]

125. Stenman, L.K.; Waget, A.; Garret, C.; Klopp, P.; Burcelin, R.; Lahtinen, S. Potential probiotic Bifidobacterium animalis ssp. lactis 420 prevents weight gain and glucose intolerance in diet-induced obese mice. *Benef. Microbes* **2014**, *5*, 437–445. [CrossRef]

126. Wickens, K.L.; Barthow, C.A.; Murphy, R.; Abels, P.R.; Maude, R.M.; Stone, P.R.; Mitchell, E.A.; Stanley, T.V.; Purdie, G.L.; Kang, J.M.; et al. Early pregnancy probiotic supplementation with *Lactobacillus rhamnosus* HN001 may reduce the prevalence of gestational diabetes mellitu: A randomised controlled trial. *Br. J. Nutr.* **2017**, *117*, 804–813. [CrossRef]

127. Chang, H.H.T. Antioxidative properties and inhibitory effect of *Bifidobacterium adolescentis* on melanogenesis. *J. Microbiol. Biotechnol.* **2012**, *28*, 2903–2912. [CrossRef]

128. Elian, S.; Souza, E.; Vieira, A.; Teixeira, M.; Arantes, R.; Nicoli, J.; Martins, F. *Bifidobacterium longum* subsp. infantis BB-02 attenuates acute murine experimental model of inflammatory bowel disease. *Benef. Microbes* **2015**, *6*, 277–286. [CrossRef]

129. Kobayashi, Y.; Kuhara, T.; Oki, M.; Xiao, J. Effects of *Bifidobacterium breve* A1 on the cognitive function of older adults with memory complaints: A randomised, double-blind, placebo-controlled trial Abstract. *Benef. Microbes* **2019**, *10*, 511–520. [CrossRef] [PubMed]

130. Khailova, L.; Petrie, B.; Baird, C.H.; Dominguez Rieg, J.A.; Wischmeyer, P.E. *Lactobacillus rhamnosus* GG and *Bifidobacterium longum* attenuate lung injury and inflammatory response in experimental sepsis. *PLoS ONE* **2014**, *9*, e97861. [CrossRef] [PubMed]

131. Kruis, W.; Chrubasik, S.; Boehm, S.; Stange, C.; Schulze, J. A double-blind placebo-controlled trial to study therapeutic effects of probiotic Escherichia coli Nissle 1917 in subgroups of patients with irritable bowel syndrome. *Int. J. Colorectal Dis.* **2012**, *27*, 467–474. [CrossRef] [PubMed]

132. Tims, S.; Roelofs, M.; Roug, C.; Rakza, T.; Chirico, G.; Roeselers, G.; Knol, J.; Christophe, J.; Turck, D. Fermented infant formula (with Bi fi dobacterium breve C50 and *Streptococcus thermophilus* O65) with prebiotic oligosaccharides is safe and modulates the gut microbiota towards a microbiota closer to that of breastfed infants. *Clin. Nutr.* **2020**, *40*, 778–787. [CrossRef]

133. Guo, M.; Wu, F.; Hao, G.; Qi, Q.; Li, R.; Li, N.; Chai, T. *Bacillus subtilis* improves immunity and disease resistance in rabbits. *Front. Immunol.* **2017**, *8*, 354. [CrossRef] [PubMed]

134. Takiishi, T.; Korf, H.; Van Belle, T.L.; Robert, S.; Grieco, F.A.; Caluwaerts, S.; Galleri, L.; Spagnuolo, I.; Steidler, L.; Van Huynegem, K.; et al. Reversal of autoimmune diabetes by restoration of antigen-specific tolerance using genetically modified *Lactococcus lactis* in mice. *J. Clin. Invest.* **2012**, *122*, 1717–1725. [CrossRef] [PubMed]

135. Joffin, N.; Jaubert, A.M.; Durant, S.; Barouki, R.; Forest, C.; Noirez, P. Citrulline counteracts overweight- and aging-related effects on adiponectin and leptin gene expression in rat white adipose tissue. *Biochim. Open* **2015**, *1*, 1–5. [CrossRef]

136. Younossi, Z.M.; Koenig, A.B.; Abdelatif, D.; Fazel, Y.; Henry, L.; Wymer, M. Global epidemiology of nonalcoholic fatty liver disease—Meta-analytic assessment of prevalence, incidence, and outcomes. *Hepatology* **2016**, *64*, 73–84. [CrossRef]

137. Wici, M. Probiotics for the treatment of overweight and obesity in humans—A review of clinical trials. *Micro-organisms* **2020**, *8*, 1148.

138. Borgeraas, H.; Johnson, L.K.; Skattebu, J.; Hertel, J.K.; Hjelmesæth, J. Effects of probiotics on body weight, body mass index, fat mass and fat percentage in subjects with overweight or obesity: A systematic review and meta-analysis of randomized controlled trials. *Etiol. Pathophysiol.* **2018**, *19*, 219–232. [CrossRef] [PubMed]

139. Howarth, G.S.; Wang, H. Role of endogenous microbiota, probiotics and their biological products in human health. *Nutrients* **2013**, *5*, 58–81. [CrossRef] [PubMed]

140. Obesity, T.; Brusaferro, A.; Cozzali, R.; Orabona, C.; Biscarini, A.; Farinelli, E.; Cavalli, E.; Grohmann, U.; Principi, N.; Esposito, S. Is it time to use probiotics to prevent or treat obesity? *Nutrients* **2018**, *10*, 1613. [CrossRef]

141. James, P.T.; Leach, R.; Kalamara, E.; Shayeghi, M. The worldwide obesity epidemic. *Obes. Res.* **2001**, *9*, 228S–233S. [CrossRef] [PubMed]

142. Kobyliak, N.; Conte, C.; Cammarota, G.; Haley, A.P.; Styriak, I.; Gaspar, L.; Fusek, J.; Rodrigo, L.; Kruzliak, P. Probiotics in prevention and treatment of obesity: A critical view. *Nutr. Metab.* **2016**, *13*, 1–13. [CrossRef] [PubMed]

143. An, H.M.; Park, S.Y.; Lee, D.K.; Kim, J.R.; Cha, M.K.; Lee, S.W.; Lim, H.T.; Kim, K.J.; Ha, N.J. Antiobesity and lipid-lowering effects of Bifidobacterium spp. in high fat diet-induced obese rats. *Lipids Health Dis.* **2011**, *10*, 116–118. [CrossRef]

144. Ranjha, M.M.A.N.; Shafique, B.; Wang, L.; Irfan, S.; Safdar, M.N.; Murtaza, M.A.; Nadeem, M.; Mahmood, S.; Mueen-ud-Din, G.; Nadeem, H.R. A comprehensive review on phytochemistry, bioactivity and medicinal value of bioactive compounds of pomegranate (*Punica granatum*). *Adv. Tradit. Med.* **2021**, 1–21. [CrossRef]

145. Mishra, S.; Wang, S.; Nagpal, R.; Miller, B.; Singh, R.; Taraphder, S.; Yadav, H. Probiotics and prebiotics for the amelioration of type 1 diabetes: Present and future perspectives. *Micro-organisms* **2019**, *7*, 67. [CrossRef]

146. Razmpoosh, E.; Javadi, M.; Ejtahed, H.; Mirmiran, P. Probiotics as beneficial agents in the management of diabetes mellitus: A systematic review. *Diabetes. Metab. Res. Rev.* **2016**, *32*, 143–168. [CrossRef]

147. Sun, Z.; Sun, X.; Li, J.; Li, Z.; Hu, Q.; Li, L. Using probiotics for type 2 diabetes mellitus intervention: Advances, questions, and potential. *Crit. Rev. Food Sci. Nutr.* **2020**, *60*, 670–683. [CrossRef] [PubMed]

148. Salgaço, M.K.; Garcia, L.; Oliveira, S.; Costa, G.N.; Bianchi, F.; Sivieri, K. Relationship between gut microbiota, probiotics, and type 2 diabetes mellitus. *Appl. Microbiol. Biotechnol.* **2019**, *103*, 9229–9238. [CrossRef]

149. Ardeshirlarijani, E.; Tabatabaei-malazy, O.; Mohseni, S.; Qorbani, M.; Larijani, B. Effect of probiotics supplementation on glucose and oxidative stress in type 2 diabetes mellitus: A meta-analysis of randomized trials. *J. Pharm. Sci.* **2019**, *27*, 827–837. [CrossRef]

150. Alokail, M.S.; Sabico, S.; Al-Saleh, Y.; Al-Daghri, N.M.; Alkharfy, K.M.; Vanhoutte, P.M.; Mcternan, P.G. Effects of probiotics in patients with diabetes mellitus type 2: Study protocol for a randomized, double-blind, placebo-controlled trial. *Tirals* **2013**, 1–8. [CrossRef] [PubMed]

151. Hulston, C.J.; Churnside, A.A.; Venables, M.C. Probiotic supplementation prevents high-fat, overfeeding-induced insulin resistance in human subjects. *Br. J. Nutr.* **2015**, *113*, 596–602. [CrossRef]

152. Callaway, L.K.; Mcintyre, H.D.; Barrett, H.L.; Foxcroft, K.; Tremellen, A.; Lingwood, B.E.; Tobin, J.M.; Wilkinson, S.; Kothari, A.; Morrison, M.; et al. Probiotics for the prevention of gestational diabetes mellitus in overweight and obese women: Findings from the SPRING double-blind randomized controlled trial. *Diabetes Care* **2019**, *42*, 364–371. [CrossRef]

153. Khraishi, M.; MacDonald, D.; Rampakakis, E.; Vaillancourt, J.; Sampalis, J.S. Prevalence of patient-reported comorbidities in early and established psoriatic arthritis cohorts. *Clin. Rheumatol.* **2011**, *30*, 877–885. [CrossRef] [PubMed]

154. Garcia-Santos, A.; Bermudez, M.G. *The Role of Probiotics and Prebiotics in the Prevention and Treatment of Obesity*; NCBI: Bethesda, MD, USA, 2019; ISBN 3462930869.

155. Watanabe, H.; Katsura, T.; Takahara, M.; Miyashita, K.; Katakami, N.; Matsuoka, T.A.; Kawamori, D.; Shimomura, I. Plasma lipopolysaccharide binding protein level statistically mediates between body mass index and chronic microinflammation in Japanese patients with type 1 diabetes. *Diabetol. Int.* **2020**, *11*, 293. [CrossRef]

156. Adeshirlarijaney, A.; Gewirtz, A.T. Considering gut microbiota in treatment of type 2 diabetes mellitus. *Gut Microbes* **2020**, *11*, 253–264. [CrossRef] [PubMed]
157. Koppe, L.; Mafra, D.; Fouque, D. Probiotics and chronic kidney disease. *Nat. Publ. Gr.* **2015**, *88*, 958–966. [CrossRef]
158. Dehghani, H.; Heidari, F.; Mozaffari-khosravi, H. Synbiotic supplementations for azotemia in patients with chronic kidney disease: A randomized controlled trial. *Iran. J. Kidney Dis.* **2016**, *10*, 351–357. [PubMed]
159. Lieske, J.C. Probiotics for prevention of urinary stones. *Ann. Transl. Med.* **2017**, *5*, 29. [CrossRef]
160. Le Bon, M.; Davies, H.E.; Glynn, C.; Thompson, C.; Madden, M.; Wiseman, J.; Dodd, C.E.; Hurdidge, L.; Payne, G.; Le Treut, Y.; et al. Influence of probiotics on gut health in the weaned pig. *Livest. Sci.* **2010**, *133*, 179–181. [CrossRef]
161. Klimesova, K.; Whittamore, J.M.; Hatch, M. Bifidobacterium animalis subsp. lactis decreases urinary oxalate excretion in a mouse model of primary hyperoxaluria. *Urolithiasis* **2014**, *43*, 107–117. [CrossRef] [PubMed]
162. Newell, D.G.; Koopmans, M.; Verhoef, L.; Duizer, E.; Aidara-Kane, A.; Sprong, H.; Opsteegh, M.; Langelaar, M.; Threfall, J.; Scheutz, F.; et al. Food-borne diseases—The challenges of 20 years ago still persist while new ones continue to emerge. *Int. J. Food Microbiol.* **2010**, *139*, S3–S15. [CrossRef]
163. Zommiti, M.; Chikindas, M.L.; Ferchichi, M. Probiotics—Live Biotherapeutics: A story of success, limitations, and future prospects—Not only for humans. *Probiotics Antimicrob. Proteins* **2019**, *12*, 1266–1289. [CrossRef]
164. Galdeano, C.M.; Perdigón, G. Role of viability of probiotic strains in their persistence in the gut and in mucosal immune stimulation. *J. Appl. Microbiol.* **2004**, *97*, 673–681. [CrossRef]
165. Lee, Y.K.; Salminen, S. *Handbook of Probiotics and Prebiotics*, 2nd ed.; John Wiley & Sons: Hoboken, NJ, USA, 2008; pp. 1–596. [CrossRef]
166. Fefana Discusses Probiotics in Animal Feed—All about Feed. Available online: https://www.allaboutfeed.net/animal-feed/feed-additives/fefana-discusses-probiotics-in-animal-feed/ (accessed on 31 October 2021).

applied sciences

MDPI

Review

Antimicrobials from Medicinal Plants: An Emergent Strategy to Control Oral Biofilms

Catarina Milho [1,†], Jani Silva [1,2,†], Rafaela Guimarães [1,†], Isabel C. F. R. Ferreira [3], Lillian Barros [3,*] and Maria José Alves [1,3,*]

[1] AquaValor—Centro de Valorização e Transferência de Tecnologia da Água—Associação, Rua Dr. Júlio Martins n.º 1, 5400-342 Chaves, Portugal; catarina.milho@aquavalor.pt (C.M.); jani.silva@aquavalor.pt (J.S.); rafaela.guimaraes@aquavalor.pt (R.G.)

[2] Molecular Oncology and Viral Pathology Group, IPO Porto Research Center (CI-IPOP), Portuguese Oncology Institute of Porto (IPO Porto), Rua Dr. António Bernardino de Almeida 865, 4200-072 Porto, Portugal

[3] Centro de Investigação de Montanha (CIMO), Instituto Politécnico de Bragança, Campus de Santa Apolónia, 5300-253 Bragança, Portugal; iferreira@ipb.pt

* Correspondence: lillian@ipb.pt (L.B.); maria.alves@ipb.pt (M.J.A.)

† These authors contributed equally to this work.

Abstract: Oral microbial biofilms, directly related to oral diseases, particularly caries and periodontitis, exhibit virulence factors that include acidification of the oral microenvironment and the formation of biofilm enriched with exopolysaccharides, characteristics and common mechanisms that, ultimately, justify the increase in antibiotics resistance. In this line, the search for natural products, mainly obtained through plants, and derived compounds with bioactive potential, endorse unique biological properties in the prevention of colonization, adhesion, and growth of oral bacteria. The present review aims to provide a critical and comprehensive view of the in vitro antibiofilm activity of various medicinal plants, revealing numerous species with antimicrobial properties, among which, twenty-four with biofilm inhibition/reduction percentages greater than 95%. In particular, the essential oils of *Cymbopogon citratus* (DC.) Stapf and *Lippia alba* (Mill.) seem to be the most promising in fighting microbial biofilm in *Streptococcus mutans*, given their high capacity to reduce biofilm at low concentrations.

Keywords: medicinal plants; oral diseases; oral biofilm; drug resistance; antibiofilm strategies

Citation: Milho, C.; Silva, J.; Guimarães, R.; Ferreira, I.C.F.R.; Barros, L.; Alves, M.J. Antimicrobials from Medicinal Plants: An Emergent Strategy to Control Oral Biofilms. *Appl. Sci.* **2021**, *11*, 4020. https://doi.org/10.3390/app11094020

Academic Editor: Marek Kieliszek

Received: 6 April 2021
Accepted: 26 April 2021
Published: 28 April 2021

1. Introduction

Oral diseases triggered by pathogenic bacteria persevere as a worldwide problem with high impact on human health. More than 750 bacteria species inhabit the oral cavity, some of which are opportunistic species capable of causing infections related to oral biofilm. Epithelial cells, dental surfaces and orthodontic prostheses are examples of oral surfaces favorable to the creation of multispecies biofilms that promote the development of infectious diseases, such as dental caries, gingivitis, and periodontitis, which represent some of the most common chronic oral diseases in adults and children [1–4]. Dental caries, a medical term for tooth decay or cavities, are part of a group of polymicrobial diseases caused by specific acid-producing bacteria, mainly Gram-positive species, such as *Streptococcus mutans*, *Streptococcus sobrinus* and *Lactobacillus* spp., responsible for the destruction of the dental enamel and its lower layer, the dentin [5,6]. These bacteria metabolize sucrose into organic acids, mainly lactic acid, which dissolve calcium phosphate from the teeth causing decalcification and possible decay [7]. On the other hand, periodontal diseases are characterized by the occurrence of severe gum infections that damage the soft tissue and the bone that supports the tooth, most of which caused by the pathogens *Aggregatibacter actinomycetemcomitans*, *Porphyromonas gingivalis* and *Prevotella intermedia* [5,8].

Several chemical compounds have been used in the control of oral infectious diseases, like chlorhexidine, fluorine and their combinations [9]. Chlorhexidine is generally accepted as a chemical antibiofilm agent, widely used in dentistry to preserve the healthy oral microbiome [10]. However, some unwanted side effects have been reported, arising from its use for prolonged periods, which may include tooth pigmentation, burning sensation in the mouth, altered taste, and restocking of the oral cavity by resistant strains [11], the latter being mainly due to an increased resistance to antibiotics and to other synthetic chemicals, with a consequent decrease in their clinical efficacy [2]. Given this, in recent years, one of the adopted strategies to overcome these and other related issues is the use of medicinal plants [12,13], which have been used as traditional treatments for thousands of years and throughout the world, given their composition in natural bioactive compounds with multiple recognized biological activities [14]. There are several reports of antibacterial and/or antibiofilm activity linked to extracts from a huge variety of plants, mostly related with the presence of secondary metabolites such as flavonoids, phenolic acids, and tannins [15], which play an important role in the resistance to various microbial pathogens and in the protection against free radicals and toxins [16,17]. Regarding phenolic compounds, several mechanisms through which antimicrobial activity is promoted have been described, since these compounds interact with bacterial proteins and cell membrane structures, damaging and reducing their fluidity, inhibiting the synthesis of nucleic acids, and interfering with the microorganisms' own energy metabolism [16,18,19]. On the other hand, the investigation of antibiofilm properties derived from phenolic compounds present in plants has revealed that, in addition to their bactericidal effect, other mechanisms can lead to biofilm suppression, namely through disturbances in its bacterial regulatory mechanisms, such as quorum sensing (QS) [20]. As an example, many catechin-based polyphenols, flavonoids, proanthocyanin oligomers, and some other plant-derived compounds, compromise the formation of biofilm through the inhibition of glucosyltransferases (GTFs) from one of the most important oral pathogens, *S. mutans* [21].

In modern medicine, natural compounds are considered valuable and with undeniable therapeutic assets, showing reduced toxicity and increased efficiency [22]. Therefore, the search for natural phytochemicals is seen as a good alternative to synthetic substances in the prevention and treatment of oral diseases. The present review focuses on the potential of plant extracts to inhibit the growth and adhesion of oral pathogens, and the development of biofilms, thus reducing the progress of oral diseases.

A literature search in PubMed and Science Direct was conducted using the search terms "oral biofilm", "dental biofilm" "medicinal plants", "aromatic plants", "natural products", "antibiofilm", "cariogenic biofilms" and "natural antimicrobials". Literature analysis included scientific papers published in the last 11 years (between 2010 and 2021). Obtained scientific papers were manually curated and selected by relevance of their findings, namely, selected by relevance of their findings and focusing on antibiofilm activity.

The inclusion criteria for the collected papers (54 papers) were as follows: (1) medicinal plants extracts, (2) oral biofilm-associated bacteria, (3) inhibition of biofilm formation and/or eradication of preformed biofilm.

2. Oral Microbiome

The oral cavity has an actual diverse microbiome, comprising bacteria, protozoa, fungi, archaea, and viruses, with the most abundant group (96%) being composed by bacteria belonging to the phyla Firmicutes, Bacteroidetes, Proteobacteria, Actinobacteria, Spirochaetes, and Fusobacteria [23,24]. The remaining microorganisms belong to the phyla Euryarchaeotic, Chlamydia, Chloroflexi, SR1, Sinergistetes, Tenericutes, and TM7. Although present in small percentages, species from the Archaea domain, such as *Methanobrevibacter oralis*, *Methanobacterium curvum/congolense* and *Methanosarcina mazeii*, are also part of the oral microbiome [25]. About fungi, *Candida* species are the most found, being present in almost 50% of the healthy world population [26]. Other frequently fungi found in the oral cavity belong to *Aspergillus*, *Saccharomycetales*, *Cryptococcus*, *Fusarium* genera [27]. When it comes to viruses, the most frequently found include those that cause sores in the oral cavity, chicken pox, herpes simplex, among others [28]. Moreover, *Tri-*

chomonas tenax and *Entamoeba gingivalis* are the main protozoa found as members of the oral microbiome, the majority of which being saprophytes [29].

Microorganisms that make up the oral microbiome can reside on two types of surfaces, the hard faces of the teeth and the soft tissues of the oral mucosa [30]. Usually, these microorganisms are present in the oral cavity in the form of a biofilm, playing an extremely important role in oral homeostasis, maintenance, and prevention of oral pathologies [31]. However, under certain conditions, changes in the composition and properties of the biofilm can lead to oral illnesses, such as tooth decay and periodontitis. Increased sugar intake, for instance, promotes the proliferation of acidogenic and aciduric bacteria, such as *S. mutans* and *Lactobacillus acidophilus*, which, in turn, create an acidic environment that stimulates the development of dental caries [32–34]. Other bacteria that are usually present in the tooth biofilm include *P. gingivalis*, *Tannerella forsythia* and *Treponema denticola*, closely related to periodontal diseases, which may result from a set of inflammatory conditions that affect the supporting tissues of the teeth [35]. Thus, the biofilm formation may represent the start of the development of different oral diseases.

3. Oral Biofilm Formation

In general, biofilm formation encompasses a series of sequential steps (Figure 1), which begins with the formation of a conditioning film on a surface. In the oral cavity, specifically, an acquired salivary surface, composed of glycoproteins and other molecules, is developed on the tooth surface [36].

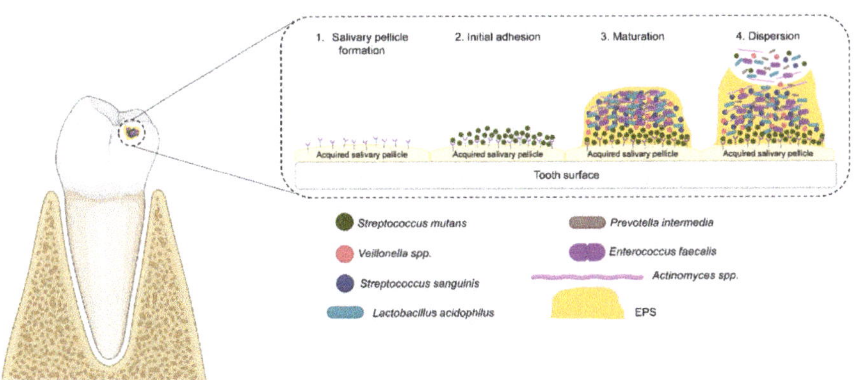

Figure 1. Sequence of pathogenic oral biofilm development. (1) Initially, a salivary pellicle is formed, derived from salivary glycoproteins attached to the tooth surface. (2) Then, initial adhesion starts, as early colonizer bacteria in saliva recognize the binding proteins in acquired pellicle and attach to them. (3) As the biofilm grows, different bacterial species attach, and the biofilm matures. (4) Finally, bacteria disperse from the biofilm and spread to colonize new surfaces. (Created with BioRender.com accessed on 27 April 2021).

Subsequently, a reversible fixation of microorganisms to the tooth occurs, whose approximation of their cell walls occurs randomly or directed through chemotaxis and motility, as well as through weak interactions such as electrostatic and van der Waals forces, and hydrophobic interactions, established between them and the surface [37]. Thus, the mechanisms that occur during the first stages of biofilm formation allow the interaction and adhesion of bacteria to proteins in the acquired salivary film, such as α-amylase and glycoproteins rich in proline [38]. The physical–chemical properties of the oral environment, the presence of nutrients, the physiological state of the bacteria and the presence of bacterial structures, such as fimbriae and flagella, also influence bacterial fixation. In dental caries, the main bacterial colonizers that stick to the tooth surface belong to the genus *Actinomyces*, *Streptococcus*, *Haemophilus*, *Capnocytophaga*, *Veillonella* and *Neisseria* [39]. At this stage, if the oral environment is unfavorable to fixation, bacteria can be easily removed from the adhered surface. However, under favorable conditions, bacteria can become irreversibly linked, with different forces involved in this process, such as dipole-dipole interactions, hydrogen, ionic and covalent bonds, and hydrophobic interactions [37]. This attachment is strengthened by bacterial surface structures such as ligands located on pili, fimbriae, and fibrillae [40,41]. The following production of extracellular polymeric substances (EPS), the most important phase of the irreversible attachment, will support the adhesion of bacteria to oral surfaces. EPS is mainly composed by polysaccharides, containing, also, nucleic and amino acids, glycoproteins and phosphoproteins, phospholipids, uronic acids, and phenolic compounds [42]. In addition to strengthening bacterial adhesion to the tooth surface, EPS is also responsible for reducing diffusional transport, causing decreased growth and metabolism rates of incorporated bacterial cells, nutrient storage, and increased resistance to antimicrobial agents [43]. At the end of this stage, if a physical or chemical procedure is not applied, the bacterial fixation becomes irreversible. *Fusobacterium nucleatum*, *Treponema* spp., *T. forsythensis*, *P. gingivalis*, and *A. actinomycetemcomitans*, are some of the main colonizing bacterial species that are late attached to the biofilm in expansion [44]. After this stage, bacterial cells proliferate, communicating with each other through chemical signals, and potentiating the production of EPS. The continuous growth of bacteria leads to the formation of biofilm that can cover the entire exposed surface, and whose complexity increases not only through the constant fixation and growth of these microorganisms, but also through the production of greater amounts of EPS, originating several layers of cells incorporated into the matrix [45]. In the biofilm itself, there are also water-filled channels responsible for transporting nutrients and removing waste products. At this step, the formation of a mature biofilm, with a complex three-dimensional structure, is completed and, as the biofilm matures, the cells become detached and dispersed, as a result of nutrient depletion, decreased pH or oxygenation, and accumulation of toxic products [46]. Additionally, enzymes that degrade EPS can also be produced by different microorganisms, further increasing the detachment of biofilm cells, which will later colonize new niches and start the formation of new biofilms.

In the oral cavity, the subsistence in a multispecies biofilm confers ecological advantages when compared to biofilms colonized by a single species. Moreover, once in the EPS matrix, oral bacteria are protected from microenvironmental damage, from the host's immune defenses and from antimicrobial agents [47]. The presence of persistent dormant cells in the biofilm is crucial, given their high tolerance to antimicrobial agents, reason why they are pointed out as the main parties responsible for the biofilm recalcitrancy to these agents [48].

4. Oral Biofilms: From Dental Caries to Systemic Diseases

Chronic and progressive oral diseases are estimated to affect more than 3.5 million people worldwide. Although preventable, this public health problem can disproportionately affect low-income people and, consequently, their longevity and quality of life, factors directly related to social and economic disparities [49]. As previously mentioned, dental caries, periodontal diseases and other problems related to the oral cavity, result from a complex interaction between the microbiome and the oral microenvironment, the pathogens, and specific characteristics of the host, sharing common risk factors between them [47,50] (Figure 2A). Both the appearance and progression of these pathologies are closely related to bacterial biofilms characteristics, which present themselves as complex microbial groups that interact with each other, triggering immunological/inflammatory responses and modulating the action of antimicrobial agents [51].

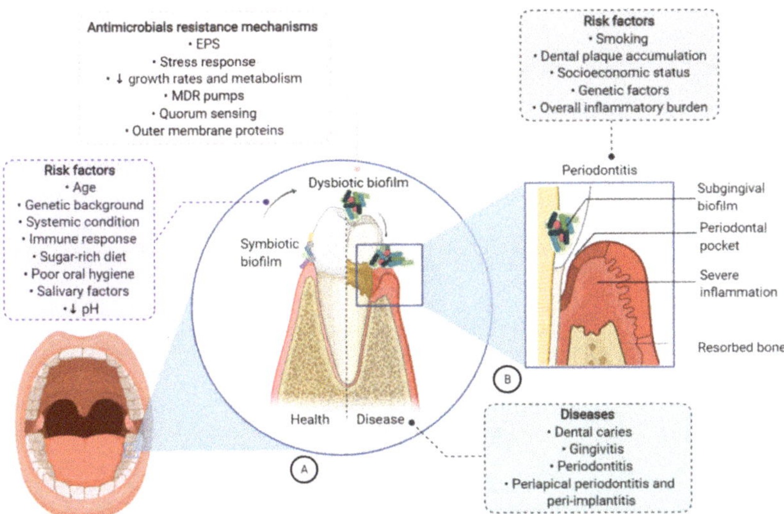

Figure 2. Dysbiosis as a trigger of oral diseases. (**A**) Development of oral biofilm from health to disease conditions, the associated risk factors, and the antimicrobial resistance mechanisms. (**B**) Development of periodontitis and related risk factors. (Created with BioRender.com accessed on 27 April 2021).

The extracellular insoluble glucans converted from the dietary sucrose that feeds the assembly of the EPS matrix are essential in bacterial adhesion and subsequent formation of dental plaque, and in the formation of the EPS matrix nucleus of the biofilm, which can be the starting point for the development of diseases such as periodontitis [47,52–54]. Cariogenic biofilms are naturally acidic and hypoxic; however, they are rich in carbohydrates, which creates an environment conducive to the growth of opportunistic microorganisms, such as the lactobacilli, *Lactobacillus casei* and *Lactobacillus reuteri*, thus accelerating the development of dental caries [47], which passes from the enamel to the dentin, and may be associated with continuous microbial dysbiosis owed to an increased bacterial diversity and to species with proteolytic capacity, such as *F. nucleatum* [55].

In contrast, endodontic infections are characterized by the presence of bacterial infections in the tooth pulp that can be caused by initial dental caries. In fact, sequencing of the 16S rRNA gene showed that polymicrobial biofilms can be identified within the infected root canals, which are mostly composed of *Firmicutes* spp. (>50%) [50,56,57]. Thus, the morphological structure of the oral biofilm may vary from case to case, in which the time of infection, the type and availability of nutrients in the oral microenvironment and the arrangement of the established microbiota are at the origin of this structure [58,59].

Periodontitis, in turn, is defined as a chronic inflammatory disease induced by biofilm, that affects the integrity of the periodontium, which consists in the periodontal ligament, gingiva and alveolar bone (Figure 2B). Destructive inflammation of the tissue, as well as increased dysbiosis, results in bone loss, that may culminate in tooth decay and, ultimately, systemic complications. This type of inflammation is associated with variations in the subgingival polymicrobial community, which changes from a predominantly aerobic Gram-positive biofilm to a Gram-negative anaerobic biofilm [50]. Chronic periodontitis has been associated with a predominance of "red complex bacteria", namely *P. gingivalis*, *T. forsythia* and *T. denticola* [50,60]. The most aggressive forms of this disease can result in faster periodontal destruction and bone loss, modulated by a set of pathogenic species that act in conjunction with *A. actinomycetemcomitans* strains [50,58].

Increasing evidence has shown that the dysbiotic oral microbiota can not only be a source of oral inflammation, but may also contribute to systemic ones, by releasing toxins or microbial by-products into the bloodstream. Thus, the synergistic effect between oral and systemic inflammation is a crucial step for the subsequent damaging effects on various organ systems, increasing the risk of developing oral and non-oral diseases [61] (Figure 3).

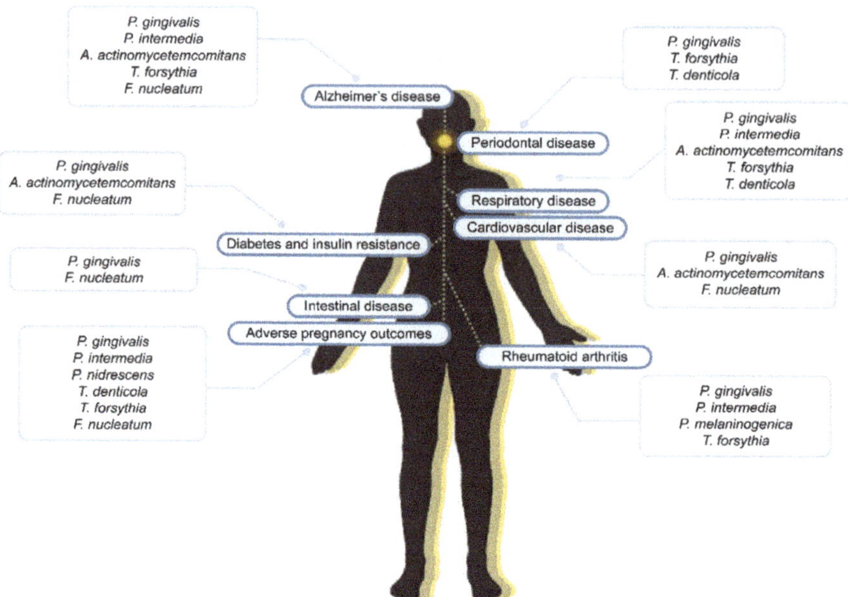

Figure 3. Schematic representation of different systemic diseases and their association with oral keystone pathogens. (Created with BioRender.com accessed on 27 April 2021).

In fact, a meta-analytical study showed that patients with periodontal disease are more susceptible to the development of oral cancer [62]. Recently, a study performed by Michaud et al. [63] also provided supporting data for a positive correlation between periodontal disease and risk of oral, lung, and pancreatic cancers. In line with these information, *P. gingivalis* was found at significantly higher levels in both oral and esophagus squamous cell carcinoma patients [64]. Moreover, other studies have shown that *F. nucleatum* can migrate from the oral cavity to the intestinal tract, promoting a pro-inflammatory microenvironment and suggesting a possible role of these bacteria in the development of colorectal cancer [65,66]. Also, Alzheimer's disease and diabetes *mellitus* seem to be bidirectionally associated with periodontal disease [61]. According to Mealey et al. [67], diabetic patients have a 3-fold increase risk of developing periodontitis when compared with healthy controls. On the other hand, Teeuw et al. [68] suggest that periodontal treatments lead to an improvement of glycemic control of type 2 diabetes, which may be related, in part, with periodontal infection and inflammatory response [69]. As so, these data show that periodontal disease management could positively control the glycemic levels in diabetic individuals. Also, Kothari et al. [70] found that individuals with acquired brain injuries had a series of unstable oral, dental and periodontal parameters, translated into generalized chronic periodontitis. In contrast, the systemic production of pro-inflammatory cytokines in response to oral bacterial infection suggests that periodontal disease can lead to cerebral inflammatory state, related to Alzheimer's disease [71,72]. Actually, higher levels of antibodies against *A. actinomycetemcomitans*, *P. gingivalis*, *T. forsythia* were found in elderly patients with Alzheimer's disease when compared to healthy individuals [72]. Likewise, the presence of periodontopathic virulence factors, namely lipopolysaccharides (LPS) from *P. gingivalis* and *T. denticola*, could figure in the development of brain inflammation and, ultimately, in Alzheimer's disease [73]. Other studies have also shown that individuals with periodontal disease had a 1.14 to 2 times greater risk of developing coronary heart disease when compared to a control group [74]. The presence of oral bacteria DNA, namely *P. gingivalis*, *A. actinomycetemcomitans*, *T. forsythia*, *Eikenella corrodens*, *F. nucleatum* and *Campylobacter. rectus* in atheromatous plaques of endarterectomy, were also detected [75]. Reinforcing the previous results, DNA from periodontal bacteria *P. gingivalis*, *A. actinomycetemcomitans*, *P. intermedia*, *T. forsythia* and cariogenic *S. mutans*, was noticed in atherosclerotic plaques, suggesting that these oral pathogens have the ability to migrate from the oral cavity to distant body sites [76,77]. In addition, the oral cavity, especially the saliva and dental plaque of individuals with periodontal disease, appears to be the source of accumulation and dissemination of pathogens to the lower respiratory tract. Several oral pathogens have already been implicated in lung infections, including *A. actinomycetemcomitans*, *Actinomyces israelii*, *Capnocytophaga* spp., *Chlamydophila pneumoniae*, *Eikenella corrodens*, *F. nucleatum*, *Fusobacterium necrophorum*, *P. gingivalis*, *P. intermedia* and *Streptococcus constellatus* [78–80]. It has been also reported that periodontal infection increases the likelihood of developing nosocomial pneumonia [81]. Genetic similarities between microorganisms isolated on dental plaque and bronchoalveolar lavage fluid suggest that the former may function as a reservoir for respiratory and/or opportunistic pathogens [82]. Porto et al. [83], in turn, showed that toothed and toothless individuals hospitalized in intensive care units and submitted to orotracheal intubation presented large amounts of *A. actinomycetemcomitans*, *P. gingivalis and T. forsythia*, suggesting that the oral microenvironment favors the pathogenic bacterial growth. Additional studies also express that systemic dissemination of periodontal-related oral pathogens, endotoxins, and inflammatory mediators can cross the placental barrier and contribute to adverse pregnancy outcomes [84,85], with *F. nucleatum* and *P. gingivalis* being detected in placental and fetal tissues [86], and in the placenta of preterm delivery patients, respectively [87,88]. An increasing number of epidemiological, serological and clinical evidences, observed between the pathogenesis of rheumatoid arthritis and periodontitis, have been presented, with *A. actinomycetemcomitans* and *P. gingivalis* being recognized as the main triggers of this disease, associating autoimmunity with periodontal diseases [89]. *P. gingivalis* is referred as the only human pathogen known to express the peptidyl arginine deiminase enzyme and to produce citrullinated epitopes that are recognized by anti-citrullinated protein antibodies, that ultimately culminate in clinical manifestations of rheumatoid arthritis [50,89].

5. Biological Properties of Plants

Traditional knowledge on medicinal plants, used for different health purposes, has attracted much attention among the scientific community due to their effectiveness in the treatment of several diseases [90]. Medicinal plants are a rich source of bioactive compounds, or bionutrients, which may be present in seeds, roots, leaves, flowers, or even in the whole plant, thus assuming themselves as important sources of compounds with characteristics of food additives, flavorings and other valences at an industrial level [91]. Bioactive compounds are secondary metabolites that can be classified based on their composition, the pathway by which they are synthesized, or through their chemical structure. A simple classification of bioactive compounds includes three main groups, consisting of phenolic compounds, terpenoids, and alkaloids, which represent about 90% of all secondary metabolites [92]. The minor groups of secondary metabolites include saponins, lipids, carbohydrates, ketones, and others [93,94]. Currently, phenolic compounds are among the most studied natural products, given their chemical and structural diversity and bioactive properties [95,96], which may include antioxidant, antimutagenic, antitumor, antiallergenic, anti-inflammatory, antiviral, antiulcer, antidiarrheal, anthelmintic, antihepatitic, and antiproliferative properties [97,98]. Medicinal plants are composed of a wide range of phenolic compounds with antimicrobial properties that provide protection against aggressive agents. Some of these compounds can deliver sustainable solutions for combating drug-resistant microorganisms [99]. These secondary metabolites are biosynthesized from the shikimate pathway, containing benzene, hydrogen and oxygen rings, the majority of which being flavonoids, the largest and most studied group of compounds in plants [100]. Flavonoids are found to be effective against a wide range of microorganisms, and their antimicrobial activity is thought to be shaped through their ability to form complexes with both extracellular and soluble proteins, as well as with bacterial membranes, increasing their permeability and disruption [101]. In addition, these compounds may also act to inhibit the activities of DNA gyrase and β-hydroxyacyl-acyl transport protein dehydratase [102]; for example, catechins belonging to this group exhibit inhibitory activity against both Gram-positive and Gram-negative bacteria [103]. In its turn, apigenin showed inhibitory activity to both GTF and fructosyltransferase proteins of *S. mutans* without major impact on bacterial viability [104]. Quercetin derivates inhibited *S. mutans* biofilm production by reducing the synthesis of both water-soluble and insoluble glucans and suppressing several virulence genes [105]. On the other hand, the antimicrobial mechanism by which terpenoid compounds act is not clearly defined, but is attributed to the rupture of microorganisms membrane [106,107]. Finally, alkaloids, which are biosynthesized from amino acids, such as tyrosine, hold an antimicrobial mechanism attributed to their ability to intercalate with DNA, thereby resulting in impaired cell division and death [108]. The plant-extracted product may exert its antimicrobial activity, not by killing the microorganism itself, but by affecting several key events in the pathogenic process [109,110].

There are several reports on plant extracts activity against a wide range of microbial pathogens from the oral cavity. Some of these studies are focused on the investigation of the ability of plant-derived products to inhibit the formation of biofilms in the oral cavity, interfering and reducing the adhesion of microbial pathogens to different oral surfaces, which constitutes the first step in the formation of dental plaque, and in the progression to cavities and periodontal diseases. It has been shown that crude plant extracts and purified phytochemicals can act as bactericides, inhibiting one or all stages of plaque formation, by interfering with biofilm adhesion/aggregation/formation, or inhibiting the production of glycolytic acid in cariogenic bacteria [111]. Several studies report the polyphenols inhibitory effects on oral biofilm formation and on dental biofilm production and accumulation. Many compounds such as catechins, flavonoids, alkaloids, terpenoids, proanthocyanin oligomers and some other plant-derived compounds, inhibit *S. mutans* GTFs, one of the crucial virulence factors of *S. mutans* with a key role in the synthesis of glucans, an important component of the biofilm matrix [21]. The use of traditional medicines clearly shows how potential biologically active compounds can suppress pathogens and prevent disease progression [94]. Thus, the use of herbal extracts and their products on a daily basis is a promising and interesting alternative to synthetic compounds in the control of oral diseases.

6. The Most Promising Medicinal Plant Extracts in the Control of Oral Biofilms

The antibiotic therapy has reached its limits regarding antimicrobial resistance, threatening the effective prevention and treatment of an increasing range of infections. Thus, new therapeutic approaches based on natural phytochemicals have been the target of several research, considering their bioactive assets, namely antimicrobial properties. Knowing that the number of medicinal plants that potential possess antimicrobial/antibiofilm properties is quite large, only published works investigating the extracts obtained from plants' aerial parts, roots and seeds were considered. Hence, Table 1 presents some of the plant species whose extracts hold compounds with antimicrobial/antibiofilm activity, among others, up to 95%, against specific microorganisms, i.e., extracts with the potential to inhibit biofilm formation and/or eradicate it, with concentrations <1 mg·mL^{-1}.

One of the described plant species with proven antimicrobial [112], antiulcerative [113] and antifungal [114] activity is *Baccharis dracunculifolia* D.C., considered to be the most important botanical source of South-Eastern Brazilian propolis [115]. The antibiofilm properties against oral cavity bacteria were studied by Galvão et al. [9], through essential oils extracted from the aerial parts of the plant in question. These authors showed that, for a concentration of 31.2 μg·mL^{-1}, *B. dracunculifolia* extracts present 95% of inhibition growth of biofilms of *S. mutans* NCTC 1091. The ability of this extract to inhibit biofilm formation seems to be related to its composition in oxygenated sesquiterpenes, such as spathulenol and trans-nerolidol, described in the literature as antibiofilm/antimicrobial mediators [116,117].

Different phenolic compounds are responsible for the wide diversity of bioactive properties of plant extracts, which, in addition to their antimicrobial assets, may also take part in the healing process. An example of this statement is the species *Camelia japonica* L., whose extract holds antioxidant, anti-inflammatory and antimicrobial properties [118–120]. Additionally, *Chelidonium majus* subsp. *asiaticum H.Hara*, commonly known as greater celandine, is a medicinal plant widely used in traditional medicine due to its anti-inflammatory and antimicrobial effects [121,122], properties also attributed to the *Thuja orientalis* L. species, a perennial conifer tree of the family Cupressaceae [123,124]. Choi et al. [125] investigated the antimicrobial and antibiofilm activities of methanolic extracts of *C. japonica*, *C. majus* subsp. *asiaticum*, *C. flagelliferum* and *T. orientalis* against oral pathogens. Notably, all these plant extracts were able to inhibit the GTF function of *S. mutans* ATCC 25175, an important virulence factor in the biofilm formation, by 99.0%, at a concentration of 1.00 mg·mL^{-1}. The total phenolic compounds concentration of these extracts is quite high, which may be the reason of their superior antimicrobial effect [126,127].

The essential oil extracted from *Cinnamomum zeylanicum* Blume, a perennial tree from which cinnamon is obtained [128], is described in the literature as an antimicrobial agent that acts against various biofilm-forming bacteria present in the oral cavity [129,130], such as *S. mutans* ATCC 25175. When used in the management of biofilms of *S. mutans*, the essential oil from *C. zeylanicum* skin was able to inhibit their formation by up to 99%, at a concentration of 0.224 mg·mL^{-1} [131]. β-linalool and (*E*)-cinnamaldehyde are the two main compounds present in this oil, and can be pointed out as responsible for its antibiofilm properties [132,133]. In addition, the essential oils of *Coriandrum sativum* L., a medicinal plant with nutritional benefits, commonly named coriander, exhibit antibacterial and antibiofilm properties against *S. mutans*, in addition to holding antioxidant and anesthetic properties [134,135]. Galvão et al. [9] used a chemical fraction of *C. sativum* essential oil as an antibacterial against *S. mutans* UA 159, at a concentration of 31.2 μg·mL^{-1}, being able to inhibit the growth of *S. mutans* biofilms by more than 95%. The fatty alcohol 1-decanol is one of the major components found in *C. sativum* essential oil, and it has been described as an antibacterial and antibiofilm agent [136,137].

Copaifera pubiflora Benth. is a flowering plant whose oleoresin is widely used in Brazilian medicine due to its anti-inflammatory, analgesic, and antimicrobial properties [138–140]. In a study performed by Moraes et al. [141], oleoresin from *C. pubiflora* was used as an alternative agent for the removal of oral pathogenic biofilms. This work showed satisfactory data regarding the antimicrobial activity of the used extract against microorganisms normally present in the oral cavity, which included *S. sanguinis* ATCC 10556 and *P. micra* clinical isolate (CI), and it was able to eliminate more than 99.9% of preformed biofilms of these species, at a concentration of 50.0 μg·mL^{-1}. The antimicrobial effect of *C. pubiflora* oleoresin was attributed to the presence of ent-hardwickiic acid, which was found to be the main compound present in this plant [142]. *Cymbopogon citratus* (DC.) Stapf, in turn, usually known as lemon grass, is a perennial aromatic plant that is cultivated in tropical and sub-tropical regions [143]. Its essential oil has been found to have many different biological properties, including anxiolytic, antibacterial and antibiofilm incomes [144–146]. When it comes to oral health, it has been shown that *C. citratus* exerts an antibiofilm effect on pathogenic bacteria from the oral cavity. As an example, 93.0% of growth inhibition of *S. mutans* biofilms was obtained at a concentration of 1.00 μg·mL^{-1} of *C. citratus* essential oil [147]. In another study, the essential oil from this plant, at a concentration of 0.100 μg·mL^{-1}, led to a reduction of more than 95% of *S. mutans* ATCC 35668 preformed biofilms [148]. In addition, *C. citratus* oil was tested against *S. mutans* ATCC 35,688 and *L. acidophilus* ATCC 4356 single-species biofilms, also inhibiting their growth by more than 95%, although at higher concentrations (26.1 mg·mL^{-1} and 13.2 mg·mL^{-1}, respectively). Regarding its chemical composition, the compounds that are usually found in *C. citra-*

tus essential oil are citral and myrcene, which are described to have good antimicrobial properties [149].

Eucalyptus globulus Labill is an evergreen tree native to Australia whose leaves have been widely used in pharmaceutical products, given their antimicrobial and antioxidant properties [150]. Tsukatani et al. [151] reported that the *E. globulus* ethanolic extract exhibited up to 99% eradication activity against *P. gingivalis* JCM 12257 at a minimum biofilm eradication concentration (MBEC) of 49.1 µg·mL^{-1}, and *S. mutans* NBRC13955 at a MBEC of 393 µg·mL^{-1}. The extracts of *Eucalyptus* species were found to equally contain antimicrobial compounds, such as macrocarpals, eucalyptine and 1,8-cineol [152,153]. Macrocarpal A, B and C are phloroglucinol derivatives referred to inhibit virulence factors of the periodontopathic bacteria *P. gingivalis*, including specific cysteine proteinases, which appear to be essential for the growth and survival of this bacterium in the periodontal pocket [152]. The presence of specific groups of bioactive compounds can also influence the progression stages of the oral bacterial biofilm's formation.

Derris reticulata Craib, a climbing medicinal plant used in folk medicine [154], has prenylated flavanones as its main constituents, whose presence has been linked to several pharmacological outcomes, particularly antibacterial activity [154,155]. According to Pulbutr et al. [155], the ethanolic extract from stems of *D. reticulata* was able to inhibit *S. mutans* DMST 1877 biofilm formation by up to 99.9%, at the highest tested concentration, 750 µg·µL^{-1}. These results are in accordance with the capability of *D. reticulata* to inhibit both sucrose-dependent and independent *S. mutans* adherence in a concentration dependent manner, at sub-MIC concentrations (<625 µg·µL^{-1}).

Dodonaea viscosa var. angustifolia leaf decoction extracts have been reported to have anti-inflammatory and antimicrobial activity [156,157], and they are traditionally used as mouthwashes for toothaches and related problems [158]. The bioactivities present in this plant can be attributed to the major compounds found in *D. viscosa* extract, namely xylopyranoside; 2,2′-methylenebis [6-(1,1-dimethyl)]-4-methyl); 2-(3-Hydroxy-4-methoxyphenyl)-3,7-dimethoxy-4H-chromen-4-one; trans-3′,4′,5′-Trimethoxy-4-(methylthio)chalcone and stigmasterol [159]. Naidoo et al. [159] studied the inhibitory effect of this plant methanolic extract against *S. mutans* NCTC 1091 biofilm, verifying that biofilm reduction was dependent on the exposure time and concentration.

The tapered roots and rhizomes of *Glycyrrhiza glabra* L. hold most of the bioactive components responsible for this plant medicinal and culinary features [160]. Its phytochemical compounds, namely glycyrrhizin, an oleanane-type triterpene saponin, stands out as its major constituent with antibiofilm activity [160,161]. Suwannakul and Chaibenjawong [162] found that the inhibition pattern and eradication of *P. gingivalis* biofilm by *G. glabra* ethanol extract were concentration-dependent. At the concentration of 500 µg·mL^{-1}, the extract exhibited up to 90% inhibition of *P. gingivalis* biofilm formation. In addition, the eradication of *P. gingivalis* biofilm was achieved at a MBEC of 62.5 µg·mL^{-1}, being higher at a concentration of 500 µg·mL^{-1}. The authors also found that the Rgp- and Kgp-proteinase activities of *P. gingivalis*, which are important virulence factors, were also reduced by approximately 50% [162]. Interestingly, Kim et al. [163] showed that *G. glabra* main bioactive compound, namely 18α-glycyrrhetinic acid, significantly inhibits the *P. gingivalis* LPS-induced endothelial permeability, both in vitro and in vivo assays.

Lippia alba (Mill.) flowering plant which essential oil presents several pharmacological properties such as sedative, analgesic, antispasmodic, anti-inflammatory and antimicrobial assets [164]. In a work published by Tofino-Rivera et al. [148], the essential oil from *L. alba* was used as an antibacterial agent against *S. mutans* ATCC 35668 biofilms, and it was shown that, at a concentration of 0.100 µg·mL^{-1}, this natural product was able to reduce the number of viable cells present in the biofilm by 95.8%. In this case, isomeric monoterpenes geraniol and citral are two of the main components found, which are known to hold antimicrobial properties [149,165].

The *Mentha* genus includes diverse aromatic herbs that are commonly used in herbal teas, flavoring agent, and as medicinal plants. Infusion, decoction, and distillate water of the aerial parts have been used for centuries as tonics, carminative, digestive, stomachic, antispasmodic, and anti-inflammatory preparations [166]. Traditionally, these plants have been also used for teeth whitening, and their distilled oils used to flavor toothpastes and chewing gum, until this day. Knowing this, several works were conducted in order to study the ability of these plants to eliminate oral pathogenic biofilms. As an example, Shafiei et al. [167] investigated the effects of *Mangifera* sp. and *Mentha* sp. aqueous extracts towards the eradication of *S. sanguinis* ATCC BAA-1455 and *S. mutans* ATCC 25175 biofilms. Both extracts showed to be more effective in reducing the biofilm population of *S. mutans* than *S. sanguinis*. At a concentration 0.50 mg·mL^{-1}, *Mangifera* sp. and *Mentha* sp. extracts reduced 99.4% and 98.5% the *S. mutans* biofilms, respectively. Previous reports showed that *Psidium* sp. and *Mentha* sp. extracts, as well as their mixtures, have antibacterial and anti-adherence activities against *S. sanguinis* and *S. mutans* in single species biofilms [168,169]. Moreover, the phenolic profile of *Mangifera* sp. extract revealed that quinic acid, benzophenone *C*-glycoside isomer, benzophenone *C*-glycoside and quercetin-3-*O*-glucoside were its main compounds, while in the *Mentha* sp. extract, methyl 2-[cyclohex-2-en-l-yl(hydroxy)-methyl]-3-hydroxy-4-(2-hydroxyethyl)-3-methyl- 5-oxoprolinate was mainly found [170], all presenting inhibiting virulence properties of *S. mutans*.

Myrtus communis L. is an important aromatic and medicinal plant species from the Mediterranean area, widely used for culinary, cosmetic, pharmaceutical, therapeutic, and industrial purposes [171]. Antimicrobial and antioxidant proprieties of *M. communis* have been reported in numerous studies [172]. Sateriale et al. [173] described the antibiofilm activity of hydroethanolic extracts from its leaves against the oral pathogens *S. mutans* ATCC 25175, *S. oralis* (CI), *S. mitis* (CI), and *R. dentocariosa* (CI). Curiously, the hydroethanolic extract of *M. communis* produced a significant ($p < 0.05$) inhibition in all tested oral pathogens biofilms. MBEC values, assigned to the lowest concentration of each antimicrobial agent that is able to eradicate preformed biofilms, ranged between 40 mg·mL^{-1} (*S. oralis* and *S. mitis*) and 120 mg·mL^{-1} (*S. mutans* and *R. dentocariosa*). In a previous study, the same authors were able to identify gallic acid derivatives, tannins, myricetin, and quercetin derivatives as the most abundant phenolic compounds in the hydroalcoholic extract of *M. communis* [174]. The authors also refer that the antibiofilm properties of *M. communis* have been directly correlated with its phenolic compounds' arrangement.

Rhodiola rosea L. is medicinal plant that has been used due to its therapeutic properties and as a potential source of antimicrobial, antioxidant, anti-inflammatory agents, among others [175,176]. This plant has been recognized to present a broad spectrum of biological activities mainly attributed to its major phytochemical compounds, which include phenylethanes and phenylpropanoids (rosavin, salidroside, rosin, cinnamyl alcohol, and tyrosol) [175,177]. In a study performed by Zhang et al. [178], the *R. rosea* root ethanolic extract inhibited the biofilm formation of *S. mutans* UA159 by 95%, at a concentration of 0.50 µg·µL^{-1}, with the highest reduction of EPS synthesis being observed at the same concentration. *R. rosea* also suppressed the expression of virulence genes and QS system as well as the enzymatic activity of GTF proteins in both 0.25 µg·µL^{-1} and 0.50 µg·µL^{-1} concentration groups [178]. At a concentration of 0.12 µg·µL^{-1}, no significant cytotoxic effect was observed, and 0.50 µg·µL^{-1} and 0.25 µg·µL^{-1} slightly inhibited the cell proliferation [178].

Rosmarinus officinalis L. is recognized as a fragrant medicinal plant, native to the Mediterranean region and cultivated worldwide. In addition to its therapeutic and prophylactic effects, this plant is extensively used as a condiment, food preservative and for ornamental purposes [179]. Extensive research has been developed regarding the characterization of the antibiofilm properties of this plant. Tsukatani et al. [151] found that the *R. officinalis* ethanolic extract was able to eliminate *P. gingivalis* JCM12257 and *S. mutans* NBRC13955 biofilm formation at a concentration of 195.5 µg·mL^{-1} and 97.8 µg·mL^{-1}, respectively. Additionally, the phytochemical profile of this medicinal herb was determined, and it showed that carnosic acid presented the lowest MBEC against *P. gingivalis* (25.0 µg·mL^{-1}) and *S. mutans* (12.5 µg·mL^{-1}). This bioactive component is a plastidial catecholic diterpene with recognized antioxidative, anti-inflammatory and antimicrobial properties [179,180]. *Syzygium aromaticum* L., another plant traditionally used as a spice and as a food preservative, holds a wide spectrum of pharmacological properties. As reported by Tsukatani et al. [151], the biofilm eradication activity of *S. aromaticum* ethanolic extract against *P. gingivalis* JCM 12,257 and *S. mutans* NBRC 13,955 was observed, respectively, at a concentration of 435.5 and 871 µg·mL^{-1}. Regarding its phytochemical signature, eugenol, eugenol acetate and β-caryophyllene are acknowledged as antimicrobial components present in this plant extract [181]. Eugenol was reported to inhibit *P. gingivalis* and *S. mutans* biofilms at higher concentrations, 800 and >800 µg·mL^{-1}, respectively [151]. In addition, other minor compounds, such as β-caryophyllene, exhibited eradication activities against *P. gingivalis* and *S. mutans* at lower concentrations (400 and 50 µg·mL^{-1}, respectively), and eugenol acetate against *S. mutans*, at 400 µg·mL^{-1} [151].

Spirostachys africana Sond., a tree originally from South America, is traditionally used as a remedy for toothache [182]. Although used as an antibacterial product, its bioactive properties have not been extensively described in the literature. As an example, the dichloromethane:methanol extract from *S. africana* leaves was used against *S. mutans* ATCC 25175 oral pathogen, being able to inhibit the growth of *S. mutans* biofilms by more than 97% at a concentration 0.25 mg·mL^{-1} [183]. Unfortunately, no studies have been found describing the chemical profile of *S. africana* leaf extracts. Another plant commonly used for its interesting properties is *Thymus vulgaris* L., commonly named thyme, whose essential oil bioactive properties include antioxidant, anti-inflammatory, antitumor, and antimicrobial effects [184]. Several studies have been conducted regarding the antibacterial activity of *T. vulgaris* essential oil against oral pathogens, namely *S. aureus* [185]. Interestingly, when used at a concentration of 0.156 mg·mL^{-1}, this essential oil was capable of inhibiting the formation of *S. aureus* biofilms by 96%. The antibiofilm effect of *T. vulgaris* essential oil may be due to the presence in its composition of different chemical compounds, in particular thymol, a monoterpenoid phenol that is extensively described as having antibiofilm properties [186].

Trachyspermum ammi L., a plant rich in thymol, is a known herb with recognized medical properties, widely cultivated in the west and northwest of Iran [187]. Its seeds hold several medicinal features, including antibacterial, antioxidants and antifungal properties, among others [188,189]. With respect to its application as an antibacterial product against oral bacteria, Khan et al. [90] found that, at a 160 µg·mL^{-1} concentration, the ethanolic extract of *T. ammi* seeds was able to reduce the number of viable cells in *S. mutans* ATCC 700610 biofilms by 89%. On the other hand, the petroleum ether fraction of the ethanol extract, applied at 40.0 µg·mL^{-1}, caused the complete inhibition of *S. mutans* biofilms growth (100%), probably due to its higher concentration in antimicrobial compounds. As mentioned before, thymol, a monoterpenoid phenol, is the major component found in ethanol extracts of *T. ammi* seeds, which is known to exert antibiofilm effects [186].

Table 1. Medicinal plants with verified antimicrobial/antibiofilm activity against oral cavity bacteria, and the respective bioactive compounds present in their extracts.

Plant Name	Plant Extract	Compound	Microorganism	Results				References
				Antimicrobial Activity		Antibiofilm Activity		
Acacia karroo Hayne	Dichloromethane: methanol (leaves)	-	*S. mutans* ATCC 25175	MIC	$0.50\ \text{mg·mL}^{-1}$	88.8% inhibition	$0.25\ \text{mg·mL}^{-1}$	[183]
Achyranthes aspera L.	Methanol	Betulin; 3,12-oleandione	*S. mutans* (CI)	MIC IZD	$125\ \mu\text{g·mL}^{-1}$ 23.0 mm ($250\ \text{mg·mL}^{-1}$)	94.9% inhibition	$125\ \mu\text{g·mL}^{-1}$	[190]
Aloysia gratissima (Aff & Hook) Tr	Essential oil (leaves) (Ag4 fraction)	-	*S. mutans* UA159	MIC MBC	31.2–$62.5\ \mu\text{g·mL}^{-1}$ 62.5–$125\ \mu\text{g·mL}^{-1}$	>90% inhibition	$62.5\ \mu\text{g·mL}^{-1}$	[9]
	Essential oil (leaves)	(*E*)-pinocamphone; β-pinene; guaiol	*F. nucleatum* ATCC 25586	MIC MBC	$0.125\ \text{mg·mL}^{-1}$ $0.250\ \text{mg·mL}^{-1}$	55.83% inhibition	$1.0\ \text{mg·mL}^{-1}$	[136]
			P. gingivalis ATCC 33277	MIC MBC	$0.125\ \text{mg·mL}^{-1}$	39.12% inhibition		
			S. sanguis ATCC 10556	MIC MBC	$0.500\ \text{mg·mL}^{-1}$ $1.0\ \text{mg·mL}^{-1}$	60.83% inhibition		
			S. mitis ATCC 903	MICMBC	$0.25\ \text{mg·mL}^{-1}$	9.00% inhibition		
Artemisia princeps Pamp.	Ethanol (leaves)	-	*S. mutans* ATCC 25175	MIC MBC	$0.40\ \text{mg·mL}^{-1}$ $3.2\ \text{mg·mL}^{-1}$	≈80.0% inhibition	$0.40\ \text{mg·mL}^{-1}$	[191]
Artocarpus lakoocha Roxb.	Aqueous	Oxyresveratrol	*S. mutans* ATCC 25175	MIC MBC IZD	$0.10\ \text{mg·mL}^{-1}$ $0.20\ \text{mg·mL}^{-1}$ 30.5 mm (10% (*w*/*v*))	≥90.0% inhibition ≥90.0% reduction	$3.12\ \text{mg·mL}^{-1}$ $6.25\ \text{mg·mL}^{-1}$	[192]
			A. actinomycetemcomitans ATCC 33384	MIC MBC IZD	$0.100\ \text{mg·mL}^{-1}$ $0.200\ \text{mg·mL}^{-1}$ 29.5 mm (10% (*w*/*v*))	≥90.0% inhibition ≥90.0% reduction	$0.39\ \text{mg·mL}^{-1}$ $3.12\ \text{mg·mL}^{-1}$	
Azadirachta indica A.Juss.	Aqueous (leaves)	-	*E. faecalis* ATCC 29212 *S. aureus* ATCC 25923		-	53.6% reduction 48.2% reduction	$30\ \text{mg·mL}^{-1}$	[193]
Baccharis dracunculifolia DC.	Essential oil (leaves)	-	*S. mutans* ATCC 35688 *S. mutans* 22 (CI) *S. mutans* 24 (CI) *S. mutans* 28 (CI)	MIC	6.0% (*v*/*v*)	39.3% reduction 78.9% reduction 90.9% reduction 91.1% reduction	6.0% (*v*/*v*)	[194]
	Essential oil (leaves)	Trans-nerolidol; spathulenol	*S. mutans* UA159	MIC MBC	15.6–$31.2\ \mu\text{g·mL}^{-1}$ 125–$250\ \mu\text{g·mL}^{-1}$	95.0% inhibition	$31.2\ \mu\text{g·mL}^{-1}$	[9]
Berula erecta (Huds.) Coville	Dichloromethane: methanol (rhizome)	-	*S. mutans* ATCC 25175	MIC	$0.50\ \text{mg·mL}^{-1}$	37.7% inhibition	$0.25\ \text{mg·mL}^{-1}$	[183]
Betula schmidtii Regel.	Methanol	-	*S mutans* ATCC 25175	MIC	$31.3\ \text{mg·mL}^{-1}$	≈46.0% reduction	$125\ \text{mg·mL}^{-1}$	[195]

Table 1. *Cont.*

Plant Name	Plant Extract	Compound	Microorganism	Results				References
				Antimicrobial Activity		**Antibiofilm Activity**		
Camellia japonica L.	Methanol (leaves)	-	*S. mutans* ATCC 25175	MIC IZD	0.5 mg·disk^{-1} 12 mm (2 mg·disk^{-1})	99.0% GTFs inhibition	1.0 mg·mL^{-1}	[125]
Chelidonium majus subsp. asiaticum H. Hara	Methanol (whole plant)	-	*S. mutans* ATCC 25175	MIC IZD	1.0 mg·disk^{-1} 8 mm (1.0 mg·disk^{-1})	99.0% GTFs inhibition	1.0 mg·mL^{-1}	[125]
Chrysosplenium flagelliferum F.Schmidt	Methanol (whole plant)	-	*S. mutans* ATCC 25175	MIC IZD	1.0 mg·disk^{-1} 12 mm (2 mg·disk^{-1})	99.0% GTFs inhibition	1.0 mg·mL^{-1}	[125]
Cinnamomum burmannii (Nees & T.Nees) Blume	Aqueous	-	*S. mutans* UA159	MIC	2.5 mg·mL^{-1}	MBIC MBBC >99% inhibition	2.5 mg·mL^{-1} 10 mg·mL^{-1} 10 mg·mL^{-1}	[196]
Cinnamomum zeylanicum Blume	Essential oil	β-linalool; (E)-cinnamaldehyde; cinnamyl acetate	*S. mutans* ATCC 25175	MIC IZD	0.056 mg·mL^{-1} 10 mm (50 mg·mL^{-1})	≈99.0% inhibition	0.224 mg·mL^{-1}	[185]
			E. faecalis ATCC 19433	MIC IZD	0.315 mg·mL^{-1} 10 mm (50 mg·mL^{-1})	≈47.0% inhibition	1.26 mg·mL^{-1}	
			S. aureus ATCC 29213	MIC IZD	0.315 mg·mL^{-1} 11 mm (50 mg·mL^{-1})	≈93.0% inhibition	0.315 mg·mL^{-1}	
	Essential oil (bark)	-	*S. mutans* KPSK2	MIC MBC IZD	0.08% (v/v) 0.08% (v/v) 32.2 mm (20% (v/v))	88.2% inhibition 81.2% reduction	0.32% (v/v) 0.32% (v/v)	[197]
	Essential oil (bark)	-	*P. gingivalis* ATCC 5397 *F. nucleatum* ATCC 25586		-	85.7% inhibition 75.3% reduction	5.0 mg·mL^{-1}	[198]
	Essential oil (bark)	Cinnamaldehyde	*P. gingivalis* ATCC 33177	MIC	6.25 µg·mL^{-1}	74.5% inhibition 33.5% reduction	4.17 µg·mL^{-1} 25 µg·mL^{-1}	[199])
Cistus creticus L.	Methanol (aerial plant parts)	Quercetin; 3-O-β-D-glucopyranoside	*S. mutans* DSM 20523	MIC MBC	5 mg·mL^{-1} 10 mg·mL^{-1}	≈80% inhibition	0.600 mg·mL^{-1}	[200]
Cistus monspeliensis L.	Methanol (aerial plant parts)	Cistodioic acid	*S. mutans* DSM 20523	MIC MBC	2.5 mg·mL^{-1} NA	≈60% inhibition	0.600 mg·mL^{-1}	[200]

Table 1. *Cont.*

Plant Name	Plant Extract	Compound	Microorganism	Antimicrobial Activity		Antibiofilm Activity		References
Copaifera pubiflora Benth.	Oleoresin	Ent-hardwickiic acid; schistochilic acid B; ent-7α-acetoxy hardwickiic acid; (13*E*)-ent-labda-7,13-dien-1-5-oic acid	*S. sanguinis* ATCC 10556	MIC MBC	12.5 µg·mL^{-1} 25.0 µg·mL^{-1}	MBIC$_{50}$ MBEC	6.25 µg·mL^{-1} 50.0 µg·mL^{-1}	[141]
			S. sanguinis (CI)	MIC MBC	25.0 µg·mL^{-1}	MBIC$_{50}$	6.25 µg·mL^{-1}	
			S. mutans ATCC 25175	MIC MBC	12.5 µg·mL^{-1}	MBIC$_{50}$	12.5 µg·mL^{-1}	
			L. paracasei (CI)	MIC MBC	12.5 µg·mL^{-1}	MBIC$_{50}$	12.5 µg·mL^{-1}	
			P. gingivalis ATCC 33277	MIC MBC	12.5 µg·mL^{-1} 50.0 µg·mL^{-1}	MBIC$_{50}$	12.5 µg·mL^{-1}	
			P. gingivalis (CI)	MIC MBC	50.0 µg·mL^{-1}	MBIC$_{50}$	100 µg·mL^{-1}	
			F. nucleatum (CI)	MIC MBC	25.0 µg·mL^{-1} 50.0 µg·mL^{-1}	MBIC$_{50}$	400 µg·mL^{-1}	
			P. micra (CI)	MIC MBC	12.5 µg·mL^{-1} 25.0 µg·mL^{-1}	MBIC$_{50}$MBEC	25.0 µg·mL^{-1} 50.0 µg·mL^{-1}	
Coriandrum sativum L.	Essential oil (leaves)	1-decanol, *E*-2-decen-1-ol; 2-dodecen-1-ol	*S. mutans* UA159	MIC MBC	15.6–31.2 µg·mL^{-1} 31.2–62.5 µg·mL^{-1}	>95% inhibition	31.2 µg·mL^{-1}	[9]
	Essential oil (leaves)	1-decanol, *E*-2-decen-1-ol; 2-dodecen-1-ol	*F. nucleatum* ATCC 25586	MIC MBC	0.015 mg·mL^{-1} 0.125 mg·mL^{-1}	55.8% inhibition	1.0 mg·mL^{-1}	[136]
			P. gingivalis ATCC 33277	MIC MBC	0.125 mg·mL^{-1}	39.7% inhibition		
			S. sanguis ATCC 10556	MIC MBC	0.250 mg·mL^{-1} 0.500 mg·mL^{-1}	58.3% inhibition		
			S. mitis ATCC 903	MIC MBC	0.062 mg·mL^{-1} 0.125 mg·mL^{-1}	1.5% inhibition		
Croton urucurana Baill.	Ethyl acetate-ethanol (stem bark)	-	*S. mutans* UA159		-	34% inhibition	0.007 mg·mL^{-1}	[201]
Curcuma longa L.	Ethanol	-	*P. gingivalis* JCM12257 *S. mutans* NBRC13955		-	99.7% reduction 99.1% reduction	5.0% (*v/v*)	[151]

Table 1. *Cont.*

Plant Name	Plant Extract	Compound	Microorganism	Results				References
				Antimicrobial Activity		Antibiofilm Activity		
	Essential oil (leaves)	Citral; myrcene	*S. mutans* ATCC UA159		-	93% inhibition	1.0 μg·mL^{-1}	[147]
Cymbopogon citratus (DC.) Stapf	Essential oil (leaves)	Geranial; neral; myrcene	*S. mutans* ATCC 35688	MIC MBC IZD	2.61 mg·mL^{-1} 10.54 mg·mL^{-1} 11 mm (100% (v/v))	95% inhibition	26.1 mg·mL^{-1}	[202]
			L. acidophilus ATCC 4356	MIC MBC IZD	1.32 mg·mL^{-1} 2.61 mg·mL^{-1} 8 mm (100% (v/v))	99.6% inhibition	13.2 mg·mL^{-1}	
	Essential oil	Geranial; neral; myrcene	*S. mutans* ATCC 35668		-	95.4% reduction	0.10 μg·mL^{-1}	[148]
Cymbopogon martinii (Roxb.) W. Watson	Essential oil	Geraniol; geranyl acetate	*S. mitis* (CI)	MIC MBC	0.25 mg·mL^{-1} >2.0 mg·mL^{-1}	28% reduction	0.25 mg·mL^{-1}	[203]
			E. faecalis (CI)	MIC MBC	0.25 mg·mL^{-1} 1.0 mg·mL^{-1}	36% reduction	1.0 mg·mL^{-1}	
			S. mitis + S. sanguinis + E. faecalis		-	20% reduction		
Cyperus articulatus L.	Essential oil (bulbs)	α-pinene; mustakone; α-bulnesene	*F. nucleatum* ATCC 25586	MIC MBC	0.250 mg·mL^{-1}	61.67% inhibition	1.0 mg·mL^{-1}	[136]
			P. gingivalis ATCC 33277	MIC MBC	0.250 mg·mL^{-1}	43.53% inhibition		
			S. sanguis ATCC 10556	MIC MBC	0.250 mg·mL^{-1} 0.500 mg·mL^{-1}	63.96% inhibition		
			S. mitis ATCC 903	MIC MBC	0.250 mg·mL^{-1} 0.500 mg·mL^{-1}	5.00% inhibition		
Derris reticulata Craib	Ethanol (stem)	-	*S. mutans* DMST 1877	MIC MBC	0.875 mg·mL^{-1} 1.75 mg·mL^{-1}	102.8% inhibition	750 μg·mL^{-1}	[155]
Dodonaea viscosa var. angustifólia (L.f.) Benth	Methanol (leaves)	Xylopyranoside; 2,2′-methylenebis[6-(1,1-dimethyl)]-4-methyl); 2-(3-Hydroxy-4-methoxyphenyl)-3,7-dimethoxy-4H-chromen-4-one; trans-3′,4′,5′-Trimethoxy-4-(methylthio)chalcone; stigmasterol	*S. mutans* NCTC 1091	MIC MBC	0.78 mg·mL^{-1} 3.125 mg·mL^{-1}	99% inhibition	0.78 mg·mL^{-1}	[159]

Table 1. *Cont.*

Plant Name	Plant Extract	Compound	Microorganism	Results				References
				Antimicrobial Activity		Antibiofilm Activity		
Englerophytum magalismontanum(Sond.) T.D.Penn.	Dichloromethane: methanol (stems)	-	*S. mutans* ATCC 25175	MIC	0.83 mg·mL^{-1}	49.28% inhibition	0.25 mg·mL^{-1}	[183]
Erythrina lysistemon Hutch.	Dichloromethane: methanol (stems)	-	*S. mutans* ATCC 25175	MIC	0.50 mg·mL^{-1}	72.54% inhibition	0.25 mg·mL^{-1}	[183]
Eucalyptus sp.	Essential oil	-	*E. faecalis* ATCC 29212		-	71.6% reduction 78.5% reduction	100% (*v*/*v*)	[204]
Eucalyptus galbie	Methanol	-	*E. faecalis* PTCC 1237	MIC IZD	12.5 mg·mL^{-1} 9.63 mm (100 mg·mL^{-1})	77.7% adherence reduction	6.25 mg·mL^{-1}	[205]
Eucalyptus globulus Labill.	Ethanol	Macrocarpal A; macrocarpal B; macrocarpal C; eucalyptin; 1,8-cineole	*P. gingivalis* JCM12257 *S. mutans* NBRC13955		-	MBEC MBEC	49.1 µg·mL^{-1} 393 µg·mL^{-1}	[151]
	Essential oil (leaves)	1,8-cineole; α-pinene	*S. mutans* ATCC 700610	MIC IZD	0.013 mg·mL^{-1} 34.7 mm (100% *v*/*v*)	81.1% reduction	2.0 mg·mL^{-1}	[206]
Eucalyptus x urograndis	Essential oil (leaves)	1,8-cineole; α-pinene	*S. mutans* ATCC 700610	MIC IZD	0.025 mg·mL^{-1} 23.0 mm (100% (*v*/*v*))	35.1% reduction	2.0 mg·mL^{-1}	[206]
Firmiana simplex (L.) W.Wight	Methanol (bark)	-	*S. mutans* ATCC 25175	MIC IZD	1.0 mg·disk^{-1} 9 mm (1.0 mg·disk^{-1})	35.7% GTFs inhibition	1.0 mg·mL^{-1}	[125]
Foeniculum vulgare Mill.	Essential oil (seeds)	-	*S. mutans* KPSK2	MIC MBC IZD	1.25% (*v*/*v*) 2.50% (*v*/*v*) 9.17 mm [20% (*v*/*v*)]	84.4% inhibition 69.7 % reduction	5.0% (*v*/*v*)	[197]
	Ethanol	-	*P. gingivalis* JCM12257 *S. mutans* NBRC13955		-	99.9% reduction 71.4% reduction	5.0% (*v*/*v*)	[151]
Geranium sibiricum L.	Methanol (whole plant)	-	*S. mutans* ATCC 25175	MIC IZD	0.5 mg·disk^{-1} 15 mm (2.0 mg·disk^{-1})	69.3 % GTFs inhibition	1.0 mg·mL^{-1}	[125]
Ginkgo biloba L.	Methanol	-	*S mutans* ATCC 25175	MIC	62.5 mg·mL^{-1}	≈38.5% reduction	125 mg·mL^{-1}	[195]
Glycyrrhiza glabra L.	Ethanol (roots)	-	*P. gingivalis* ATCC 33277	MIC MBC	62.5 µg·mL^{-1} 125 µg·mL^{-1}	92.3% inhibition MBEC	500 µg·mL^{-1} 62.5 µg·mL^{-1}	[162]

Table 1. *Cont.*

Plant Name	Plant Extract	Compound	Microorganism	Results				References
				Antimicrobial Activity		**Antibiofilm Activity**		
Hibiscu sabdariffa L.	Ethanol (calices)	-	*S. mutans* Ingbritt	MIC / MBC	7.2 mg·mL^{-1} / 57.6 mg·mL^{-1}	99.0% inhibition	3.60 mg·mL^{-1}	[207]
			S. sanguinis ATCC 10556T	MIC / MBC	28.8 mg·mL^{-1} / 57.6 mg·mL^{-1}	97.0% inhibition	14.4 mg·mL^{-1}	
			L. casei ATCC 4646	MIC / MBC	28.8 mg·mL^{-1} / >57.6 mg·mL^{-1}	92.0% inhibition	28.8 mg·mL^{-1}	
			A. naeslundii ATCC 12104T	MIC / MBC	14.4 mg·mL^{-1} / >57.6 mg·mL^{-1}	97.0% inhibition	14.4 mg·mL^{-1}	
			A. actinomycetemcomitans ATCC 29522	MIC / MBC	28.8 mg·mL^{-1} / 57.6 mg·mL^{-1}	97.0% inhibition	28.8 mg·mL^{-1}	
			F. nucleatum JCM 6328	MIC / MBC	7.2 mg·mL^{-1} / 14.4 mg·mL^{-1}	83.0% inhibition	1.8 mg·mL^{-1}	
			P. gingivalis ATCC 33277T	MIC / MBC	7.2 mg·mL^{-1} / 28.8 mg·mL^{-1}	98.0% inhibition	14.4 mg·mL^{-1}	
			P. intermedia ATCC 25611T	MIC / MBC	14.4 mg·mL^{-1} / 28.8 mg·mL^{-1}	89.0% inhibition	7.2 mg·mL^{-1}	
Houttuynia cordata Thunb.	Ethanol (leaves)	-	*S. mutans* MT8148	MIC	1.09 µg·mL^{-1}	≈80.0% reduction	10% (*v/v*)	[208]
			F. nucleatum JCM8532	MIC	0.543 µg·mL^{-1}	≈90.0% reduction		
Hypericum perforatum L.	Ethanol	-	*S. mutans* ATCC 21752	MIC	64.0 µg·mL^{-1}	MBIC$_{50}$	18.0 µg·mL^{-1}	[209]
			S. sobrinus ATCC 6715	MIC	16.0 µg·mL^{-1}		7.23 µg·mL^{-1}	
			L. plantarum ATCC 80141	MIC	32.0 µg·mL^{-1}		29.9 µg·mL^{-1}	
			E. faecalis ATCC 29912	MIC	32.0 µg·mL^{-1}		18.1 µg·mL^{-1}	
	Hexane		*S. mutans* ATCC 21752	MIC	64.0 µg·mL^{-1}		16.3 µg·mL^{-1}	
			S. sobrinus ATCC 6715	MIC	16.0 µg·mL^{-1}		7.29 µg·mL^{-1}	
			L. plantarum ATCC 80141	MIC	32.0 µg·mL^{-1}		27.6 µg·mL^{-1}	
			E. faecalis ATCC 29912	MIC	32.0 µg·mL^{-1}		16.1 µg·mL^{-1}	
	Chloroform		*S. mutans* ATCC 21752	MIC	32.0 µg·mL^{-1}		17.2 µg·mL^{-1}	
			S. sobrinus ATCC 6715	MIC	16.0 µg·mL^{-1}		8.03 µg·mL^{-1}	
			L. plantarum ATCC 80141	MIC	32.0 µg·mL^{-1}		27.5 µg·mL^{-1}	
			E. faecalis ATCC 29912	MIC	32.0 µg·mL^{-1}		17.2 µg·mL^{-1}	
	Acetic acid		*S. mutans* ATCC 21752	MIC	32.0 µg·mL^{-1}		17.7 µg·mL^{-1}	
			S. sobrinus ATCC 6715	MIC	16.0 µg·mL^{-1}		7.52 µg·mL^{-1}	
			L. plantarum ATCC 80141	MIC	16.0 µg·mL^{-1}		25.1 µg·mL^{-1}	
			E. faecalis ATCC 29912	MIC	16.0 µg·mL^{-1}		17.7 µg·mL^{-1}	

Table 1. *Cont.*

Plant Name	Plant Extract	Compound	Microorganism		Antimicrobial Activity	Antibiofilm Activity		References
						Results		
					Antimicrobial Activity		**Antibiofilm Activity**	
Hypericum perforatum L.	Butanol	-	*S. mutans* ATCC 21752	MIC	32.0 µg·mL^{-1}		17.5 µg·mL^{-1}	[209]
			S. sobrinus ATCC 6715	MIC	16.0 µg·mL^{-1}		7.25 µg·mL^{-1}	
			L. plantarum ATCC 80141	MIC	16.0 µg·mL^{-1}	MBIC$_{50}$	27.3 µg·mL^{-1}	
			E. faecalis ATCC 29912	MIC	32.0 µg·mL^{-1}		17.5 µg·mL^{-1}	
	Aqueous		*S. mutans* ATCC 21752	MIC	32.0 µg·mL^{-1}		20.1 µg·mL^{-1}	
			S. sobrinus ATCC 6715	MIC	8.00 µg·mL^{-1}		7.60 µg·mL^{-1}	
			L. plantarum ATCC 80141	MIC	8.00 µg·mL^{-1}		24.9 µg·mL^{-1}	
			E. faecalis ATCC 29912	MIC	16.0 µg·mL^{-1}		20.1 µg·mL^{-1}	
Laurus nobilis L.	Essential oil (from Gafsa)	1,8-cineole; methyl eugenol; α-terpinyl acetate; linalool	*S. aureus* ATCC 6538	MIC	31.25 mg·mL^{-1}	95.35% inhibition	31.25 mg·mL^{-1}	[210]
				MBC	62.5 mg·mL^{-1}	78.4 % eradication	100 mg mL^{-1}	
				IZD	7.75 mm (100% (v/v))			
			S. aureus L36 (CI)	MIC	15.625 mg·mL^{-1}	78.92% inhibition	15.625 mg·mL^{-1}	
				MBC	125 mg·mL^{-1}	27.2% eradication	100 mg·mL^{-1}	
				IZD	8.0 mm (100% (v/v))			
			S. aureus L37 (CI)	MIC	15.625 mg·mL^{-1}	86.07% inhibition	15.625 mg·mL^{-1}	
				MBC	125 mg·mL^{-1}	30.0% eradication	100 mg·mL^{-1}	
				IZD	9.5 mm (100% (v/v))			
	Essential oil (from Sousse)	1,8-cineole; methyl eugenol; α-terpinyl acetate; linalool	*S. aureus* ATCC 6538	MIC	31.25 mg·mL^{-1}	96.61% inhibition	15.625 mg·mL^{-1}	
				MBC	62.5 mg·mL^{-1}	78.0 % eradication	100 mg·mL^{-1}	
				IZD	10.5 mm (100% (v/v))			
			S. aureus L36 (CI)	MIC	3.91 mg·mL^{-1}	88.22% inhibition	3.91 mg·mL^{-1}	
				MBC	31.25 mg·mL^{-1}	45.0% eradication	100 mg·mL^{-1}	
				IZD	11.5 mm (100% (v/v))			
			S. aureus L37 (CI)	MIC	3.91 mg·mL^{-1}	93.00% inhibition	3.91 mg·mL^{-1}	
				MBC	62.5 mg·mL^{-1}	31.0% eradication	100 mg·mL^{-1}	
				IZD	10.75 mm (100% (v/v))			

Table 1. *Cont.*

Plant Name	Plant Extract	Compound	Microorganism	Results				References
				Antimicrobial Activity		Antibiofilm Activity		
Ligustrum robustum (Roxb.)	Methanol (roots)	Ligurobustoside B, N, J and C	*S. mutans* UA159	MIC MBC	0.40% (*w/v*) 0.80% (*w/v*)	≈35.0% inhibition MBIC-0.8% (*w/v*)	0.10% (*w/v*)	[211]
			S. mutans C1 (CI)	MIC MBC	0.25% (*w/v*) 0.50% (*w/v*)	≈61.0% inhibition MBIC-0.5% (*w/v*)		
			S. mutans C2 (CI)	MIC MBC	0.25% (*w/v*) 0.50% (*w/v*)	≈62.0% inhibition MBIC-0.5% (*w/v*)		
			S. mutans C3 (CI)	MIC MBC	0.35% (*w/v*) 0.70% (*w/v*)	≈41.0% inhibition MBIC-0.7% (*w/v*)		
			S. mutans C4 (CI)	MIC MBC	0.25% (*w/v*) 0.50% (*w/v*)	≈54.0% inhibition MBIC-0.5% (*w/v*)		
			S. mutans C5 (CI)	MIC MBC	0.25% (*w/v*) 0.50% (*w/v*)	≈57.0% inhibition MBIC-0.25% (*w/v*)		
			S. mutans C6 (CI)	MIC MBC	0.25% (*w/v*) 0.50% (*w/v*)	≈47.0% inhibition MBIC-0.5% (*w/v*)		
			S. mutans C7 (CI)	MIC MBC	0.30% (*w/v*) 0.60% (*w/v*)	≈42.0% inhibition MBIC-0.6% (*w/v*)		
			S. mutans C8 (CI)	MIC MBC	0.35% (*w/v*) 0.70% (*w/v*)	≈40.0% inhibition MBIC-0.7% (*w/v*)		
Lindera glauca (Siebold & Zucc.) Blume	Methanol (leaves)	-	*S. mutans* ATCC 25175	MIC IZD	1.0 mg·disk^{-1} 8 mm (1.0 mg·disk^{-1})	85.9% GTFs inhibition	1.0 mg·mL^{-1}	[125]
Lippia alba (Mill.) N.E.Br. ex Britton & P.Wilson	Essential oil	Geraniol; citral	*S. mutans* ATCC 35668		-	95.8% reduction	0.10 µg·mL^{-1}	[148]
	Essential oil (leaves)	Thymol	*S. mutans* UA159	MIC MBC	62.5–125 µg·mL^{-1} 125–250 µg·mL^{-1}	>90% inhibition	125 µg·mL^{-1}	[9]
Lippia sidoides Cham.	Essential oil (leaves)	Thymol; *p*-cymene; α-caryophyllene	*F. nucleatum* ATCC 25586	MIC MBC	0.125 mg·mL^{-1}	58.33% inhibition	1.0 mg·mL^{-1}	[136]
			P. gingivalis ATCC 33277	MIC MBC	0.250 mg·mL^{-1}	12.94% inhibition		
			S. sanguis ATCC 10556	MIC MBC	0.125 mg·mL^{-1} 0.500 mg·mL^{-1}	58.13% inhibition		
			S. mitis ATCC 903	MIC MBC	0.250 mg·mL^{-1} >1.0 mg·mL^{-1}	5.50% inhibition		

Table 1. *Cont.*

Plant Name	Plant Extract	Compound	Microorganism	Results				References
				Antimicrobial Activity		Antibiofilm Activity		
Mangifera sp.	Aqueous (leaves)	Quinic acid; benzophenone C-glycoside isomer; benzophenone C-glycoside; quercetin-3-O-glucoside	*S. mutans* ATCC 25175 *S. sanguinis* ATCC BAA-1455		-	99.4% reduction 61.4% reduction	0.50 mg·mL^{-1}	[167]
Mangifera indica L.	Aqueous (leaves)	-	*E. faecalis* ATCC 29212 *S. aureus* ATCC 25923		-	37.1% reduction 30.3% reduction	30 mg·mL^{-1}	[193]
Matricaria aurea (Loefl.) Sch.Bip.	Ethanol (flowers)	-	*P. gingivalis* (CI)	MIC MBC IZD	0.78 mg·mL^{-1} 3.12 mg·mL^{-1} 20 mm (0.2 g·mL^{-1})	78% inhibition	1.56 mg·mL^{-1}	[212]
			T. denticola (CI)	MIC MBC IZD	1.56 mg·mL^{-1} 3.12 mg·mL^{-1} 15 mm (0.2 g·mL^{-1})	74% inhibition	3.12 mg·mL^{-1}	
			T. forsythia (CI)	MIC MBC IZD	0.39 mg·mL^{-1} 1.56 mg·mL^{-1} 23 mm (0.2 g·mL^{-1})	86% inhibition	0.78 mg·mL^{-1}	
			A. actinomycetemcomitans (CI)	MIC MBC IZD	1.56 mg·mL^{-1} 6.25 mg·mL^{-1} 13 mm (0.2 g·mL^{-1})	62% inhibition	3.12 mg·mL^{-1}	
Matricaria recutita L.	Essential oil (leaves)	Alkaloid; saponin; tannin, phenolic; flavonoid; triterpenoid and glycoside compounds	*A. actinomycetemcomitans* ATCC 29522 *T. denticola* ATCC 35405		-	87.6% reduction 99.1% reduction	100% (v/v) 50% (v/v)	[213]
Melaleuca alternifolia (Maiden & Betche) Cheel Myrtaceae	Essential oil	-	*S. mutans* ATCC 25175	MIC MBC IZD	0.125% (v/v) 0.25% (v/v) 20.3 mm [20% (v/v)]	≈94% reduction (metabolic activity) ≈85% reduction (biomass)	1.0% (v/v) 0.5% (v/v)	[214]
Mentha sp.	Aqueous (leaves)	Methyl-2-[cyclohex-2-en-1-yl(hydroxy)methyl]-3-hydroxy-4-(2-hydroxyethyl)-3-methyl-5-oxoprolinate	*S. mutans* ATCC 25175 *S. sanguinis* ATCC BAA-1455		-	98.5% reduction 85.5% reduction	0.50 mg·mL^{-1}	[167]

Table 1. *Cont.*

Plant Name	Plant Extract	Compound	Microorganism	Results				References
				Antimicrobial Activity		**Antibiofilm Activity**		
Mentha × *piperita* L.	Essential oil (leaves)	-	*S. mutans* KPSK2	MIC MBC IZD	1.25% (*v/v*) 1.25% (*v/v*) 11.33 mm (20% (*v/v*))	83.42% inhibition 71.01% reduction	5% (*v/v*)	[197]
	Essential oil (flowers)	Menthol; menthone	*F. nucleatum* ATCC 25586	MIC MBC	0.25% (*v/v*) 0.5% (*v/v*)	>90.0% inhibition	0.25% (*v/v*)	[215]
Mentha spicata L.	Methanol (leaves)	-	*S. mutans* KPSK2	MIC MBC IZD	1.25% (*v/v*) 1.25% (*v/v*) 9.50 mm (20% (*v/v*))	82.7% inhibition 71.5% reduction	5.0% (*v/v*)	[197]
Mikania glomerata Spreng	Essential oil (leaves)	Germacrene D; α-caryophyllene; bicyclogermacrene	*F. nucleatum* ATCC 25586	MIC MBC	0.25 mg·mL^{-1} 0.50 mg·mL^{-1}	58.96% inhibition	1.0 mg·mL^{-1}	[136]
			P. gingivalis ATCC 33277	MIC MBC	0.50 mg·mL^{-1} >1.00 mg·mL^{-1}	40.00% inhibition		
			S. sanguis ATCC 10556	MIC MBC	0.062 mg·mL^{-1} 0.125 mg·mL^{-1}	54.79% inhibition		
			S. mitis ATCC 903	MIC MBC	0.125 mg·mL^{-1}	1.00% inhibition		
Myracrodruon urundeuva All.	Hydroalcoholic (leaves)	-	*S. mutans* ATCC 25175	MIC MBC	2.5 mg·mL^{-1}	MBIC MBEC	1.25 mg·mL^{-1} 2.5 mg·mL^{-1}	[216]
Myrtus communis L.	Hydroethanolic (leaves)	-	*S. mutans* ATCC 25175			MBIC MBEC	40 µg·µL^{-1} 120 µg·µL^{-1}	[173]
			S. oralis (CI)		-	MBIC MBEC	20 µg·µL^{-1} 40 µg·µL^{-1}	
			S. mitis (CI)			MBIC MBEC	20 µg·µL^{-1} 40 µg·µL^{-1}	
			R. dentocariosa (CI)			MBIC MBEC	40 µg·µL^{-1} 120 µg·µL^{-1}	
Ocimum basilicum L.	Essential oil (leave)	-	*S. mutans* KPSK2	MIC MBC IZD	0.31% (*v/v*) 0.31% (*v/v*) 14.7 mm (20% (*v/v*))	86.8% inhibition 73.3% reduction	1.25% (*v/v*)	[197]
	Ethanol (seeds)	-	*P. gingivalis* JCM12257 *S. mutans* NBRC13955		-	99.9% reduction 90.9% reduction	5.0% (*v/v*)	[151]

Table 1. *Cont.*

Plant Name	Plant Extract	Compound	Microorganism	Results				References
				Antimicrobial Activity		Antibiofilm Activity		
	Methanol (aerial plant parts)	Rosmarinic acid	*S. mutans* DSM 20523	MIC MBC	2.5 mg·mL^{-1} 5.0 mg·mL^{-1}	≈60.0% inhibition	5.0 mg·mL^{-1}	[200]
Origanum vulgare L.	Essential oil	*O*-cymene; carvacrol	*S. mutans* ATCC 25175	MIC IZD	0.315 mg·mL^{-1} 10.0 mm (50 mg·mL^{-1})	≈97.0% inhibition	1.26 mg·mL^{-1}	[185]
			E. faecalis ATCC 19433	MIC IZD	0.625 mg·mL^{-1} 7.0 mm (50 mg·mL^{-1})	≈60.0% inhibition	2.5 mg·mL^{-1}	
			S. aureus ATCC 29213	MIC IZD	0.625 mg·mL^{-1} 8.0 mm (50 mg·mL^{-1})	≈94% inhibition	0.625 mg·mL^{-1}	
Phellodendron amurense Rupr.	Ethanol (bark)	-	*S. mutans* UA159	MIC	0.313 mg·mL^{-1}	100% inhibition	1.25 mg·mL^{-1}	[217]
Piper betle L.	Aqueous (leaves)	-	*E. faecalis* ATCC 29212 *S. aureus* ATCC 25923		-	53.6% reduction 32.7% reduction	30.0 mg·mL^{-1}	[193]
	Ethanol (leaves)	4-chromanol	*S. mutans* ATCC 25175	MIC MBC	1.56 mg·mL^{-1} 3.17 mg·mL^{-1}	90% inhibition 90% reduction	1.56 mg·mL^{-1} 6.25 mg·mL^{-1}	[10]
			A. actinomycetemcomitans ATCC 33384	MIC MBC	1.04 mg·mL^{-1} 2.08 mg·mL^{-1}	90% inhibition 90 reduction	0.39 mg·mL^{-1} 3.13 mg·mL^{-1}	
Piper nigrum L.	Methanol (seed)	-	*S. mutans* KPSK2	MIC MBC IZD	1.25% (*v*/*v*) 2.50% (*v*/*v*) 14.0 mm (20% (*v*/*v*))	82.7% inhibition 67.8% reduction	5.0% (*v*/*v*)	[197]
	Aqueous (leaves)	-	*E. faecalis* ATCC 29212 *S. aureus* ATCC 25923		-	49.1% reduction 44.2% reduction	30.0 mg·mL^{-1}	[193]
Pistacia lentiscus L.	Essential oil (berry)	Phenols; free unsaturated-saturated fatty acids	*S. salivarius* K12	MIC MBC	>50.0% (*v*/*v*)	MBIC	> 50% (*v*/*v*)	[218]
			S. salivarius M18	MIC MBC	>50.0% (*v*/*v*)	MBIC	>50% (*v*/*v*)	
			S. pyogenes (CI)	MIC MBC	50.0% (*v*/*v*)	MBIC	25.0–3.0% (*v*/*v*)	
			S. agalactiae (CI)	MIC MBC	50.0% (*v*/*v*)	MBIC	25.0–3.0% (*v*/*v*)	
			S. mutans CIP103220	MIC MBC	50.0% (*v*/*v*)	MBIC	25.0–3.0% (*v*/*v*)	
			S. intermedius DSM 20573	MIC MBC	6.0–3.0% (*v*/*v*) 50.0% (*v*/*v*)	MBIC	<4.0% (*v*/*v*)	
			S. mitis (CI)	MIC MBC	6.0–3.0% (*v*/*v*) 50.0% (*v*/*v*)	MBIC	<4.0% (*v*/*v*)	

Table 1. *Cont.*

Plant Name	Plant Extract	Compound	Microorganism	Antimicrobial Activity		Antibiofilm Activity		References
Pistacia vera L.	Purified oleoresin	α-pinene; β-pinene	*S. mutans* ATCC 25175 *S. sanguinis* ATCC 10556	MBC MBC	>2048 µg·mL^{-1} 1024 µg·mL^{-1}	49.4% inhibition 71.2% inhibition	256 µg·mL^{-1}	[219]
Psidium sp.	Aqueous (leaves)	Methyl quercetin sulfate; 2,6-dihydroxy-3-methyl-4-*O*-(6″-*O*-galloyl-β-*D*-glucopyranosyl)-benzophenone	*S. mutans* ATCC 25175 *S. sanguinis* ATCC BAA-1455		-	93.8% reduction 48.6% reduction	0.5 mg·mL^{-1}	[167]
Psidium cattleianum Sabine	Aqueous (leaves)	-	*E. faecalis* ATCC 51299	MIC MBC	0.5 mg·mL^{-1}	>99.9% reduction	5.0 mg·mL^{-1}	[220]
			P. aeruginosa ATCC 15442	MIC MBC	4.0 mg·mL^{-1}	-	-	
	Hydroethanol (leaves)		*E. faecalis* ATCC 51299	MICMBC	0.5 mg·mL^{-1}	>99.9% reduction	2.5 mg·mL^{-1}	
			P. aeruginosa ATCC 15442	MIC MBC	4.0 mg·mL^{-1}	>99.9% reduction	40 mg·mL^{-1}	
Punica granatum L.	Flowers infusion	-	*S. mutans* ATCC 35608	MIC MBC	50.0 mg·mL^{-1}	84.4% reduction	25 mg·mL^{-1}	[221]
			S. sanguinis ATCC 10556	MIC MBC	6.25 mg·mL^{-1} 25.0 mg·mL^{-1}	100% reduction	1.56 mg·mL^{-1}	
			S. sobrinus ATCC 27607	MIC MBC	25.0 mg·mL^{-1}	99.9% reduction	12.5 mg·mL^{-1}	
			S. salivarius ATCC 9222	MIC MBC	25.0 mg·mL^{-1} 100 mg·mL^{-1}	86.5% reduction	12.5 mg·mL^{-1}	
			E. faecalis CIP 55142	MIC MBC	50.0 mg·mL^{-1}	56.3% reduction	50 mg·mL^{-1}	
Pyrostegia venusta (Ker Gawl.) Miers	Hydroalcoholic (flowers)	-	*S. mutans* ATCC 25175	MIC MBC	500 mg·mL^{-1} 1000 mg·mL^{-1}	68.9% inhibition	500 mg·mL^{-1}	[222]
Qualea grandiflora Mart.	Hydroalcoholic (leaves)	-	*S. mutans* ATCC 25175	MIC	5.0 mg·mL^{-1}	MBIC MBEC	5.0 mg·mL^{-1} 10.0 mg·mL^{-1}	[216]
Rhodiola rosea L.	Ethanol (root)	-	*S. mutans* UA159		-	≈95.0% inhibition ≈48.6% reduction	0.50 µg·µL^{-1}	[178]
Rosa rugosa Thunb.	Methanol (leaves)	-	*S. mutans* ATCC 25175	MIC IZD	1.0 mg·disk^{-1} 12 mm (2.0 mg·disk^{-1})	64.9% GTFs inhibition	1.0 mg·mL^{-1}	[125]

Table 1. *Cont.*

Plant Name	Plant Extract	Compound	Microorganism	Results				References
				Antimicrobial Activity		**Antibiofilm Activity**		
	Ethanol	Carnosic acid; carnosol; rosmarinic acid	*S. mutans* NBRC13955 *P. gingivalis* JCM12257		-	MBEC MBEC	97.8 µg·mL^{-1} 195.5 µg·mL^{-1}	[151]
Rosmarinus officinalis L.	Extract (leaves)	-	*S. aureus* ATCC 6538	MIC MMC	25 mg·mL^{-1} >50 mg·mL^{-1}	99% reduction	200 mg·mL^{-1}	[223]
			E. faecalis ATCC 4083	MIC MMC	50 mg·mL^{-1} >50 mg·mL^{-1}	80% reduction		
			S. mutans ATCC 35688	MIC MMC	25 mg·mL^{-1} >50 mg·mL^{-1}	80% reduction		
			P. aeruginosa ATCC 15442	MIC MMC	6.25 mg·mL^{-1}	70% reduction		
	Methanol (aerial plant parts)	7-methoxyrosmanol, rosmanol isomers; rosmarinic acid	*S. mutans* DSM 20523	MIC MBC	0.60 mg·mL^{-1} 2.50 mg·mL^{-1}	>90% inhibition	1.25 mg·mL^{-1}	[200]
	Extract (leaves)		*S. mutans* ATCC 35688		-	32% reduction of total biofilm protein	200 mg·mL^{-1}	[224]
Salvadora persica L.	Methanol	Benzyl (6Z,9Z,12Z)-6,9,12-octadecatrienoate; 3-benzyloxy-1-nitro-butan-2-ol; 1,3-cyclohexane dicarbohydrazide	*S. mutans* CI	MIC IZD	2.60 mg·mL^{-1} 24.0 mm (2.5 g. mL^{-1})	87.9% inhibition	2.6 mg·mL^{-1}	[225]
Salvia sclarea L.	Methanol (aerial plant parts)	Rosmarinic acid	*S. mutans* DSM 20523	MIC MBC	1.25 mg·mL^{-1} 5.0 mg·mL^{-1}	60% inhibition	2.5 mg·mL^{-1}	[200]
Schinus terebinthifolia Raddi.	Methanol (leaves)	*p*-anisaldehyde	*S. mutans* UA159		-	41% inhibition	0.0035 mg·mL^{-1}	[201]
Sophora flavescens Aiton.	Methyl-alcohol	-	*S mutans* ATCC 25175	MIC	62.5 mg·mL^{-1}	≈88.5% reduction	125 mg·mL^{-1}	[195]
Spirostachys africana Sond.	Dichloromethane: methanol (leaves)	-	*S. mutans* ATCC 25175	MIC	0.50 mg·mL^{-1}	97.56% inhibition	0.25 mg·mL^{-1}	[183]

Table 1. *Cont.*

Plant Name	Plant Extract	Compound	Microorganism	Results				References
				Antimicrobial Activity		Antibiofilm Activity		
	Essential oil (flower buds)	-	*P. gingivalis* ATCC 53978 *F. nucleatum* ATCC 25586		-	78.6% inhibition 76.2% reduction	5.0 mg·mL^{-1} 1.25 mg·mL^{-1}	[198]
	Ethanol	Eugenol; eugenol acetate; β-caryophyllene	*P. gingivalis* JCM12257 *S. mutans* NBRC13955		-	MBEC MBEC	435.5 µg·mL^{-1} 871 µg·mL^{-1}	[151]
Syzygium aromaticum (L.) Merr. & L.M.Perry	Essential oil	Eugenol; β-caryophyllene	*S. mutans* ATCC 25175	MIC IZD	0.315 mg·mL^{-1} 5 mm (50 mg·mL^{-1})	96% reduction	1.26 mg·mL^{-1}	[185]
			S. aureus ATCC 29213	MIC IZD	0.625 mg·mL^{-1} 6 mm (50 mg·mL^{-1})	94% reduction	2.5 mg·mL^{-1}	
			E. faecalis ATCC 19433	MIC IZD	1.25 mg·mL^{-1} 1 mm (50 mg·mL^{-1})	45% reduction	5.0 mg·mL^{-1}	
	Hydroethanol (buds)	Eugenol; β-caryophyllene; eugenol acetate	*E. faecalis* ATCC 29212	MBC IZD	10% (*w/v*) 5.467 mm (20% (*w/v*))	100% reduction	10% (*w/v*)	[226]
Tarchonanthus camphoratus L.	Dichloromethane: methanol (bark)	-	*S. mutans* ATCC 25175	MIC	0.67 mg·mL^{-1}	86.37% inhibition	0.08 mg·mL^{-1}	[183]
Tecoma capensis (Thunb.) Lindl.	Dichloromethane: methanol (leaves)	-	*S. mutans* ATCC 25175	MIC	0.67 mg·mL^{-1}	53.10% inhibition	0.25 mg·mL^{-1}	[183]
Thuja orientalis L.	Methanol (leaves and stems)	-	*S. mutans* ATCC 25175	MIC IZD	0.1 mg·disk^{-1} 13 mm (2.0 mg·disk^{-1})	99.0% GTFs inhibition	1.0 mg·mL^{-1}	[125]
Thymus longicaulis C.Presl.	Methanol (aerial parts)	Rosmarinic acids; flavonoids; triterpenic acids	*S. mutans* DSM 20523	MIC MBC	0.60 mg·mL^{-1} 1.25 mg·mL^{-1}	95% inhibition	5.0 mg·mL^{-1}	[200]

Table 1. *Cont.*

Plant Name	Plant Extract	Compound	Microorganism	Antimicrobial Activity		Antibiofilm Activity		References
				Results				
Thymus vulgaris L.	Extract (leaves)	-	*S. aureus* ATCC 6538	MIC MMC	>50.0 mg·mL^{-1}	66% reduction	200 mg·mL^{-1}	[227]
			E. faecalis ATCC 4083	MIC MMC	>50.0 mg·mL^{-1}	82% reduction		
			S. mutans ATCC 35688	MIC MMC	>50.0 mg·mL^{-1}	64% reduction		
			P. aeruginosa ATCC 15442	MIC MMC	>50.0 mg·mL^{-1}	88% reduction		
	Essential oil	*O*-cymene; thymol	*S. mutans* ATCC 25175	MIC IZD	0.625 mg·mL^{-1} 10 mm (50 mg·mL^{-1})	96% inhibition	2.5 mg·mL^{-1}	[185]
			S. aureus ATCC 29213	MIC IZD	0.625 mg·mL^{-1} 8 mm (50 mg·mL^{-1})	96% inhibition	0.156 mg·mL^{-1}	
			E. faecalis ATCC 19433	MIC IZD	0.625 mg·mL^{-1} 9 mm (50 mg·mL^{-1})	55% inhibition	2.5 mg·mL^{-1}	
Thymus zygis L.	Essential oil	Thymol; *p*-cymene; terpinene; linalool; carvacrol	*S. mitis* (CI) *S. sanguinis* (CI) *E. faecalis* (CI)	MIC MBC	1 mg·mL^{-1} 2 mg·mL^{-1}	50% inhibition 16% inhibition 22% inhibition	0.5 mg·mL^{-1} 4 mg·mL^{-1} 4 mg·mL^{-1}	[203]
			S. mitis + *S. sanguinis* + *E. faecalis*		-	16% inhibition	0.5 mg·mL^{-1}	
Trachyspermum ammi L.	Ethanol (seeds)	2-isopropyl-5-methyl-phenol; oleic acid; octadecanoic acid	*S. mutans* ATCC 700610	MIC	320 µg·mL^{-1}	61% inhibition 89% reduction	160 µg·mL^{-1}	[90]
	Petroleum ether fraction (seeds)	2-isopropyl-5-methyl-phenol; oleic acid	*S. mutans* ATCC 700610	MIC	40 µg·mL^{-1}	100% inhibition 100% reduction	40 µg·mL^{-1} 80 µg·mL^{-1}	
Zingiber officinale Roscoe.	Ethanol	-	*P. gingivalis* JCM12257 *S. mutans* NBRC13955		-	99.9% reduction 97.6% reduction	5.0% (*v/v*)	[151]

CI, clinical isolate; CSH, cell surface hydrophobicity; IZD, inhibition zone diameter; GTFs, glucosyltransferase; MIC, minimum inhibitory concentration; MBC, minimum bactericidal concentration; MFC, minimum fungicidal concentration; MMC, minimum microbial concentration; MBIC, minimum biofilm inhibitory concentration; MBBC, minimum biofilm bactericidal concentration; MBEC, minimum biofilm eradication concentration; NA, no activity observed; -, not tested.

7. Conclusions

Medicinal plants are still a greatly unexplored source of powerful natural products with antibiofilm potential, especially when antibiotic resistances continue to rise. The focus of this review was to emphasize the potential of extracted products from medicinal plants such as essential oils and plant extracts, to treat common oral diseases, like dental caries and periodontitis, which are mainly caused by the formation of bacterial oral biofilms. Although the extracts of many medicinal plants have shown promising results in the control of oral biofilms, the two most promising extracts exerting this activity were found to be the essential oils extracted from two aromatic plants, namely *C. citratus* and *L. alba*. Interestingly, the terpenoid citral is one of the main components found in both plants, which is in accordance with several studies that point the powerful antibiofilm effect of this compound. The use of essential oils from *C. citratus* and *L. alba* could be a great alternative to antibiotics in the treatment of oral diseases since they show low cytotoxicity levels and do not induce resistance in bacterial pathogens. Nonetheless, research regarding the use of medicinal plants on the treatment of oral ailments continues to be an extremely interesting topic, mainly due to the extensive variety of unscreened plants that potentially have antimicrobial and antibiofilm properties.

Author Contributions: Conceptualization C.M., J.S., R.G.; methodology C.M., J.S., R.G.; data collection C.M., J.S., R.G.; writing—original draft preparations; C.M., J.S., R.G., writing—review and editing I.C.F.R.F., L.B.; supervision, M.J.A. All authors have read and agreed to the published version of the manuscript.

Funding: The authors wish to acknowledge financial support from the project "AquaValor—Centro de Valorização e Transferência de Tecnologia da Água" (NORTE-01-0246-FEDER-000053), supported by Norte Portugal Regional Operational Programme (NORTE 2020), under the PORTUGAL 2020 Partnership Agreement, through the European Regional Development Fund (ERDF); to the Foundation for Science and Technology (FCT, Portugal) for financial support through national funds FCT/MCTES to CIMO (UIDB/00690/2020); and L. Barros thanks the national funding by FCT, P.I., through the institutional scientific employment program-contract for her contract.

Institutional Review Board Statement: Not applicable.

Informed Consent Statement: Not applicable.

Data Availability Statement: Not applicable.

Conflicts of Interest: The authors declare no conflict of interest.

References

1. Marsh, P. Dental plaque as a microbial biofilm. *Caries Res.* **2004**, *38*, 204–211. [CrossRef]
2. Chinsembu, K.C. Plants and other natural products used in the management of oral infections and improvement of oral health. *Acta Trop.* **2016**, *154*, 6–18. [CrossRef] [PubMed]
3. Hwang, G.; Klein, M.I.; Koo, H. Analysis of the mechanical stability and surface detachment of mature Streptococcus mutans biofilms by applying a range of external shear forces. *Biofouling* **2014**, *30*, 1079–1091. [CrossRef]
4. Nishikawara, F.; Nomura, Y.; Imai, S.; Senda, A.; Hanada, N. Evaluation of cariogenic bacteria. *Eur. J. Dent.* **2007**, *1*, 31–39. [CrossRef] [PubMed]
5. Allaker, R.P.; Douglas, C.I. Novel anti-microbial therapies for dental plaque-related diseases. *Int. J. Antimicrob. Agents* **2009**, *33*, 8–13. [CrossRef] [PubMed]
6. Maeda, H.; Hirai, K.; Mineshiba, J.; Yamamoto, T.; Kokeguchi, S.; Takashiba, S. Medical microbiological approach to Archaea in oral infectious diseases. *Jpn. Dent. Sci. Rev.* **2013**, *49*, 72–78. [CrossRef]
7. Loesche, W. Dental caries and periodontitis: Contrasting two infections that have medical implications. *Infect. Dis. Clin. N. Am.* **2007**, *21*, 471–502. [CrossRef] [PubMed]
8. Alireza, R.G.; Afsaneh, R.; Hosein, M.S.; Siamak, Y.; Afshin, K.; Zeinab, K.; Mahvash, M.J.; Reza, R.A. Inhibitory activity of Salvadora persica extracts against oral bacterial strains associated with periodontitis: An in-vitro study. *J. Oral Biol. Craniofac. Res.* **2014**, *4*, 19–23. [CrossRef]
9. Galvão, L.C.; Furletti, V.F.; Bersan, S.M.; da Cunha, M.G.; Ruiz, A.L.; Carvalho, J.E.; Sartoratto, A.; Rehder, V.L.; Figueira, G.M.; Teixeira Duarte, M.C.; et al. Antimicrobial activity of essential oils against Streptococcus mutans and their antiproliferative effects. *Evid.-Based Complement. Altern. Med.* **2012**, *2012*, 751435. [CrossRef]

10. Teanpaisan, R.; Kawsud, P.; Pahumunto, N.; Puripattanavong, J. Screening for antibacterial and antibiofilm activity in Thai medicinal plant extracts against oral micro-organisms. *J. Tradit. Complement. Med.* **2017**, *7*, 172–177. [CrossRef]

11. Vieira, D.R.; Amaral, F.M.; Maciel, M.C.; Nascimento, F.R.; Libério, S.A.; Rodrigues, V.P. Plant species used in dental diseases: Ethnopharmacology aspects and antimicrobial activity evaluation. *J. Ethnopharmacol.* **2014**, *155*, 1441–1449. [CrossRef]

12. Brown, D. Antibiotic resistance breakers: Can repurposed drugs fill the antibiotic discovery void? *Nat. Rev. Drug Discov.* **2015**, *14*, 821–832. [CrossRef]

13. Rana, R.; Sharma, R.; Kumar, A. Repurposing of existing statin drugs for treatment of microbial infections: How much promising? *Infect. Disord. Drug Targets* **2019**, *19*, 224–237. [CrossRef]

14. Palombo, E.A. Traditional medicinal plant extracts and natural products with activity against oral bacteria: Potential applica-tion in the prevention and treatment of oral diseases. *Evid.-Based Complement. Altern. Med.* **2011**, *2011*, 680354. [CrossRef] [PubMed]

15. Slobodníková, L.; Fialová, S.; Rendeková, K.; Kováč, J.; Mučaji, P. Antibiofilm activity of plant polyphenols. *Molecules* **2016**, *21*, 1717. [CrossRef] [PubMed]

16. Daglia, M. Polyphenols as antimicrobial agents. *Curr. Opin. Biotechnol.* **2012**, *23*, 174–181. [CrossRef] [PubMed]

17. Quideau, S.; Deffieux, D.; Douat-Casassus, C.; Pouységu, L. Plant polyphenols: Chemical properties, biological activities, and synthesis. *Angew. Chem. Int. Ed.* **2011**, *50*, 586–621. [CrossRef]

18. Cushnie, T.T.; Lamb, A.J. Recent advances in understanding the antibacterial properties of flavonoids. *Int. J. Antimicrob. Agents* **2011**, *38*, 99–107. [CrossRef] [PubMed]

19. Gyawali, R.; Ibrahim, S.A. Natural products as antimicrobial agents. *Food Control* **2014**, *46*, 412–429. [CrossRef]

20. Silva, L.N.; Zimmer, K.R.; Macedo, A.J.; Trentin, D.S. Plant natural products targeting bacterial virulence factors. *Chem. Rev.* **2016**, *116*, 9162–9236. [CrossRef]

21. Ren, Z.; Chen, L.; Li, J.; Li, Y. Inhibition of Streptococcus mutans polysaccharide synthesis by molecules targeting glycosyltrans-ferase activity. *J. Oral Microbiol.* **2016**, *8*, 31095. [CrossRef] [PubMed]

22. Koparde, A.A.; Doijad, R.C.; Magdum, C.S. Natural products in drug discovery. In *Pharmacognosy—Medicinal Plants*; IntechOpen: London, UK, 2019.

23. Wade, W.G. Characterisation of the human oral microbiome. *J. Oral Biosci.* **2013**, *55*, 143–148. [CrossRef]

24. Dewhirst, F.E.; Chen, T.; Izard, J.; Paster, B.J.; Tanner, A.C.R.; Yu, W.-H.; Lakshmanan, A.; Wade, W.G. The human oral microbiome. *J. Bacteriol.* **2010**, *192*, 5002–5017. [CrossRef] [PubMed]

25. Lurie-Weinberger, M.N.; Gophna, U. Archaea in and on the human body: Health implications and future directions. *PLoS Pathog.* **2015**, *11*, e1004833. [CrossRef] [PubMed]

26. Agrawal, A.; Singh, A.; Verma, R.; Murari, A. Oral candidiasis: An overview. *J. Oral Maxillofac. Pathol.* **2014**, *18*, 81–85. [CrossRef]

27. Ghannoum, M.A.; Jurevic, R.J.; Mukherjee, P.K.; Cui, F.; Sikaroodi, M.; Naqvi, A.; Gillevet, P.M. Characterization of the oral fungal microbiome (Mycobiome) in healthy individuals. *PLoS Pathog.* **2010**, *6*, e1000713. [CrossRef]

28. Santosh, A.B.R.; Muddana, K. Viral infections of oral cavity. *J. Fam. Med. Prim. Care* **2020**, *9*, 36–42. [CrossRef]

29. Sharma, N.; Bhatia, S.; Sodhi, A.S.; Batra, N. Oral microbiome and health. *AIMS Microbiol.* **2018**, *4*, 42–66. [CrossRef]

30. Deo, P.N.; Deshmukh, R. Oral microbiome: Unveiling the fundamentals. *J. Oral Maxillofac. Pathol.* **2019**, *23*, 122–128. [CrossRef]

31. Lamont, R.J.; Koo, H.; Hajishengallis, G. The oral microbiota: Dynamic communities and host interactions. *Nat. Rev. Genet.* **2018**, *16*, 745–759. [CrossRef]

32. Loesche, W.J. Role of Streptococcus mutans in human dental decay. *Microbiol. Rev.* **1986**, *50*, 353. [CrossRef]

33. Marsh, P.D. Microbiology of dental plaque biofilms and their role in oral health and caries. *Dent. Clin. N. Am.* **2010**, *54*, 441–454. [CrossRef] [PubMed]

34. McNeill, K.; Hamilton, I. Acid tolerance response of biofilm cells of Streptococcus mutans. *FEMS Microbiol. Lett.* **2003**, *221*, 25–30. [CrossRef]

35. Dahlen, G.; Basic, A.; Bylund, J. Importance of virulence factors for the persistence of oral bacteria in the inflamed gingival crevice and in the pathogenesis of periodontal disease. *J. Clin. Med.* **2019**, *8*, 1339. [CrossRef]

36. Hannig, M.; Joiner, A. The structure, function and properties of the acquired pellicle. *Monogr. Oral Sci.* **2005**, *19*, 29–64. [CrossRef]

37. Bos, R.; Van der Mei, H.C.; Busscher, H.J. Physico-chemistry of initial microbial adhesive interactions—Its mechanisms and methods for study. *FEMS Microbiol. Rev.* **1999**, *23*, 179–230. [CrossRef]

38. Scannapieco, F.A.; Torres, G.; Levine, M.J. Salivary α-amylase: Role in dental plaque and caries formation. *Crit. Rev. Oral Biol. Med.* **1993**, *4*, 301–307. [CrossRef]

39. Ihara, Y.; Takeshita, T.; Kageyama, S.; Matsumi, R.; Asakawa, M.; Shibata, Y.; Sugiura, Y.; Ishikawa, K.; Takahashi, I.; Yamashita, Y. Identification of initial colonizing bacteria in dental plaques from young adults using full-length 16S rRNA gene sequencing. *mSystems* **2019**, *4*, e00360-19. [CrossRef]

40. Vitkov, L.; Krautgartner, W.D.; Hannig, M.; Fuchs, K. Fimbria-mediated bacterial adhesion to human oral epithelium. *FEMS Microbiol. Lett.* **2001**, *202*, 25–30. [CrossRef]

41. Okahashi, N.; Nakata, M.; Terao, Y.; Isoda, R.; Sakurai, A.; Sumitomo, T.; Yamaguchi, M.; Kimura, R.K.; Oiki, E.; Kawabata, S.; et al. Pili of oral Streptococcus sanguinis bind to salivary amylase and promote the biofilm formation. *Microb. Pathog.* **2011**, *50*, 148–154. [CrossRef]

42. Karygianni, L.; Ren, Z.; Koo, H.; Thurnheer, T. Biofilm matrixome: Extracellular components in structured microbial communities. *Trends Microbiol.* **2020**, *28*, 668–681. [CrossRef] [PubMed]

43. Costa, O.Y.A.; Raaijmakers, J.M.; Kuramae, E.E. Microbial extracellular polymeric substances: Ecological function and impact on soil aggregation. *Front. Microbiol.* **2018**, *9*, 1636. [CrossRef]
44. Rickard, A.H.; Gilbert, P.; High, N.J.; E Kolenbrander, P.; Handley, P.S. Bacterial coaggregation: An integral process in the development of multi-species biofilms. *Trends Microbiol.* **2003**, *11*, 94–100. [CrossRef]
45. Rabin, N.; Zheng, Y.; Opoku-Temeng, C.; Du, Y.; Bonsu, E.; O Sintim, H. Biofilm formation mechanisms and targets for developing antibiofilm agents. *Future Med. Chem.* **2015**, *7*, 493–512. [CrossRef] [PubMed]
46. Kaplan, J. Biofilm dispersal: Mechanisms, clinical implications, and potential therapeutic uses. *J. Dent. Res.* **2010**, *89*, 205–218. [CrossRef] [PubMed]
47. Bowen, W.H.; Burne, R.A.; Wu, H.; Koo, H. Oral biofilms: Pathogens, matrix, and polymicrobial interactions in microenvironments. *Trends Microbiol.* **2018**, *26*, 229–242. [CrossRef] [PubMed]
48. Suppiger, S.; Astasov-Frauenhoffer, M.; Schweizer, I.; Waltimo, T.; Kulik, E.M. Tolerance and persister formation in oral streptococci. *Antibiotics* **2020**, *9*, 167. [CrossRef] [PubMed]
49. Peres, M.A.; Macpherson, L.M.D.; Weyant, R.J.; Daly, B.; Venturelli, R.; Mathur, M.R.; Listl, S.; Celeste, R.K.; Guarnizo-Herreño, C.C.; Kearns, C.; et al. Oral diseases: A global public health challenge. *Lancet* **2019**, *394*, 249–260. [CrossRef]
50. Maddi, A.; Scannapieco, F.A. Oral biofilms, oral and periodontal infections, and systemic disease. *Am. J. Dent.* **2013**, *26*, 249–254.
51. Kouidhi, B.; Al Qurashi, Y.M.A.; Chaieb, K. Drug resistance of bacterial dental biofilm and the potential use of natural compounds as alternative for prevention and treatment. *Microb. Pathog.* **2015**, *80*, 39–49. [CrossRef]
52. Matsumoto-Nakano, M. Role of Streptococcus mutans surface proteins for biofilm formation. *Jpn. Dent. Sci. Rev.* **2018**, *54*, 22–29. [CrossRef] [PubMed]
53. Bowen, W.; Koo, H. Biology of Streptococcus mutans-derived glucosyltransferases: Role in extracellular matrix formation of cariogenic biofilms. *Caries Res.* **2011**, *45*, 69–86. [CrossRef] [PubMed]
54. Miller, M.B.; Bassler, B.L. Quorum sensing in bacteria. *Annu. Rev. Microbiol.* **2001**, *55*, 165–199. [CrossRef] [PubMed]
55. Colombo, A.; Tanner, A. The role of bacterial biofilms in dental caries and periodontal and peri-implant diseases: A historical perspective. *J. Dent. Res.* **2019**, *98*, 373–385. [CrossRef] [PubMed]
56. Mosaddad, S.A.; Tahmasebi, E.; Yazdanian, A.; Rezvani, M.B.; Seifalian, A.; Yazdanian, M.; Tebyanian, H. Oral microbial biofilms: An update. *Eur. J. Clin. Microbiol. Infect. Dis.* **2019**, *38*, 2005–2019. [CrossRef]
57. Santos, A.L.; Siqueira, J.F., Jr.; Rôças, I.N.; Jesus, E.C.; Rosado, A.S.; Tiedje, J.M. Comparing the bacterial diversity of acute and chronic dental root canal infections. *PLoS ONE* **2011**, *6*, e28088. [CrossRef]
58. Colombo, A.P.; do Souto, R.M.; da Silva-Boghossian, C.M.; Miranda, R.; Lourenço, T.G. Microbiology of oral biofilm-dependent diseases: Have we made significant progress to understand and treat these diseases? *Curr. Oral Health Rep.* **2015**, *2*, 37–47. [CrossRef]
59. Ricucci, D.; Siqueira, J.F., Jr. Biofilms and apical periodontitis: Study of prevalence and association with clinical and histopathologic findings. *J. Endod.* **2010**, *36*, 1277–1288. [CrossRef]
60. Socransky, S.S.; Haffajee, A.D.; Cugini, M.A.; Smith, C.; Kent, R.L., Jr. Microbial complexes in subgingival plaque. *J. Clin. Periodontol.* **1998**, *25*, 134–144. [CrossRef]
61. Bui, F.Q.; Almeida-Da-Silva, C.L.C.; Huynh, B.; Trinh, A.; Liu, J.; Woodward, J.; Asadi, H.; Ojcius, D.M. Association between periodontal pathogens and systemic disease. *Biomed. J.* **2019**, *42*, 27–35. [CrossRef]
62. Yao, Q.-W.; Zhou, D.-S.; Peng, H.-J.; Ji, P.; Liu, D.-S. Association of periodontal disease with oral cancer: A meta-analysis. *Tumor Biol.* **2014**, *35*, 7073–7077. [CrossRef] [PubMed]
63. Michaud, D.S.; Fu, Z.; Shi, J.; Chung, M. Periodontal disease, tooth loss, and cancer risk. *Epidemiol. Rev.* **2017**, *39*, 49–58. [CrossRef]
64. Inaba, H.; Sugita, H.; Kuboniwa, M.; Iwai, S.; Hamada, M.; Noda, T.; Morisaki, I.; Lamont, R.J.; Amano, A. Porphyromonas gingivalis promotes invasion of oral squamous cell carcinoma through induction of pro MMP 9 and its activation. *Cell. Microbiol.* **2014**, *16*, 131–145. [CrossRef] [PubMed]
65. Fukugaiti, M.H.; Ignacio, A.; Fernandes, M.R.; Júnior, U.R.; Nakano, V.; AvilaCampos, M.J. High occurrence of Fusobacterium nucleatum and Clostridium difficile in the intestinal microbiota of colorectal carcinoma patients. *Braz. J. Microbiol.* **2015**, *46*, 1135–1140. [CrossRef] [PubMed]
66. Castellarin, M.; Warren, R.L.; Freeman, J.D.; Dreolini, L.; Krzywinski, M.; Strauss, J.; Barnes, R.; Watson, P.; Allen-Vercoe, E.; Moore, R.A.; et al. Fusobacterium nucleatum infection is prevalent in human colorectal carcinoma. *Genome Res.* **2011**, *22*, 299–306. [CrossRef]
67. Mealey, B.L.; Ocampo, G.L. Diabetes mellitus and periodontal disease. *Periodontol. 2000* **2007**, *44*, 127–153. [CrossRef] [PubMed]
68. Teeuw, W.J.; Gerdes, V.E.A.; Loos, B.G. Effect of periodontal treatment on glycemic control of diabetic patients: A systematic review and meta-analysis. *Diabetes Care* **2010**, *33*, 421–427. [CrossRef] [PubMed]
69. Nishimura, F.; Iwamoto, Y.; Mineshiba, J.; Shimizu, A.; Soga, Y.; Murayama, Y. Periodontal Disease and diabetes mellitus: The role of tumor necrosis factor-α in a 2-way relationship. *J. Periodontol.* **2003**, *74*, 97–102. [CrossRef]
70. Kothari, M.; Spin-Neto, R.; Nielsen, J.F. Comprehensive oral-health assessment of individuals with acquired brain-injury in neuro-rehabilitation setting. *Brain Inj.* **2016**, *30*, 1103–1108. [CrossRef]
71. Akiyama, H.; Barger, S.; Barnum, S.; Bradt, B.; Bauer, J.; Cole, G.M.; Cooper, N.R.; Eikelenboom, P.; Emmerling, M.; Fiebich, B.L.; et al. Inflammation and Alzheimer's disease. *Neurobiol. Aging* **2000**, *21*, 383–421. [CrossRef]

72. Kamer, A.R.; Craig, R.G.; Pirraglia, E.; Dasanayake, A.P.; Norman, R.G.; Boylan, R.J.; Nehorayoff, A.; Glodzik, L.; Brys, M.; de Leon, M.J. TNF-α and antibodies to periodontal bacteria discriminate between Alzheimer's disease patients and normal subjects. *J. Neuroimmunol.* **2009**, *216*, 92–97. [CrossRef] [PubMed]

73. Poole, S.; Singhrao, S.K.; Kesavalu, L.; Curtis, M.A.; Crean, S. Determining the presence of periodontopathic virulence factors in short-term postmortem Alzheimer's disease brain tissue. *J. Alzheimers Dis.* **2013**, *36*, 665–677. [CrossRef] [PubMed]

74. Bahekar, A.A.; Singh, S.; Saha, S.; Molnar, J.; Arora, R. The prevalence and incidence of coronary heart disease is significantly increased in periodontitis: A meta-analysis. *Am. Heart J.* **2007**, *154*, 830–837. [CrossRef]

75. Figuero, E.; Sánchez-Beltrán, M.; Cuesta-Frechoso, S.; Tejerina, J.M.; Del Castro, J.A.; Gutiérrez, J.M.; Herrera, D.; Sanz, M. Detection of periodontal bacteria in atheromatous plaque by nested polymerase chain reaction. *J. Periodontol.* **2011**, *82*, 1469–1477. [CrossRef] [PubMed]

76. Haraszthy, V.; Zambon, J.; Trevisan, M.; Zeid, M.; Genco, R. Identification of periodontal pathogens in atheromatous plaques. *J. Periodontol.* **2000**, *71*, 1554–1560. [CrossRef] [PubMed]

77. Nakano, K.; Inaba, H.; Nomura, R.; Nemoto, H.; Takeda, M.; Yoshioka, H.; Matsue, H.; Takahashi, T.; Taniguchi, K.; Amano, A.; et al. Detection of cariogenic streptococcus mutans in extirpated heart valve and atheromatous plaque specimens. *J. Clin. Microbiol.* **2006**, *44*, 3313–3317. [CrossRef] [PubMed]

78. Hajishengallis, G.; Wang, M.; Bagby, G.J.; Nelson, S. Importance of TLR2 in early innate immune response to acute pulmonary infection with Porphy-romonas gingivalis in mice. *J. Immunol.* **2008**, *181*, 4141–4149. [CrossRef]

79. Sonti, R.; Fleury, C. Fusobacterium necrophorum presenting as isolated lung nodules. *Respir. Med. Case Rep.* **2015**, *15*, 80–82. [CrossRef]

80. Williams, D.M.; Kerber, C.A.; Tergin, H.F. Unusual presentation of Lemierre's syndrome due to Fusobacterium nucleatum. *J. Clin. Microbiol.* **2003**, *41*, 3445–3448. [CrossRef]

81. Gomes-Filho, I.S.; De Oliveira, T.F.L.; Da Cruz, S.S.; Passos-Soares, J.D.S.; Trindade, S.C.; Oliveira, M.T.; Souza-Machado, A.; Cruz, A.A.; Barreto, M.L.; Seymour, G.J. Influence of periodontitis in the development of nosocomial pneumonia: A case control study. *J. Periodontol.* **2014**, *85*, e82–e90. [CrossRef]

82. Heo, S.M.; Sung, R.S.; Scannapieco, F.A.; Haase, E.M. Genetic relationships between Candida albicans strains isolated from dental plaque, trachea, and bron-choalveolar lavage fluid from mechanically ventilated intensive care unit patients. *J. Oral Microbiol.* **2011**, *3*, 6362. [CrossRef]

83. Tonetto, M.R.; Rocatto, G.; Matos, F.Z.; Pedro, F.M.; Lima, S.L.; Aranha, A.F.; Porto, A.N.; Borges, A.H.; Borba, A.M.; Patil, S.; et al. Periodontal and microbiological profile of intensive care unit inpatients. *J. Contemp. Dent. Pract.* **2016**, *17*, 807–814. [CrossRef]

84. Kaur, M.; Geisinger, M.L.; Geurs, N.C.; Griffin, R.; Vassilopoulos, P.J.; Vermeulen, L.; Haigh, S.; Reddy, M.S. Effect of intensive oral hygiene regimen during pregnancy on periodontal health, cytokine levels, and pregnancy outcomes: A pilot study. *J. Periodontol.* **2014**, *85*, 1684–1692. [CrossRef] [PubMed]

85. Lin, D.; Moss, K.; Beck, J.D.; Hefti, A.; Offenbacher, S. Persistently high levels of periodontal pathogens associated with preterm pregnancy outcome. *J. Periodontol.* **2007**, *78*, 833–841. [CrossRef] [PubMed]

86. Han, Y.; Wang, X. Mobile microbiome: Oral bacteria in extra-oral infections and inflammation. *J. Dent. Res.* **2013**, *92*, 485–491. [CrossRef] [PubMed]

87. Katz, J.; Chegini, N.; Shiverick, K.; Lamont, R. Localization of P. gingivalis in preterm delivery placenta. *J. Dent. Res.* **2009**, *88*, 575–578. [CrossRef] [PubMed]

88. Reyes, L.; Phillips, P.; Wolfe, B.; Golos, T.G.; Walkenhorst, M.; Progulske-Fox, A.; Brown, M. Porphyromonas gingivalis and adverse pregnancy outcome. *J. Oral Microbiol.* **2017**, *9*, 1374153. [CrossRef]

89. Potempa, J.; Mydel, P.; Koziel, J. The case for periodontitis in the pathogenesis of rheumatoid arthritis. *Nat. Rev. Rheumatol.* **2017**, *13*, 606–620. [CrossRef]

90. Khan, R.; Adil, M.; Danishuddin, M.; Verma, P.K.; Khan, A.U. In vitro and in vivo inhibition of Streptococcus mutans biofilm by Trachyspermum ammi seeds: An approach of alternative medicine. *Phytomedicine* **2012**, *19*, 747–755. [CrossRef]

91. Tiwari, R.; Rana, C. Plant secondary metabolites: A review. *Int. J. Eng. Res. Gen. Sci.* **2015**, *3*, 661–670.

92. Gorlenko, C.L.; Kiselev, H.Y.; Budanova, E.V.; Zamyatnin, A.A.; Ikryannikova, L.N. Plant secondary metabolites in the battle of drugs and drug-resistant bacteria: New heroes or worse clones of antibiotics? *Antibiotics* **2020**, *9*, 170. [CrossRef] [PubMed]

93. Hussein, R.A.; El-Anssary, A.A. Plants secondary metabolites: The key drivers of the pharmacological actions of medicinal plants. *Herb. Med.* **2019**, *1*, 13. [CrossRef]

94. Anand, U.; Jacobo-Herrera, N.; Altemimi, A.; Lakhssassi, N. A comprehensive review on medicinal plants as antimicrobial therapeutics: Potential avenues of biocompatible drug discovery. *Metabolites* **2019**, *9*, 258. [CrossRef]

95. Boudet, A.-M. Evolution and current status of research in phenolic compounds. *Phytochemistry* **2007**, *68*, 2722–2735. [CrossRef]

96. Cohen, S.D.; Kennedy, J.A. Plant metabolism and the environment: Implications for managing phenolics. *Crit. Rev. Food Sci. Nutr.* **2010**, *50*, 620–643. [CrossRef]

97. Zahin, M.; Aqil, F.; Khan, M.S.; Ahmad, I. Ethnomedicinal plants derived antibacterials and their prospects. In *Ethnomedicine: A Source of Complementary Therapeutics*; Research Signpost: Thiruvananthapuram, India, 2010; pp. 149–178.

98. Carocho, M.; Barreiro, M.F.; Morales, P.; Ferreira, I.C. Adding molecules to food, pros and cons: A review on synthetic and natural food additives. *Compr. Rev. Food Sci. Food Saf.* **2014**, *13*, 377–399. [CrossRef]

99.   Silva, N.; Júnior, A.F. Biological properties of medicinal plants: A review of their antimicrobial activity. *J. Venom. Anim. Toxins Incl. Trop. Dis.* **2010**, *16*, 402–413. [CrossRef]

100. Hassan, B.A.R. Medicinal plants (importance and uses). *Pharm. Anal. Acta* **2012**, *3*, 2153–2435. [CrossRef]

101. Górniak, I.; Bartoszewski, R.; Króliczewski, J. Comprehensive review of antimicrobial activities of plant flavonoids. *Phytochem. Rev.* **2019**, *18*, 241–272. [CrossRef]

102. Zhang, L.; Kong, Y.; Wu, D.; Zhang, H.; Wu, J.; Chen, J.; Ding, J.; Hu, L.; Jiang, H.; Shen, X. Three flavonoids targeting the β-hydroxyacyl-acyl carrier protein dehydratase from Helicobacter pylori: Crystal structure characterization with enzymatic inhibition assay. *Protein Sci.* **2008**, *17*, 1971–1978. [CrossRef] [PubMed]

103. Taylor, P.W.; Hamilton-Miller, J.M.T.; Stapleton, P.D. Antimicrobial properties of green tea catechins. *Food Sci. Technol. Bull. Funct. Foods* **2005**, *2*, 71–81. [CrossRef] [PubMed]

104. Koo, H.; Hayacibara, M.F.; Schobel, B.D.; Cury, J.A.; Rosalen, P.L.; Park, Y.K.; Vacca-Smith, A.M.; Bowen, W.H. Inhibition of Streptococcus mutans biofilm accumulation and polysaccharide production by apigenin and tt-farnesol. *J. Antimicrob. Chemother.* **2003**, *52*, 782–789. [CrossRef] [PubMed]

105. Hasan, S.; Singh, K.; Danisuddin, M.; Verma, P.K.; Khan, A.U. Inhibition of major virulence pathways of Streptococcus mutans by quercitrin and deoxynojirimycin: A synergistic approach of infection control. *PLoS ONE* **2014**, *9*, e91736. [CrossRef]

106. Pagare, S.; Bhatia, M.; Tripathi, N.; Pagare, S.; Bansal, Y.K. Secondary metabolites of plants and their role: Overview. *Curr. Trends Biotechnol. Pharm.* **2015**, *9*, 293–304.

107. Fokialakis, N.; Skaltsounis, A.L. Natural resins and bioactive natural products thereof as potential anitimicrobial agents. *Curr. Pharm. Des.* **2011**, *17*, 1267–1290. [CrossRef]

108. Savoia, D. Plant-derived antimicrobial compounds: Alternatives to antibiotics. *Futur. Microbiol.* **2012**, *7*, 979–990. [CrossRef]

109. Bazaka, K.; Jacob, M.V.; Chrzanowski, W.; Ostrikov, K. Anti-bacterial surfaces: Natural agents, mechanisms of action, and plasma surface modification. *RSC Adv.* **2015**, *5*, 48739–48759. [CrossRef]

110. Sen, T.; Samanta, S.K. Medicinal plants, human health and biodiversity: A broad review. *Adv. Biochem. Eng. Biotechnol.* **2015**, *147*, 59–110.

111. Patel, D.M.; Chauhan, J.B.; Ishnava, K.B. Studies on the anticariogenic potential of medicinal plant seed and fruit extracts. In *Natural Oral Care in Dental Therapy*; Wiley: Hoboken, NJ, USA, 2020; pp. 81–96.

112. Cazella, L.N.; Glamoclija, J.; Soković, M.; Gonçalves, J.E.; Linde, G.A.; Colauto, N.B.; Gazim, Z.C. Antimicrobial activity of essential oil of Baccharis dracunculifolia DC (Asteraceae) aerial parts at flowering period. *Front. Plant Sci.* **2019**, *10*, 27. [CrossRef]

113. Boeing, T.; Costa, P.; Venzon, L.; Meurer, M.; Mariano, L.N.B.; França, T.C.S.; Gouveia, L.; De Bassi, A.C.; Steimbach, V.; De Souza, P.; et al. Gastric healing effect of p-coumaric acid isolated from Baccharis dracunculifolia DC on animal model. *Naunyn-Schmiedebergs Arch. Pharmacol.* **2021**, *394*, 49–57. [CrossRef]

114. Luchesi, L.A.; Paulus, D.; Busso, C.; Frata, M.T.; Oliveira, J.B. Chemical composition, antifungal and antioxidant activity of essential oils from Baccharis dracunculifolia and Pogostemon cablin against Fusarium graminearum. *Nat. Prod. Res.* **2020**, 1–4. [CrossRef] [PubMed]

115. Lemos, M.; De Barros, M.P.; Sousa, J.P.B.; Filho, A.A.D.S.; Bastos, J.K.; De Andrade, S.F. Baccharis dracunculifolia, the main botanical source of Brazilian green propolis, displays antiulcer activity[†]. *J. Pharm. Pharmacol.* **2010**, *59*, 603–608. [CrossRef] [PubMed]

116. Gilabert, M.; Ramos, A.N.; Schiavone, M.M.; Arena, M.E.; Bardón, A. Bioactive sesqui- and diterpenoids from the Argentine Liverwort Porella chilensis. *J. Nat. Prod.* **2011**, *74*, 574–579. [CrossRef] [PubMed]

117. Lee, K.; Lee, J.-H.; Kim, S.-I.; Cho, M.H.; Lee, J. Anti-biofilm, anti-hemolysis, and anti-virulence activities of black pepper, cananga, myrrh oils, and nerolidol against Staphylococcus aureus. *Appl. Microbiol. Biotechnol.* **2014**, *98*, 9447–9457. [CrossRef]

118. Piao, M.J.; Yoo, E.S.; Koh, Y.S.; Kang, H.K.; Kim, J.; Kim, Y.J.; Kang, H.H.; Hyun, J.W. Antioxidant effects of the ethanol extract from flower of Camellia japonica via scavenging of reactive oxygen species and induction of antioxidant enzymes. *Int. J. Mol. Sci.* **2011**, *12*, 2618–2630. [CrossRef]

119. Kim, K.Y.; Davidson, P.M.; Chung, H.J. Antibacterial activity in extracts of Camellia japonica L. Petals and its application to a model food system. *J. Food Prot.* **2001**, *64*, 1255–1260. [CrossRef] [PubMed]

120. Nam, H.H.; Nan, L.; Choo, B.K. Inhibitory effects of Camellia japonica on cell inflammation and acute rat reflux esopha-gitis. *Chin. Med.* **2021**, *16*, 1–12. [CrossRef]

121. Zhang, Z.H.; Mi, C.; Wang, K.S.; Wang, Z.; Li, M.Y.; Zuo, H.X.; Xu, G.H.; Li, X.; Piao, L.X.; Ma, J.; et al. Chelidonine inhibits TNF-α-induced inflammation by suppressing the NF-κB pathways in HCT116 cells. *Phytother. Res.* **2018**, *32*, 65–75. [CrossRef]

122. Dobrucka, R.; Dlugaszewska, J.; Kaczmarek, M. Cytotoxic and antimicrobial effects of biosynthesized ZnO nanoparticles using of Chelidonium majus extract. *Biomed. Microdevices* **2018**, *20*, 1–13. [CrossRef] [PubMed]

123. Kim, T.-H.; Li, H.; Wu, Q.; Lee, H.J.; Ryu, J.-H. A new labdane diterpenoid with anti-inflammatory activity from Thuja orientalis. *J. Ethnopharmacol.* **2013**, *146*, 760–767. [CrossRef]

124. Chakraborty, S.; Afaq, N.; Singh, N.; Majumdar, S. Antimicrobial activity of Cannabis sativa, Thuja orientalis and Psidium guajava leaf extracts against methicillin-resistant Staphylococcus aureus. *J. Integr. Med.* **2018**, *16*, 350–357. [CrossRef]

125. Choi, H.-A.; Cheong, D.-E.; Lim, H.-D.; Kim, W.-H.; Ham, M.-H.; Oh, M.-H.; Wu, Y.; Shin, H.-J.; Kim, G.-J. Antimicrobial and anti-biofilm activities of the methanol extracts of medicinal plants against dental pathogens Streptococcus mutans and Candida albicans. *J. Microbiol. Biotechnol.* **2017**, *27*, 1242–1248. [CrossRef]

126. Cushnie, T.T.; Lamb, A.J. Antimicrobial activity of flavonoids. *Int. J. Antimicrob. Agents* **2005**, *26*, 343–356. [CrossRef] [PubMed]
127. Xie, Y.; Yang, W.; Tang, F.; Chen, X.; Ren, L. Antibacterial activities of flavonoids: Structure-activity relationship and mechanism. *Curr. Med. Chem.* **2014**, *22*, 132–149. [CrossRef] [PubMed]
128. Dorri, M.; Hashemitabar, S.; Hosseinzadeh, H. Cinnamon (Cinnamomum zeylanicum) as an antidote or a protective agent against natural or chemical toxicities: A review. *Drug Chem. Toxicol.* **2018**, *41*, 338–351. [CrossRef]
129. Kerekes, E.B.; Vidács, A.; Takó, M.; Petkovits, T.; Vágvölgyi, C.; Horváth, G.; Balázs, V.L.; Krisch, J. Anti-biofilm effect of selected essential oils and main components on mono- and polymicrobic bacterial cultures. *Microorganisms* **2019**, *7*, 345. [CrossRef] [PubMed]
130. Bardají, D.; Reis, E.; Medeiros, T.; Lucarini, R.; Crotti, A.; Martins, C. Antibacterial activity of commercially available plant-derived essential oils against oral pathogenic bacteria. *Nat. Prod. Res.* **2015**, *30*, 1178–1181. [CrossRef]
131. Taguchi, Y.; Takizawa, T.; Ishibashi, H.; Sagawa, T.; Arai, R.; Inoue, S.; Yamaguchi, H.; Abe, S. Therapeutic effects on murine oral candidiasis by oral administration of Cassia (Cinnamomum cassia) preparation. *Nippon. Ishinkin Gakkai Zasshi* **2010**, *51*, 13–21. [CrossRef]
132. Ali, I.A.; Cheung, B.P.; Matinlinna, J.; Lévesque, C.M.; Neelakantan, P. Trans-cinnamaldehyde potently kills Enterococcus faecalis biofilm cells and prevents biofilm recovery. *Microb. Pathog.* **2020**, *149*, 104482. [CrossRef] [PubMed]
133. Durgadevi, R.; Ravi, A.V.; Alexpandi, R.; Swetha, T.K.; Abirami, G.; Vishnu, S.; Pandian, S.K. Virulence targeted inhibitory effect of linalool against the exclusive uropathogen Proteus mirabilis. *Biofouling* **2019**, *35*, 508–525. [CrossRef]
134. Kačániová, M.; Galovičová, L.; Ivanišová, E.; Vukovic, N.L.; Štefániková, J.; Valková, V.; Borotová, P.; Žiarovská, J.; Terentjeva, M.; Felšöciová, S.; et al. Antioxidant, antimicrobial and antibiofilm activity of coriander (*Coriandrum sativum* L.) essential oil for its application in foods. *Foods* **2020**, *9*, 282. [CrossRef]
135. Can, E.; Kizak, V.; Can, Ş.S.; Özçiçek, E. Anesthetic efficiency of three medicinal plant oils for aquatic species: Coriander Coriandrum sativum, linaloe tree Bursera delpechiana, and lavender Lavandula hybrida. *J. Aquat. Anim. Health* **2019**, *31*, 266–273. [CrossRef] [PubMed]
136. Bersan, S.M.F.; Galvão, L.C.C.; Goes, V.F.F.; Sartoratto, A.; Figueira, G.M.; Rehder, V.L.G.; Alencar, S.M.; Duarte, R.M.T.; Rosalen, P.L.; Duarte, M.C.T. Action of essential oils from Brazilian native and exotic medicinal species on oral biofilms. *BMC Complement. Altern. Med.* **2014**, *14*, 451. [CrossRef]
137. Mukherjee, K.; Tribedi, P.; Mukhopadhyay, B.; Sil, A.K. Antibacterial activity of long-chain fatty alcohols against mycobacteria. *FEMS Microbiol. Lett.* **2013**, *338*, 177–183. [CrossRef] [PubMed]
138. Ames-Sibin, A.P.; Barizão, C.L.; Castro-Ghizoni, C.V.; Silva, F.M.S.; Sá-Nakanishi, A.B.; Bracht, L.; Bersani-Amado, C.A.; Marçal-Natali, M.R.; Bracht, A.; Comar, J.F. β-Caryophyllene, the major constituent of copaiba oil, reduces systemic inflammation and oxidative stress in arthritic rats. *J. Cell. Biochem.* **2018**, *119*, 10262–10277. [CrossRef] [PubMed]
139. Ribeiro, V.P.; Arruda, C.; Da Silva, J.J.M.; Mejia, J.A.A.; Furtado, N.A.J.C.; Bastos, J.K. Use of spinning band distillation equipment for fractionation of volatile compounds of Copaifera oleoresins for developing a validated gas chromatographic method and evaluating antimicrobial activity. *Biomed. Chromatogr.* **2019**, *33*, e4412. [CrossRef]
140. Símaro, G.V.; Lemos, M.; da Silva, J.J.M.; Ribeiro, V.P.; Arruda, C.; Schneider, A.H.; Wanderley, C.W.D.S.; Carneiro, L.J.; Mariano, R.L.; Ambrósio, S.R.; et al. Antinociceptive and anti-inflammatory activities of Copaifera pubiflora Benth oleoresin and its major metabolite ent-hardwickiic acid. *J. Ethnopharmacol.* **2021**, *271*, 113883. [CrossRef]
141. Moraes, T.D.S.; Leandro, L.F.; Santiago, M.B.; Silva, L.D.O.; Bianchi, T.C.; Veneziani, R.C.S.; Ambrósio, S.R.; Ramos, S.B.; Bastos, J.K.; Martins, C.H.G. Assessment of the antibacterial, antivirulence, and action mechanism of Copaifera pubiflora oleoresin and isolated compounds against oral bacteria. *Biomed. Pharmacother.* **2020**, *129*, 110467. [CrossRef]
142. Carneiro, L.J.; Tasso, T.O.; Santos, M.F.; Goulart, M.O.; Santos, R.A.; Bastos, J.K.; da Silva, J.J.; Crotti, A.E.; Parreira, R.L.; Orenha, R.P.; et al. Copaifera multijuga, Copaifera pubiflora and Copaifera trapezifolia Oleoresins: Chemical characterization and in vitro cytotoxic potential against tumoral cell lines. *J. Braz. Chem. Soc.* **2020**, *31*, 1679–1689. [CrossRef]
143. Ekpenyong, C.E.; Akpan, E.E. Use of Cymbopogon citratus essential oil in food preservation: Recent advances and future perspectives. *Crit. Rev. Food Sci. Nutr.* **2017**, *57*, 2541–2559. [CrossRef]
144. Hacke, A.C.M.; Miyoshi, E.; Marques, J.A.; Pereira, R.P. Anxiolytic properties of Cymbopogon citratus (DC.) stapf extract, essential oil and its constituents in zebrafish (Danio rerio). *J. Ethnopharmacol.* **2020**, *260*, 113036. [CrossRef]
145. Oliveira, J.B.; Teixeira, M.A.; Paiva, L.F.; Oliveira, R.F.; Mendonça, A.R.; Brito, M.J. In vitro and in vivo antimicrobial activity of Cymbopogon citratus (DC.) Stapf. against Staphylococcus spp. isolated from newborn babies in an intensive care unit. *Microb. Drug Resist.* **2019**, *25*, 1490–1496. [CrossRef]
146. Ortega-Ramirez, L.A.; Gutiérrez-Pacheco, M.M.; Vargas-Arispuro, I.; González-Aguilar, G.A.; Martínez-Téllez, M.A.; Ayala-Zavala, J.F. Inhibition of glucosyltransferase activity and glucan production as an antibiofilm mechanism of lemongrass essential oil against Escherichia coli O157:H7. *Antibiotics* **2020**, *9*, 102. [CrossRef]
147. Ortega-Cuadros, M.; Tofiño-Rivera, A.P.; Merini, L.J.; Martínez-Pabón, M.C. Antimicrobial activity of Cymbopogon citratus (Poaceae) on Streptococcus mutans biofilm and its cytotoxic effects. *Rev. Biol. Trop.* **2018**, *66*, 1519–1529. [CrossRef]
148. Tofiño-Rivera, A.; Ortega-Cuadros, M.; Galvis-Pareja, D.; Jiménez-Rios, H.; Merini, L.; Martínez-Pabón, M. Effect of Lippia alba and Cymbopogon citratus essential oils on biofilms of Streptococcus mutans and cytotoxicity in CHO cells. *J. Ethnopharmacol.* **2016**, *194*, 749–754. [CrossRef] [PubMed]

149. Chaves-Quirós, C.; Usuga-Usuga, J.-S.; Morales-Uchima, S.-M.; Tofiño-Rivera, A.-P.; Tobón-Arroyave, S.-I.; Martínez-Pabón, M.-C.; Rivera, A.T. Assessment of cytotoxic and antimicrobial activities of two components of Cymbopogon citratus essential oil. *J. Clin. Exp. Dent.* **2020**, *12*, e749–e754. [CrossRef]

150. González-Burgos, E.; Liaudanskas, M.; Viškelis, J.; Žvikas, V.; Janulis, V.; Gómez-Serranillos, M.P. Antioxidant activity, neuroprotective properties and bioactive constituents analysis of varying polarity extracts from Eucalyptus globulus leaves. *J. Food Drug Anal.* **2018**, *26*, 1293–1302. [CrossRef] [PubMed]

151. Tsukatani, T.; Sakata, F.; Kuroda, R.; Akao, T. Biofilm eradication activity of herb and spice extracts alone and in combination against oral and food-borne pathogenic bacteria. *Curr. Microbiol.* **2020**, *77*, 2486–2495. [CrossRef]

152. Nagata, H.; Inagaki, Y.; Yamamoto, Y.; Maeda, K.; Kataoka, K.; Osawa, K.; Shizukuishi, S. Inhibitory effects of macrocarpals on the biological activity of Porphyromonas gingivalis and other periodontopathic bacteria. *Oral Microbiol. Immunol.* **2006**, *21*, 159–163. [CrossRef]

153. Sebei, K.; Sakouhi, F.; Herchi, W.; Khouja, M.L.; Boukhchina, S. Chemical composition and antibacterial activities of seven Eucalyptus species essential oils leaves. *Biol. Res.* **2015**, *48*, 1–5. [CrossRef] [PubMed]

154. Issarachot, P.; Sangkaew, W.; Sianglum, W.; Saeloh, D.; Limsuwan, S.; Voravuthikunchai, S.P.; Joycharat, N. α-glucosidase inhibitory, antibacterial, and antioxidant activities of natural substances from the wood of Derris reticulata Craib. *Nat. Prod. Res.* **2019**, 1–8. [CrossRef] [PubMed]

155. Pulbutr, P.; Rattanakiat, S.; Phetsaardeiam, N.; Modtaku, P.; Denchai, R.; Jaruchotikamol, A.; Khunawattanakul, W. Anticariogenic activities of Derris reticulata ethanolic stem extract against Streptococcus mutans. *Pak. J. Biol. Sci.* **2018**, *21*, 300–306. [CrossRef] [PubMed]

156. Getie, M.; Gebre-Mariam, T.; Rietz, R.; Höhne, C.; Huschka, C.; Schmidtke, M.; Abate, A.; Neubert, R. Evaluation of the antimicrobial and anti-inflammatory activities of the medicinal plants Dodonaea viscosa, Rumex nervosus and Rumex abyssinicus. *Fitoterapia* **2003**, *74*, 139–143. [CrossRef]

157. Khalil, N.; Sperotto, J.; Manfron, M. Antiinflammatory activity and acute toxicity of Dodonaea viscosa. *Fitoterapia* **2006**, *77*, 478–480. [CrossRef] [PubMed]

158. Qureshi, S.; Khan, M.; Ahmad, M. A survey of useful medicinal plants of Abbottabad in northern Pakistan. *Trakia J. Sci.* **2008**, *6*, 39–51.

159. Naidoo, R.; Patel, M.; Gulube, Z.; Fenyvesi, I. Inhibitory activity of Dodonaea viscosa var. angustifolia extract against Streptococcus mutans and its biofilm. *J. Ethnopharmacol.* **2012**, *144*, 171–174. [CrossRef] [PubMed]

160. Karkanis, A.; Martins, N.; Petropoulos, S.; Ferreira, I. Phytochemical composition, health effects, and crop management of liquorice (*Glycyrrhiza glabra* L.): A medicinal plant. *Food Rev. Int.* **2016**, *34*, 182–203. [CrossRef]

161. Chakotiya, A.S.; Tanwar, A.; Narula, A.; Sharma, R.K. Alternative to antibiotics against Pseudomonas aeruginosa: Effects of Glycyrrhiza glabra on membrane permeability and inhibition of efflux activity and biofilm formation in Pseudomonas aeruginosa and its in vitro time-kill activity. *Microb. Pathog.* **2016**, *98*, 98–105. [CrossRef]

162. Suwannakul, S.; Chaibenjawong, P. Antibacterial activities of Glycyrrhiza gabra Linn. (licorice) root extract against Porphyromonas gingivalis rand its inhibitory effects on cysteine proteases and biofilms. *J. Dent. Indones.* **2017**, *24*, 85–92. [CrossRef]

163. Kim, S.-R.; Jeon, H.-J.; Park, H.-J.; Kim, M.-K.; Choi, W.-S.; Jang, H.-O.; Bae, S.-K.; Jeong, C.-H.; Bae, M.-K. Glycyrrhetinic acid inhibits Porphyromonas gingivalis lipopolysaccharide-induced vascular permeability via the suppression of interleukin-8. *Inflamm. Res.* **2012**, *62*, 145–154. [CrossRef]

164. Hennebelle, T.; Sahpaz, S.; Joseph, H.; Bailleul, F. Ethnopharmacology of Lippia alba. *J. Ethnopharmacol.* **2008**, *116*, 211–222. [CrossRef]

165. Mączka, W.; Wińska, K.; Grabarczyk, M. One hundred faces of geraniol. *Molecules* **2020**, *25*, 3303. [CrossRef]

166. Nikavar, B.; Ali, N.A.; Kamalnezhad, M. Evaluation of the antioxidant properties of five Mentha species. *Iran J. Pharm. Res.* **2008**. [CrossRef]

167. Shafiei, Z.; Rahim, Z.H.; Philip, K.; Thurairajah, N.; Yaacob, H. Potential effects of Psidium sp., Mangifera sp., Mentha sp. and its mixture (PEM) in reducing bacterial populations in biofilms, adherence and acid production of S. sanguinis and S. mutans. *Arch. Oral Biol.* **2020**, *109*, 104554. [CrossRef] [PubMed]

168. Wi, W.N.; Fathilah, A.; Rahim, Z. Plant extracts of Psidium guajava, Mangifera and Mentha sp. inhibit the growth of the population of single-species oral biofilm. *Altern. Integr. Med.* **2013**, *31*, 1–6.

169. Rahim, Z.H.A.; Shaikh, S.; Ismail, W.N.H.W.; Harun, W.H.-A.W.; Razak, F.A. The effect of selected plant extracts on the development of single-species dental biofilms. *J. Coll. Physicians Surg. Pak.* **2014**, *24*, 796–801. [PubMed]

170. Shafiei, Z.; Rahim, Z.H.; Philip, K.; Thurairajah, N. Antibacterial and anti-adherence effects of a plant extract mixture (PEM) and its individual constituent ex-tracts (Psidium sp., Mangifera sp., and Mentha sp.) on single- and dual-species biofilms. *PeerJ* **2016**, *4*, e2519. [CrossRef]

171. Aleksic, V.; Knezevic, P. Antimicrobial and antioxidative activity of extracts and essential oils of *Myrtus communis* L. *Microbiol. Res.* **2014**, *169*, 240–254. [CrossRef]

172. Kaya, D.A.; Ghica, M.V.; Dănilă, E.; Öztürk, Ş.; Türkmen, M.; Kaya, M.G.A.; Dinu-Pîrvu, C.-E. Selection of optimal operating conditions for extraction of Myrtus Communis L. essential oil by the steam distillation method. *Molecules* **2020**, *25*, 2399. [CrossRef]

173. Sateriale, D.; Imperatore, R.; Colicchio, R.; Pagliuca, C.; Varricchio, E.; Volpe, M.G.; Salvatore, P.; Paolucci, M.; Pagliarulo, C. Phytocompounds vs. dental plaque bacteria: In vitro effects of myrtle and pomegranate polyphenolic extracts against single-species and multispecies oral biofilms. *Front. Microbiol.* **2020**, *11*, 11. [CrossRef]

174. Sateriale, D.; Facchiano, S.; Colicchio, R.; Pagliuca, C.; Varricchio, E.; Paolucci, M.; Volpe, M.G.; Salvatore, P.; Pagliarulo, C. In vitro synergy of polyphenolic extracts from honey, myrtle and pomegranate against oral pathogens, S. mutans and R. dentocariosa. *Front. Microbiol.* **2020**, *11*, 1465. [CrossRef] [PubMed]

175. Panossian, A.; Wikman, G.; Sarris, J. Rosenroot (Rhodiola rosea): Traditional use, chemical composition, pharmacology and clinical efficacy. *Phytomedicine* **2010**, *17*, 481–493. [CrossRef] [PubMed]

176. Chiang, H.-M.; Chen, H.-C.; Wu, C.-S.; Wu, P.-Y.; Wen, K.-C. Rhodiola plants: Chemistry and biological activity. *J. Food Drug Anal.* **2015**, *23*, 359–369. [CrossRef] [PubMed]

177. Elameen, A.; Kosman, V.M.; Thomsen, M.; Pozharitskaya, O.N.; Shikov, A.N. Variability of major phenyletanes and phenylpropanoids in 16-year-old Rhodiola rosea L. clones in Norway. *Molecules* **2020**, *25*, 3463. [CrossRef]

178. Zhang, Z.; Liu, Y.; Lu, M.; Lyu, X.; Gong, T.; Tang, B.; Wang, L.; Zeng, J.; Li, Y. Rhodiola rosea extract inhibits the biofilm formation and the expression of virulence genes of cariogenic oral pathogen Streptococcus mutans. *Arch. Oral Biol.* **2020**, *116*, 104762. [CrossRef]

179. De Oliveira, J.R.; Camargo, S.E.A.; De Oliveira, L.D. Rosmarinus officinalis L. (rosemary) as therapeutic and prophylactic agent. *J. Biomed. Sci.* **2019**, *26*, 1–22. [CrossRef]

180. Birtić, S.; Dussort, P.; Pierre, F.-X.; Bily, A.C.; Roller, M. Carnosic acid. *Phytochemistry* **2015**, *115*, 9–19. [CrossRef]

181. Mirpour, M.; Siahmazgi, Z.G.; Kiasaraie, M.S. Antibacterial activity of clove, gall nut methanolic and ethanolic extracts on Streptococcus mutans PTCC 1683 and Streptococcus salivarius PTCC 1448. *J. Oral Biol. Craniofacial Res.* **2015**, *5*, 7–10. [CrossRef]

182. Philander, L.A. An ethnobotany of Western Cape Rasta bush medicine. *J. Ethnopharmacol.* **2011**, *138*, 578–594. [CrossRef]

183. Akhalwaya, S.; van Vuuren, S.; Patel, M. An in vitro investigation of indigenous South African medicinal plants used to treat oral infections. *J. Ethnopharmacol.* **2018**, *210*, 359–371. [CrossRef] [PubMed]

184. Nikolić, M.; Glamočlija, J.; Ferreira, I.C.; Calhelha, R.C.; Fernandes, Â.; Marković, T.; Marković, D.; Giweli, A.; Soković, M. Chemical composition, antimicrobial, antioxidant and antitumor activity of Thymus serpyllum L., Thymus algeriensis Boiss. and Reut and Thymus vulgaris L. essential oils. *Ind. Crops Prod.* **2014**, *52*, 183–190. [CrossRef]

185. De Oliveira Carvalho, I.; Purgato, G.A.; Píccolo, M.S.; Pizziolo, V.R.; Coelho, R.R.; Diaz-Muñoz, G.; Diaz, M.A.N. In vitro anticariogenic and antibiofilm activities of toothpastes formulated with essential oils. *Arch. Oral Biol.* **2020**, *117*, 104834. [CrossRef] [PubMed]

186. Dong, J.; Zhang, L.; Liu, Y.; Xu, N.; Zhou, S.; Yang, Q.; Yang, Y.; Ai, X. Thymol protects channel catfish from Aeromonas hydrophila infection by inhibiting aerolysin expression and biofilm formation. *Microorganisms* **2020**, *8*, 636. [CrossRef] [PubMed]

187. Hajiaghapour, M.; Rezaeipour, V. Comparison of two herbal essential oils, probiotic, and mannan-oligosaccharides on egg production, hatchability, serum metabolites, intestinal morphology, and microbiota activity of quail breeders. *Livest. Sci.* **2018**, *210*, 93–98. [CrossRef]

188. Vitali, L.A.; Beghelli, D.; Nya, P.C.B.; Bistoni, O.; Cappellacci, L.; Damiano, S.; Lupidi, G.; Maggi, F.; Orsomando, G.; Papa, F.; et al. Diverse biological effects of the essential oil from Iranian Trachyspermum ammi. *Arab. J. Chem.* **2016**, *9*, 775–786. [CrossRef]

189. Dahake, P.; Dadpe, M.; Dhore, S.; Kale, Y.; Kendre, S.; Siddiqui, A. Evaluation of antimicrobial efficacy of Trachyspermum ammi (Ajwain) oil and chlorhexidine against oral bacteria: An in vitro study. *J. Indian Soc. Pedod. Prev. Dent.* **2018**, *36*, 357–363. [CrossRef]

190. Murugan, K.; Sekar, K.; Sangeetha, S.; Ranjitha, S.; Sohaibani, S.A. Antibiofilm and quorum sensing inhibitory activity of Achyranthes aspera on cariogenic Streptococcus mutans: An in vitro and in silico study. *Pharm. Biol.* **2013**, *51*, 728–736. [CrossRef]

191. Yang, Y.; Hwang, E.-H.; Park, B.-I.; Choi, N.-Y.; Kim, K.-J.; You, Y.-O. Artemisia princeps inhibits growth, biofilm formation, and virulence factor expression of Streptococcus mutans. *J. Med. Food* **2019**, *22*, 623–630. [CrossRef]

192. Teanpaisan, R.; Senapong, S.; Puripattanavong, J. In vitro antimicrobial and antibiofilm activity of Artocarpus Lakoocha (Moraceae) extract against some oral pathogens. *Trop. J. Pharm. Res.* **2014**, *13*, 1149. [CrossRef]

193. Geethashri, A.; Manikandan, R.; Ravishankar, B.; Shetty, A.V. Comparative evaluation of biofilm suppression by plant extracts on oral pathogenic bacteria. *J. Appl. Pharm. Sci.* **2014**, *4*, 20.

194. Pereira, C.A.; Costa, A.C.B.P.; Liporoni, P.C.S.; Rego, M.A.; Jorge, A.O.C. Antibacterial activity of Baccharis dracunculifolia in planktonic cultures and biofilms of Streptococcus mutans. *J. Infect. Public Health* **2016**, *9*, 324–330. [CrossRef] [PubMed]

195. Lee, S.-H. Antimicrobial effects of herbal extracts on Streptococcus mutans and normal oral streptococci. *J. Microbiol.* **2013**, *51*, 484–489. [CrossRef] [PubMed]

196. Alshahrani, A.M.; Gregory, R.L. In vitro Cariostatic effects of cinnamon water extract on nicotine-induced Streptococcus mutans biofilm. *BMC Complement. Med. Ther.* **2020**, *20*, 1–9. [CrossRef] [PubMed]

197. Wiwattanarattanabut, K.; Choonharuangdej, S.; Srithavaj, T. In vitro anti-cariogenic plaque effects of essential oils extracted from culinary herbs. *J. Clin. Diagn. Res.* **2017**, *11*, DC30–DC35. [CrossRef]

198. Azizan, N.; Mohd-Said, S.; Mazlan, M.K.; Chelvan, K.T.; Hanafiah, R.M.; Zainal-Abidin, Z. In-vitro inhibitory effect of Cinnamomum zeylanicum and Eugenia caryophyllata oils on multispecies anaerobic oral biofilm. *J. Int. Dent. Med. Res.* **2019**, *12*, 411–417.

199. Wang, Y.; Zhang, Y.; Shi, Y.-Q.; Pan, X.-H.; Lu, Y.-H.; Cao, P. Antibacterial effects of cinnamon (Cinnamomum zeylanicum) bark essential oil on Porphyromonas gingivalis. *Microb. Pathog.* **2018**, *116*, 26–32. [CrossRef]

200. Hickl, J.; Argyropoulou, A.; Sakavitsi, M.E.; Halabalaki, M.; Al-Ahmad, A.; Hellwig, E.; Aligiannis, N.; Skaltsounis, A.L.; Wittmer, A.; Vach, K.; et al. Mediterranean herb extracts inhibit microbial growth of representative oral microorganisms and biofilm formation of Streptococcus mutans. *PLoS ONE* **2018**, *13*, e0207574. [CrossRef]

201. Barbieri, D.S.; Tonial, F.; Lopez, P.V.; Maia, B.H.; Santos, G.D.; Ribas, M.O.; Glienke, C.; Vicente, V.A. Antiadherent activity of Schinus terebinthifolius and Croton urucurana extracts on in vitro biofilm formation of Candida albicans and Streptococcus mutans. *Arch. Oral Biol.* **2014**, *59*, 887–896. [CrossRef]

202. Oliveira, M.A.; Borges, A.C.; Brighenti, F.L.; Salvador, M.J.; Gontijo, A.V.; Koga-Ito, C.Y. Cymbopogon citratus essential oil: Effect on polymicrobial caries-related biofilm with low cytotoxicity. *Braz. Oral Res.* **2017**, *31*, e89. [CrossRef] [PubMed]

203. Marinković, J.; Ćulafić, D.M.; Nikolić, B.; Đukanović, S.; Marković, T.; Tasić, G.; Ćirić, A.; Marković, D. Antimicrobial potential of irrigants based on essential oils of Cymbopogon martinii and Thymus zygis towards in vitro multispecies biofilm cultured in ex vivo root canals. *Arch. Oral Biol.* **2020**, *117*, 104842. [CrossRef] [PubMed]

204. Martos, J.; Luque, C.M.; González-Rodríguez, M.P.; Arias-Moliz, M.T.; Baca, P. Antimicrobial activity of essential oils and chloroform alone and combinated with cetrimide against Enterococcus faecalis biofilm. *Eur. J. Microbiol. Immunol.* **2013**, *3*, 44–48. [CrossRef] [PubMed]

205. Raoof, M.; Khaleghi, M.; Siasar, N.; Mohannadalizadeh, S.; Haghani, J.; Amanpour, S. Antimicrobial activity of methanolic extracts of Myrtus Communis L. and Eucalyptus Galbie and their combination with calcium hydroxide powder against Enterococcus Faecalis. *J. Dent. Shiraz* **2019**, *20*, 195–202.

206. Goldbeck, J.C.; Nascimento, J.E.D.; Jacob, R.G.; Fiorentini, Â.M.; da Silva, W.P. Bioactivity of essential oils from Eucalyptus globulus and Eucalyptus urograndis against planktonic cells and biofilms of Streptococcus mutans. *Ind. Crops Prod.* **2014**, *60*, 304–309. [CrossRef]

207. Sulistyani, H.; Fujita, M.; Miyakawa, H.; Nakazawa, F. Effect of roselle calyx extract on in vitro viability and biofilm formation ability of oral pathogenic bacteria. *Asian Pac. J. Trop. Med.* **2016**, *9*, 119–124. [CrossRef] [PubMed]

208. Sekita, Y.; Murakami, K.; Yumoto, H.; Amoh, T.; Fujiwara, N.; Ogata, S.; Matsuo, T.; Miyake, Y.; Kashiwada, Y. Preventive effects of Houttuynia cordata extract for oral infectious diseases. *BioMed Res. Int.* **2016**, *2016*, 2581876. [CrossRef]

209. Süntar, I.; Oyardı, O.; Akkol, E.K.; Ozçelik, B. Antimicrobial effect of the extracts from Hypericum perforatum against oral bacteria and biofilm formation. *Pharm. Biol.* **2016**, *54*, 1065–1070. [CrossRef]

210. Merghni, A.; Marzouki, H.; Hentati, H.; Aouni, M.; Mastouri, M. Antibacterial and antibiofilm activities of Laurus nobilis L. essential oil against Staphylococcus aureus strains associated with oral infections. *Curr. Res. Transl. Med.* **2016**, *64*, 29–34. [CrossRef]

211. Zhang, Z.; Zeng, J.; Zhou, X.; Xu, Q.; Li, C.; Liu, Y.; Zhang, C.; Wang, L.; Zeng, W.; Li, Y. Activity of Ligustrum robustum (Roxb.) Blume extract against the biofilm formation and exopolysaccharide synthesis of Streptococcus mutans. *Mol. Oral Microbiol.* **2021**, *36*, 67–79. [CrossRef]

212. Ahmad, I.; Wahab, S.; Nisar, N.; Dera, A.A.; Alshahrani, M.Y.; Abullias, S.S.; Irfan, S.; Alam, M.M.; Srivastava, S. Evaluation of antibacterial properties of Matricaria aurea on clinical isolates of periodontitis patients with special reference to red complex bacteria. *Saudi Pharm. J.* **2020**, *28*, 1203–1209. [CrossRef]

213. Nurrahman, H.F.; Widyarman, A.S. Effectiveness of Matricaria chamomilla essential oil on Aggregatibacter actinomy-cetemcomitans and Treponema denticola biofilms. *J. Indones. Dent. Assoc.* **2020**, *3*, 77–82.

214. Song, Y.-M.; Zhou, H.-Y.; Wu, Y.; Wang, J.; Liu, Q.; Mei, Y.-F. In vitro evaluation of the antibacterial properties of tea tree oil on planktonic and biofilm-forming Streptococcus mutans. *AAPS PharmSciTech* **2020**, *21*, 1–12. [CrossRef]

215. Ben Lagha, A.; Vaillancourt, K.; Huacho, P.M.; Grenier, D. Effects of labrador tea, peppermint, and winter savory essential oils on Fusobacterium nucleatum. *Antibiotics* **2020**, *9*, 794. [CrossRef]

216. Pires, J.G.; Zabini, S.S.; Braga, A.S.; de Cássia Fabris, R.; de Andrade, F.B.; de Oliveira, R.C.; Magalhães, A.C. Hydroalcoholic extracts of Myracrodruon urundeuva All. and Qualea grandiflora Mart. leaves on Streptococcus mutans biofilm and tooth demineralization. *Arch. Oral Biol.* **2018**, *91*, 17–22. [CrossRef] [PubMed]

217. Tsujii, T.; Kawada-Matsuo, M.; Migita, H.; Ohta, K.; Oogai, Y.; Yamasaki, Y.; Komatsuzawa, H. Antibacterial activity of phellodendron bark against Streptococcus mutans. *Microbiol. Immunol.* **2020**, *64*, 424–434. [CrossRef] [PubMed]

218. Orrù, G.; Demontis, C.; Mameli, A.; Tuveri, E.; Coni, P.; Pichiri, G.; Coghe, F.; Rosa, A.; Rossi, P.; D'Hallewin, G. The selective interaction of Pistacia lentiscus oil vs. human Streptococci, an old functional food revisited with new tools. *Front. Microbiol.* **2017**, *8*, 2067. [CrossRef] [PubMed]

219. Magi, G.; Marini, E.; Brenciani, A.; Di Lodovico, S.; Gentile, D.; Ruberto, G.; Cellini, L.; Nostro, A.; Facinelli, B.; Napoli, E. Chemical composition of Pistacia vera L. oleoresin and its antibacterial, anti-virulence and anti-biofilm activities against oral streptococci, including Streptococcus mutans. *Arch. Oral Biol.* **2018**, *96*, 208–215. [CrossRef] [PubMed]

220. Massunari, L.; Novais, R.Z.; Oliveira, M.T.; Valentim, D.; Junior, E.D.; Duque, C. Antimicrobial activity and biocompatibility of the Psidium cattleianum extracts for endodontic purposes. *Braz. Dent. J.* **2017**, *28*, 372–379. [CrossRef]

221. Dastjerdi, E.V.; Abdolazimi, Z.; Ghazanfarian, M.; Amdjadi, P.; Kamalinejad, M.; Mahboubi, A. Effect of Punica granatum L. flower water extract on five common oral bacteria and bacterial biofilm formation on orthodontic wire. *Iran. J. Public Health* **2014**, *43*, 1688–1694.

222. De Sousa, M.B.; Júnior, J.O.; Barbosa, W.L.; da Silva Valério, E.; da Mata Lima, A.; de Araújo, M.H.; Muzitano, M.F.; Nakamura, C.V.; de Mello, J.C.; Teixeira, F.M. Pyrostegia venusta (Ker Gawl.) Miers crude extract and fractions: Prevention of dental biofilm formation and immunomodulatory capacity. *Pharmacogn. Mag.* **2016**, *12*, S218–S222.

223. De Oliveira, J.R.; de Jesus, D.; Figueira, L.W.; de Oliveira, F.E.; Pacheco Soares, C.; Camargo, S.E.; Jorge, A.O.; de Oliveira, L.D. Biological activities of Rosmarinus officinalis L.(rosemary) extract as analyzed in microorganisms and cells. *Exp. Biol. Med.* **2017**, *242*, 625–634. [CrossRef] [PubMed]

224. Oliveira, J.R.; Santana-Melo, G.D.; Camargo, S.E.; Vasconcellos, L.M.; Oliveira, L.D. Total protein level reduction of odontopathogens biofilms by Rosmarinus officinalis L. (rosemary) extract: An analysis on Candida albicans and Streptococcus mutans. *Braz. Dent. Sci.* **2019**, *22*, 260–266. [CrossRef]

225. Al-Sohaibani, S.; Murugan, K. Anti-biofilm activity of Salvadora persica on cariogenic isolates of Streptococcus mutans: In vitro and molecular docking studies. *Biofouling* **2012**, *28*, 29–38. [CrossRef] [PubMed]

226. Gupta, A.; Duhan, J.; Tewari, S.; Sangwan, P.; Yadav, A.; Singh, G.; Juneja, R.; Saini, H. Comparative evaluation of antimicrobial efficacy of Syzygium aromaticum, Ocimum sanctum and Cinnamomum zeylanicum plant extracts against Enterococcus faecalis: A preliminary study. *Int. Endod. J.* **2013**, *46*, 775–783. [CrossRef]

227. De Oliveira, J.R.; de Jesus Viegas, D.; Martins, A.P.; Carvalho, C.A.; Soares, C.P.; Camargo, S.E.; Jorge, A.O.; de Oliveira, L.D. Thymus vulgaris L. extract has antimicrobial and anti-inflammatory effects in the absence of cytotoxicity and genotoxicity. *Arch. Oral Biol.* **2017**, *82*, 271–279. [CrossRef] [PubMed]

Article

The Influence of Chestnut Flour on the Quality of Gluten-Free Bread

Katarzyna Marciniak-Lukasiak [1,*], Patrycja Lesniewska [1], Dorota Zielińska [2], Michal Sowinski [1], Katarzyna Zbikowska [3], Piotr Lukasiak [4] and Anna Zbikowska [1]

[1] Institute of Food Sciences, Faculty of Food Assessment and Technology, Warsaw University of Life Sciences (WULS-SGGW), Nowoursynowska St. 159c, 02-776 Warsaw, Poland
[2] Department of Food Gastronomy and Food Hygiene, Institute of Human Nutrition Sciences, Warsaw University of Life Sciences (WULS-SGGW), Nowoursynowska 159c, 02-776 Warsaw, Poland
[3] Faculty of Medicine, Medical University of Warsaw, Zwirki i Wigury St. 61, 02-091 Warsaw, Poland
[4] Institute of Computing Science, Faculty of Computing and Telecommunications, Poznan University of Technology, Piotrowo 2, 60-965 Poznan, Poland
*   Correspondence: katarzyna_marciniak_lukasiak@sggw.edu.pl; Tel.: +48-22-59-37548

Abstract: Gluten-free bread is the basis of an elimination diet in the case of many glucose-related diseases. The quality of this bread differs significantly from traditional products; therefore, it is necessary to conduct research aimed at improving the quality of this type of product. The aim of the study was to determine the effect of the addition of chestnut flour and the method of packaging on the quality of gluten-free bread. The addition of chestnut flour (partially replacing corn starch) was used in the amount of 5, 10, 15 and 20% of the total weight of the concentrate. The influence of the storage method on the quality of the tested bread was examined after 7, 14 and 21 days from baking. The refrigerated breads were packed using PA/PE barrier foil with air and vacuum (58%) and were stored in room temperature (22 $\pm$ 2 °C). Water content, texture and color were determined, and sensory evaluation and microbiological analysis were performed. As a result of the conducted research, we observed that the addition of chestnut flour to the recipe affects significantly ($p < 0.05$) the texture of the finished product, reducing the hardness and increasing the elasticity and cohesiveness of the bread crumb. The use of chestnut flour in an amount of up to 10% increases significantly ($p < 0.05$) the volume of the resulting loaves. Microbiological research has indicated vacuum packaging as a better way to protect and store gluten-free bread. For practical use in future production, it is recommended to replace corn starch in gluten-free breads by no more than 10% by chestnut flour.

Keywords: celiac disease; gluten-free bread; chestnut flour; texture; sensory analysis; vacuum packaging

Citation: Marciniak-Lukasiak, K.; Lesniewska, P.; Zielińska, D.; Sowinski, M.; Zbikowska, K.; Lukasiak, P.; Zbikowska, A. The Influence of Chestnut Flour on the Quality of Gluten-Free Bread. *Appl. Sci.* **2022**, *12*, 8340. https://doi.org/10.3390/app12168340

Academic Editor: Akikazu Sakudo

Received: 3 August 2022
Accepted: 18 August 2022
Published: 20 August 2022

1. Introduction

Celiac disease (CD), defined as chronic autoimmune gluten intolerance, is becoming an increasingly important health problem in modern medicine in developed countries [1,2]. The occurrence of CD is strongly related to genetic factors, among which HLA class II plays a major role—HLA-DQ2 heterodimers are expressed in over 90% of patients, where the remnant express HLA-DQ8 [3,4]. So far, the only effective treatment option is to follow a gluten-free diet throughout the entire life [5,6].

Gluten-free products are the basis of the elimination diet of patients suffering from gluten-dependent diseases, which include, among others: celiac disease, nonceliac hypersensitivity to gluten, Duhring's disease and wheat allergy [7,8]. According to the research, celiac disease affects about 1% of the population, and the incidence of this disease is gradually increasing. Failure to follow a strict diet ultimately leads to the disappearance of intestinal villi, which in turn results in the malabsorption of nutrients from food. The limited absorption of ingredients necessary for the proper functioning of the body may cause various clinical symptoms [2].

Bread is one of the basic ingredients in the daily human diet. It is the main source of carbohydrates and provides many valuable nutrients, including B vitamins and fiber to support the proper functioning of the intestines. Due to the changing eating habits and the increase in various types of food allergies and intolerances, newer recipes and technological solutions are sought to expand the range of products available on the market [9–11].

According to the Codex Standard [12], gluten is the protein fraction of wheat, rye, barley, oats or their hybrids and derivatives that some people cannot tolerate. The water-insoluble prolamins and glutelins (collectively referred to as gluten) typically make up 70–80% of the cereal grain protein. They are the most important cereal proteins from a technological point of view. In this sense, gluten is a specific structure, a viscoelastic gel, which is responsible for creating the correct structure of the finished product with unique properties. As a result of absorbing large amounts of water, gluten swells, giving the dough its appropriate characteristics, such as flexibility, plasticity, stickiness, cohesiveness and elasticity. It allows dough to obtain a thin and spongy structure with fine pores, which is fixed during baking [10].

Therefore, the production of bread from gluten-free raw materials is a major techno-logical challenge. The lack of gluten causes many problems in the preparation of the dough, significantly affecting its rheological properties and the quality of the final product. The obtained products are often characterized by a poorer structure, color and porosity of the crumb, smaller volume, stickiness and considerable brittleness in comparison to the regular one. They are also characterized by a short shelf life and worse sensory features compared to traditional baking [13].

In order to improve the rheological properties of the dough, which will allow for the proper forming of the billets and obtaining the appropriate properties of the finished product, many studies are carried out to improve the recipes of gluten-free bread. In order to replace gluten, which is a structure-forming factor, various substances are used to support the proper development of the dough, emulsifiers, as well as texturing and thickening substances. The most commonly used ingredients are hydrocolloids [14,15], which include, inter alia, guar gum, xanthan gum, Locust bean gum, pectin, carboxymethyl cellulose and hydroxyethyl cellulose. These substances also have the ability to bind water, gel and act as stabilizers. Enzymes (transglutaminases, amylases and proteases) and exopolysaccharides, which are produced by lactic acid bacteria, are also increasingly used, which allows for the elimination of chemical substances forming the structure [16–20].

Due to their specific recipe composition, gluten-free products have a lower nutritional value and a higher glycemic index compared to their traditional counterparts. In order to improve the properties of gluten-free bread, enrichment with the addition of raw materials of plant and animal origin is used. This will increase its nutritional value, obtain a high-quality product and gain consumer recognition [21,22]. An example of such an additive is chestnut flour, which also positively influences the physical and chemical properties of food products like cookies [23], pasta [24] or bread [23,25–28]. Chestnuts are a good source of antioxidants and minerals such as potassium and magnesium, which are associated with reducing the risk of cardiovascular disease and stroke [26]. Chestnut flour, on the other hand, is a good functional ingredient and may increase the content of some nutrients, positively affecting the physical and nutritional properties of cereal products [23] and the quality features of the finished product [27,28]. In practice, however, the amount of the additive used often has to be limited, because too much chestnut flour may reduce the quality of the finished product (the addition of chestnut flour creates a darker and often harder product, but, especially for gluten-free bread, improved dough workability, texture, color and flavor) [9,19,23,29].

Demand for gluten-free products continues to increase, with a global market of USD 21.61 billion in 2019 and is projected to reach nearly USD 24 billion by 2027 [30].

Taking into account the upward trend in the value of the gluten-free products market, consumer interest and the increasing availability and variety of these products, it is justified to undertake research aimed at determining the impact of the addition of chestnut flour on the texture of bread baked from gluten-free bread concentrates.

The novelty of the provided study is to provide the gluten-free products that can be addressed to the gluten-free community taking into consideration the storage time as well as the way of packaging.

2. Materials and Methods

Gluten-free bread concentrates were used with following ingredients:

- corn starch, Bogutyn Młyn, Poland;
- potato starch, Melvit, Poland;
- corn flour, Kupiec, Poland;
- chestnut flour, ViVio, Poland;
- instant yeast, Lesaffre, Poland;
- sugar, Suedzucker Polska Sp. z o.o, Poland;
- salt, Kłodawa S.A. Poland;
- hydrocypropyl methylcellulose (HPMC), J. Rettenmaier & Söhne, Poland

The following ingredients were added to the concentrates:

- milk, "Łaciate", Mlekpol, Poland, fat content 3.2%;
- potable water.

The composition of the chestnut flour:

- nutritional value w 100 g;
- energy value 1563 kJ/371 kcal—19% (Nutrient Reference Values);
- Fat 3.4 g—5% (saturated fatty acids 0 g—0%);
- carbohydrates 70 g—27% (sugar 18 g—20%);
- protein 7 g—14%;
- dietary fiber 16 g;
- salt 0.03 g < 1%.

2.1. Preparation of Gluten-Free Bread

The basis of gluten-free bread concentrates was corn starch. The same amount of potato starch, corn flour, instant yeast, sugar, salt and hydroxypropyl methylcellulose (HPMC) was present in each of the sample. Only the amount of corn starch and chestnut flour were changed. The concentrate recipes were established based on our preliminary findings, in which the amount of ingredients needed to prepare the dough was determined. The loose ingredients, which were used to prepare the mixture, were weighed in accordance with the recipes on technical scales with a precision of 0.01. The recipe composition is given in Table 1.

Table 1. The recipe composition of gluten-free bread concentrates.

Ingredients	Raw Material Content [%]				
Corn starch	63.0	58.0	53.0	48.0	43.0
Potato starch			19.0		
Corn flour			7.0		
Instant yeast			2.4		
Sugar			5.1		
Salt			1.5		
HPMC			2.0		
Chestnut flour	0.0	5.0	10.0	15.0	20.0

All ingredients were mixed in mixer for 5 min; next, the mixture was stored in a plastic bowl for 30 min at a temp. of 40 °C. After 30 min, the dough was placed in the shaped bowls, where the process of fermentation was continued for the next 10 min until the optimum volume was reached. Baking proceeded in the combi-steamer oven by UNOX (type XBC, model XBC 404) at a temp. of 175 °C for 23 min on the third level of vaporization. The refrigerated breads were packed using the packing machine EMPRA (VP 370) in a packaging made of PA/PE barrier foil (multilayer vacuum films laminated with polyamide) with air and vacuum (58%), and they were stored in room temperature (22 $\pm$ 2 °C). The temperature was set and controlled using air conditioning infrastructure. All measurements were made in 2 repetitions.

2.2. Determination of Volume, Moisture and Porosity

Bread volume was measured by the rapeseed displacement method 10-05. Bread moisture was determined according to the approved method 44-15A [31]. The porosity of the crumb was assessed by the differences between the volume of a bread crumb cylinder and the volume of a compressed crumb cylinder measured by oil displacement with a graduated cylinder. Moreover, the H/D ratio was determined (height/diameter).

2.3. Determination of Water Activity

The water activity determination was carried out on a Pre AquaLab Water Activity Analyzer (METER Group, Pullman, WA, USA). The samples were placed in airtight water activity cups and then placed in the cells of the apparatus. The measurements were made at the temperature of 20 $\pm$ 1 °C. After placing the samples in the apparatus, we waited about 3–5 min until the measured value stabilized. Measurements were made in five parallel replications. The final result was considered as the arithmetic mean of the measurements, after excluding the results differing from the others by 5%, considered as an error of the apparatus. In the tested products, both the initial water activity at zero point (baking time) and the water activity after 7, 14 and 21 days of storage were determined.

2.4. Determination of Crumb Hardness

Measurements of crumb hardness were done by using texture Analyzer TA-XTplus (Stable Micro Systems, Godalming, UK). The slices with a thickness of 20 mm were cut from the center of the analyzed loafs. Next, the slices were squeezed and relaxed. As before, we used a cylindrical head with an attachment in the shape of a cylinder of a diameter of 36 mm. Measurements were done at the speed of movement of the head of 1 mm/s, penetrating the sample to a depth of 10 mm with a charge cell of 250 N. The analysis of the moisture and the hardness of the crumb was made after 24 and 48 h after baking.

2.5. Sensory Analysis

Sensory analysis was carried out by 10 well-trained panelists. A nine-point hedonic scale was used to evaluate the overall acceptability of the bread formulations, ranging from 1 (dislike extremely) to 9 (like extremely) [32]. Samples of the same size were prepared for each panelist. Each assessor received bread samples in an identical form, which were marked in a coded manner and impossible to identify by the panelists.

2.6. Determination of Color (L*, a*, b*)

Characterization of the bread color was performed using the L*a*b* system proposed by the International Commission on Illumination (CIE) in the work of Papadakis [33]. L* refers to the luminosity or lightness component, and a* (intensity of red (+) and green (−)) and b* (intensity of yellow (+) and blue (−)) are the chromaticity coordinates. All sampled breads were analyzed in terms of the referred parameters using a Minolta CR-310 colorimeter (Konica-Minolta, Osaka, Japan), which was calibrated a priori with a white standard tile. Five repetitions were performed for each measurement.

2.7. Microbiological Quality

In the study, the microbiological quality of developed bread samples was investigated. Samples were evaluated immediately after production (zero point) and after 7, 14 and

21 days of storage at room temperature ($22 \pm 2°$ C) with lightening. Storage trials were carried out for 21 days (based on preliminary studies), because after the mentioned period, microbiological changes were observed, and the stored samples did not contain any added preservatives. Briefly, bread samples (10 g) were transferred to 90 mL peptone water (Lab M, Heywood, UK) and homogenized, serially diluted in sterile peptone water and surface spread on duplicate plates with the appropriate medium.

Nutrient agar (Biokar Diagnostic, Noack, Poland) was used for the enumeration of total viable counts (TVC), while MRS agar (LabM, Heywood, UK) was used for the enumeration of lactic acid bacteria (LAB). The plates were incubated at 30 °C for 72 h [34]. Chloramphenicol glucose agar (Biokar Diagnostic, Noack, Poland) and incubation at 25 °C for 120 h were used for enumeration of yeast and molds [35,36].

Bacillus spp. was investigated on agar PEMBA (LabM, Heywood, UK) [37].

2.8. Statistical Analysis

The statistical analysis of the obtained results was performed applying Statgraphics Plus 4.1., and the differences between the averages were estimated using multivariate regression analysis. The significance level (α) was set to 0.05 and the smallest statistically significant difference was chosen using Tukey's test. PCA analysis was done using the software Statistica 13.3. The analyzed features were done in 3 repetitions.

3. Results and Discussion

3.1. Physical Properties

The breads with the chestnut flour in the amounts of 5, 10, 15 and 20% were analyzed, as well as the control sample, which was bread without the addition of chestnut flour. By analyzing the obtained results of the average weight value after baking (Table 2), it was found that the lowest weight characterized bread with a 10% addition of chestnut flour (153.88 g) followed by the control sample (160.97 g). No statistically significant differences were noticed in the case of the control sample and the bread with the 5% addition of chestnut flour. Moreover, no statistically significant differences were found for the weight of the bread with the 15% addition of chestnut flour and the weight of bread with the 20% addition of chestnut flour (166.69 g), which was characterized by the highest weight among all variants.

Table 2. Weight, volume, specific weight and porosity of gluten-free bread with the addition of chestnut flour in the amount of 0, 5, 10, 15 and 20%.

Sample	Finished Product Weight [g]	Volume [cm^3/100 g]	Specific Mass [g/cm^3]	Porosity [%]
Control	160.97 ± 1.64 b	207.00 ± 7.35 a	0.27 ± 0.02 c	73.46 ± 1.02 c
5%	162.11 ± 2.13 b	283.8 ± 8.47 b	0.20 ± 0.03 b	69.75 ± 1.07 b
10%	153.88 ± 1.63 a	310.4 ± 6.69 b	0.17 ± 0.03 a	74.07 ± 1.00 c
15%	165.73 ± 1.78 c	215.22 ± 7.41 a	0.20 ± 0.02 b	72.22 ± 1.65 c
20%	166.69 ± 1.65 c	224.00 ± 5.65 a	0.25 ± 0.03 c	64.81 ± 1.83 a

Values in the same column marked with the same symbols mean no statistically significant differences ($\alpha = 0.05$).

The highest volume (310.4 cm^3/100 g) among the gluten-free breads was characteristic for the bread with the 10% addition of chestnut flour (Table 2). No statistically significant differences were found between the bread with the 10% addition of chestnut flour and the bread with the 5% addition of chestnut flour (283.8 cm^3/100 g). Comparing the variant without the addition of chestnut flour, with the breads with the 15% and 20% flour additions, no statistically significant differences were found in the values of this parameter. The value of this characteristic for breads with a variable addition of chestnut flour ranged from 207 cm^3/100 g to 310.4 cm^3/100 g. According to the standard for gluten-free bread [38], the volume of 100 g of bread should be no less than 190 cc. All the values obtained in the experiment are within the optimum range. Taking into account the obtained results, it was found that the addition of 5 and 10% chestnut flour to the recipe causes an increase in the volume of the tested breads. Increasing the percentage of the addition of this flour causes a decrease in the baking volume. Similar relationships were obtained by Aguilar [39] in a study of the effect of chestnut flour leaven on the properties of gluten-free bread. The breads contained 15, 20 and 25% chestnut flour, and the basic raw material was corn starch. The increase in the concentration of chestnut flour caused a decrease in the volume of the bread. The research conducted by Demirkesen [40] on an attempt to replace rice flour with chestnut flour in bread showed that the volume of the bread increases with increasing the ratio of chestnut flour to rice flour. These differences may be due to the use of flour with different dietary fiber content. In this study, flour with the amount of fiber of 16% was used, whereas the amount of fiber in the flour used by Aguilar [39] was 15%, but Demirkesen [40] used chestnut flour with a fiber content of 9.5%. Due to its gas retention and viscoelastic properties, fiber can increase the volume of gluten-free bread. Too much fiber reduces the volume of the bread. These differences may also be caused by the use of various types of technological additives, e.g., emulsifiers [39].

The highest specific mass (Table 1) was found in the control sample ($0.267 \, \text{g/cm}^3$), the second in terms of this parameter was the bread with the 20% addition of chestnut flour ($0.252 \, \text{g/cm}^3$). The values of the specific weight of the breads with the 5% ($0.197 \, \text{g/cm}^3$) and 15% additions of chestnut flour ($0.198 \, \text{g/cm}^3$) did not differ statistically significantly. The performed statistical analysis showed a significant influence of the recipe composition on the specific weight of the tested bread. The largest statistically significant difference occurred between the control sample and the bread with a 10% addition of chestnut flour, which was characterized by the lowest value of the specific weight ($0.174 \, \text{g/cm}^3$).

The porosity is the ratio of the volume occupied by the pores to the total volume of the bread. High-quality bread is distinguished by a crumb with thin-walled and evenly spaced pores. For wheat bread, the porosity should be from 73 to 83%, and for rye bread, from 55 to 70% [41]. The lowest crumb porosity was characteristic for bread with the 20% addition of chestnut flour (64.8%), which was also characterized by the lowest volume. No statistically significant differences were found between the control sample (73.5%) and the bread with the 10% addition of chestnut flour (74.1%). These breads were characterized by the highest porosity among all the tested variants. Taking into account the obtained results and comparing them with the requirements concerning the porosity of the crumb of wheat bread, it was found that, apart from the control sample, the bread with the 10% addition of chestnut flour met the requirements specified in the standards for traditional bread. Similar results were obtained by Marciniak-Łukasiak and Skrzypacz [42], who investigated the effect of the addition of amaranth flour on the physicochemical and sensory properties of gluten-free bread. Amaranth flour was added (partially replacing the corn flour) in the amounts of 2.5, 5, 7.5, 10 and 12.5% of the total weight of the concentrate. The authors observed that the breads with the addition of amaranth flour in the amount of 5, 7.5 and 10% showed the highest porosity among all those baked in the series. Amaranth flour, added to the recipe in the amount of 12.5%, reduced the porosity of the gluten-free bread.

3.2. Color

One of the most important parameters in assessing the quality of food products and raw materials is color. The degree of the color intensity affects the positive or negative attitudes of consumers to a given product.

The L* parameter values for gluten-free breads ranged from 84.73 to 94.63 (Table 3). The highest value of this parameter was found in the control sample (94.63). Along with increasing the amount of the addition of chestnut flour, the value of the L* parameter decreased. No statistically significant differences were found between the variant of the breads with the 15% and 20% additions of this flour. These results are consistent with the results obtained by Aguliar [39], who, in the study of the properties of gluten-free bread, replaced rice flour with chestnut flour (15, 20, 25% chestnut flour). The L* brightness parameter showed that the greater the proportion of chestnut flour in the recipe, the darker the color of the crumb. Similar observations were made by Rinaldi [43] when introducing chestnut flour to the bread recipe at the level of 20% (in relation to the amount of wheat flour). The addition of chestnut flour resulted in lower L* values (69.0 for bread made from wheat flour; 61.7 for bread with a 20% addition of chestnut flour). According to the authors, the cause of this phenomenon is not only the darker color of the raw material, but also the intense caramelization and Maillard reactions taking place due to the high sugar content in chestnut flour.

Table 3. The values of the color parameters of the crumb of gluten-free bread with the addition of chestnut flour in the amounts of 0, 5, 10, 15 and 20%.

Sample	L*	a*	b*	Browning Index
Control	94.63 ± 0.85 d	−0.22 ± 0.07 a	14.54 ± 0.70 b	16.31 ± 0.80 a,b
5%	90.37 ± 1.61 c	0.11 ± 0.03 b	12.65 ± 0.76 a	15.01 ± 0.96 a
10%	87.77 ± 2.13 b	0.23 ± 0.05 c	12.86 ± 1.04 a	16.42 ± 1.59 b
15%	85.08 ± 1.93 a	0.34 ± 0.03 d	14.73 ± 0.67 b,c	18.45 ± 1.16 c
20%	84.73 ± 0.97 a	0.45 ± 0.02 e	15.54 ± 0.31 c	20.40 ± 0.51 d

Values in the same column marked with the same symbols (a–e) mean no statistically significant differences ($\alpha = 0.05$).

The value of the color chromaticity index (a*) of the crumb ranged from −0.22 (for the control sample) to 0.45 (for bread with the 20% addition of chestnut flour) and was statistically significantly differentiated (Table 3). The control sample was characterized by the predominance of a green color, while the highest share of a red color was the bread with the 20% addition of chestnut flour. Along with increasing the amount of the addition of chestnut flour, the value of the a* parameter increased. The analysis of the a* parameter value showed that higher amount of chestnut flour in the dough causes a larger proportion of the red color. In gluten-free breads studied by Aguilar [39], this tendency was also noticed; however, the values of this parameter were much higher, as they amounted to 4.87, 5.82 and 6.58, respectively, for bread with 15, 20 and 25% addition of chestnut flour, which may result from a different recipe composition.

The values of the b* parameter of gluten-free bread crumb ranged from 12.65 (for bread with a 5% addition of chestnut flour) to 15.55 (for bread with a 20% addition of chestnut flour) (Table 3). The lowest share of the shade of yellow in the color of the crumb was found in bread with the 5% addition of chestnut flour (12.65), and no statistically significant differences were found between bread with the 5% and 10% addition of chestnut flour (12.86). The b* parameter values for the control sample and the bread with the 15% addition of chestnut flour were 14.54 and 14.73, respectively. Comparing the obtained values, it was found that, with the increase in the concentration of chestnut flour, the share of the yellow color in the tested breads increases, which was also observed in [39], where the values of the b* parameter were 17.76, 18.47 and 19.14 for loaves with the 15, 20 and 25% addition of chestnut flour, respectively.

3.3. Browning Index

One of the parameters that proves the color change is the browning index (BI). It represents the purity of a brown color and is reported as an important parameter in processes where enzymatic and nonenzymatic browning occurs [44].

The obtained browning index values for gluten-free bread are presented in Table 3 and were statistically significantly differentiated. The mean value of the browning index for the control sample was 16.31. With the addition of chestnut flour to the recipe, it was observed that the browning index values gradually increase. However, the browning index value for bread with the 5% addition of chestnut flour is lower compared to the control sample.

3.4. Moisture

Moisture is one of the most important physical and chemical parameters of the bread crumb. It mainly shows the degree of the freshness of the bread and influences the staling process, which adversely affects changes in the sensory characteristics of the bread [45].

The purpose of determining the humidity of the crumb of gluten-free bread at point zero after 7, 14 and 21 days of storage was to compare the changes occurring during their storage. After 7, 14 and 21 days, the samples were stored and packed in two ways: in a barrier foil with air access and in a barrier foil with a vacuum (58%) at a temperature of $22 \pm 2\ °C$.

According to the standard [38], the humidity of gluten-free bread should not exceed 53%. The tested gluten-free breads packed in a barrier foil with air had a humidity ranging from 38.89% to 47.26%, so it was within the norm (Figure 1). Research conducted by Cacak-Pietrzak [20] also confirms that the humidity of gluten-free bread ranges from 34.3% to 49.7%. The control sample (47.26%) was characterized by the highest humidity of the crumb at the zero point. With the increase of the storage time, the humidity of the crumb in the control sample gradually decreased to the level of 42.03%. All breads had the highest humidity at the zero point. The obtained relationships are consistent with the observations of Demirkesen [46], who examined the moisture content of gluten-free bread baked with rice flour and the addition of chestnut flour. They showed that, during the storage of bread, the process of staling occurs due to the migration of moisture from the crumb to the crust. As a result of this process, the ability to bind water by the crumb decreases, so the lowest moisture losses in the bread are observed in the samples stored for the shortest time. In the case of breads with the addition of chestnut flour, a decrease in humidity was observed after 7 days of storage. The lowest humidity was observed after 21 days of storage in all variants of gluten-free bread, except for bread with the 5% addition of chestnut flour, the humidity of which was 39.10%.

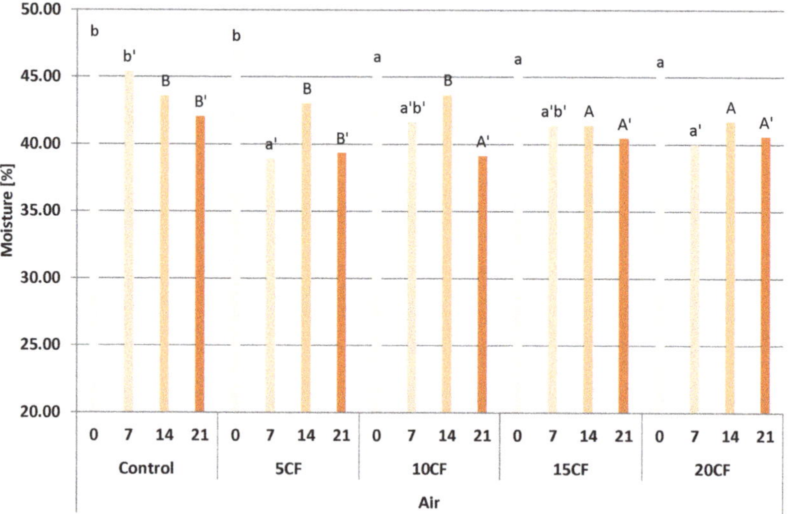

Figure 1. Moisture of the crumb of gluten-free breads with the addition of chestnut flour in the amount of 5% (5CF), 10% (10CF), 15% (15CF) and 20% (20CF) after 7, 14 and 21 days of storage, packed in a barrier film with air. Values marked with the same symbols mean no statistically significant differences ($\alpha = 0.05$).

The moisture content of the vacuum-packed gluten-free bread ranges from 38.56% to 47.26%, so it complies with the standard [38] (Figure 2). As the bread storage time increased, the crumb moisture decreased successively, except for the breads with the 5% and 10% addition of chestnut flour. In both cases, the humidity increased on the 21st day of storage.

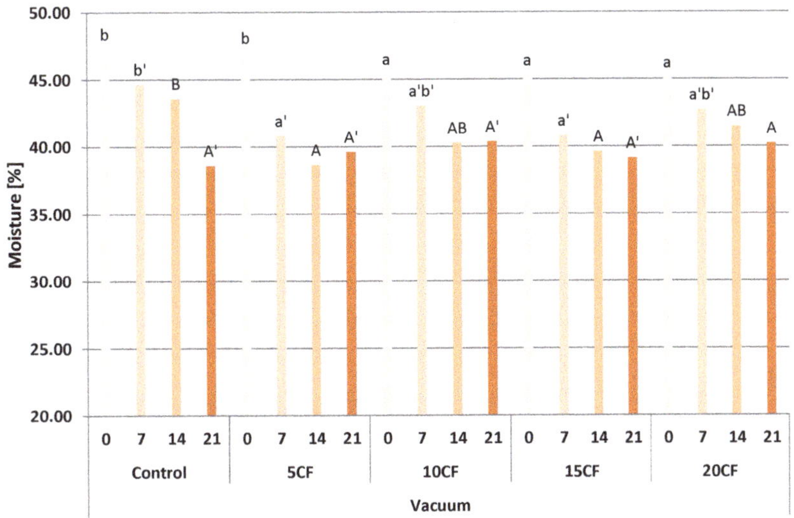

Figure 2. Moisture of the crumb of gluten-free breads with the addition of chestnut flour in the amount of 5% (5CF), 10% (10CF), 15% (15CF) and 20% (20CF) after 7, 14 and 21 days of storage, packed in a barrier film with a vacuum. Values marked with the same symbols mean no statistically significant differences ($\alpha = 0.05$).

Comparing the method of packing the tested breads, higher humidity was observed in the case of the breads with the 5, 10 and 20% addition of chestnut flour, vacuum-packed, after 7 days of storage. The control sample and the bread with 15% addition of chestnut flour were characterized by lower humidity compared to the bread packed with air access. After 14 days of storage, the humidity of the crumb was lower in the vacuum-packed bread, with the exception of the control sample, in which the humidity of the bread packed with air was 43.57%, while in vacuum-packed bread it was 43.58%. However, these differences are not statistically significant. After 21 days of storage, in the control and the breads with the 15 and 20% addition of chestnut flour lower humidity in the vacuum-packed samples was observed.

3.5. Water Activity

On the basis of the conducted research, it was found that the fresh bread at the zero point was characterized by a higher water activity compared to the bread tested after the given storage time. Water activity in the breads with the 5% and 20% addition of chestnut flour gradually decreased until they were stored. In the case of the control sample, the bread with the 15% and the bread with the 20% addition of chestnut flour, water activity increased after 14 days compared to its level after 7 days of storage. It can be assumed that the reason for this is the separation of free water in the product, which contributed to its evaporation into the environment [38]. The water activity in the analyzed breads packed in a barrier foil with air ranges from 0.956 to 0.983 (Figure 3).

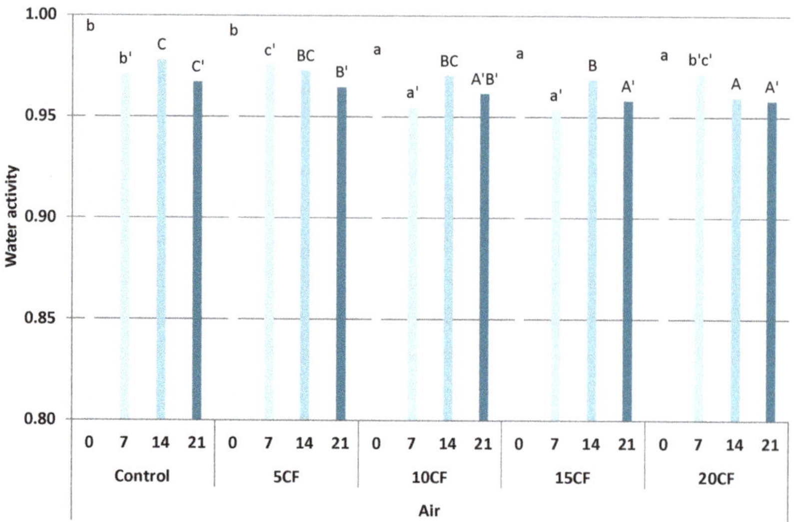

Figure 3. Water activity of gluten-free breads with the addition of chestnut flour in the amount of 5% (5CF), 10% (10CF), 15% (15CF) and 20% (20CF) after 7, 14 and 21 days of storage, packed in a barrier film with air. Values marked with the same symbols mean no statistically significant differences ($\alpha = 0.05$).

These results are on a similar level to the results obtained by Pałacha and Makarewicz [47], who examined the crumb of various types of bread, the water activity of which ranged from 0.926 to 0.975. Similar results were obtained by Aguilar [39], who tested the water activity in gluten-free bread with the addition of 15, 20 and 25% chestnut flour at the zero point and after 7 days of storage. It was shown that the water activity after 7 days of storage decreased compared to the water activity at the zero point. The water activity of the breads with 15, 20 and 25% chestnut flour added at the zero point was 0.975, respectively, 0.975 and 0.973, and after 7 days 0.972, 0.970 and 0.970. As in the case of air-packed bread, the water activity was highest in fresh bread—at the zero point. Water activity in the control sample and the breads with 5 and 15% addition of chestnut flour increased after 14 days compared to its level after 7 days of storage, and the breads with the 10 and 20% addition of chestnut flour was characterized by higher water activity after 21 days after baking compared to its level after 14 days of storage. The range of water activity in the analyzed vacuum-packed breads ranged from 0.949 to 0.983 (Figure 4). The factors that prolong the shelf life of products and ensure their microbiological stability, despite the high activity of water, could be contributed by: high baking temperature and the presence of a crust on the surface of the bread, which is a natural protection against external factors [47].

Comparing the method of packing the tested gluten-free breads, higher water activity was observed in the control sample and in the vacuum-packed bread with the 10% addition of chestnut flour after 7 days of storage. After 14 days of storage, the water activity was lower in the vacuum-packed bread, except for the control sample, in which the water activity of the air-packed bread was 0.974, and in the vacuum was 0.978. After 21 days of storage in the control sample and the breads with the 5% and 15% addition of chestnut flour, lower water activity was observed in the samples packed under vacuum.

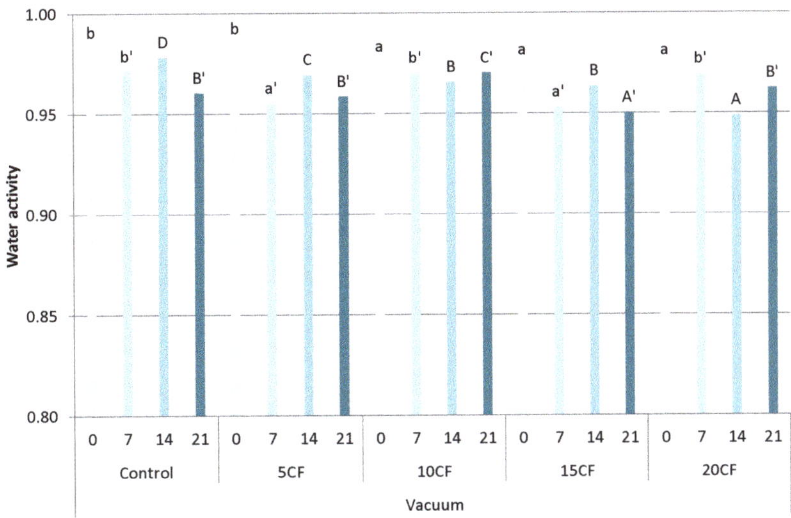

Figure 4. Water activity of gluten-free breads with the addition of chestnut flour in the amount of 5% (5CF), 10% (10CF), 15% (15CF) and 20% (20CF) after 7, 14 and 21 days of storage, packed in a barrier film with a vacuum. Values marked with the same symbols mean no statistically significant differences ($\alpha = 0.05$).

3.6. Texture

3.6.1. Hardness

Hardness is defined as the peak force during the first compression cycle [48].

The crumb hardness of gluten-free breads packed in a zero point with an air barrier film was from 1.25 N (for the bread with the 5% addition of chestnut flour) to 7.32 N (for the control) (Table 4). No statistically significant differences were found between the variants with the 10, 15 and 20% addition of chestnut flour. The addition of chestnut flour significantly reduced the hardness of the crumb. The 5% share of the chestnut flour in the tested breads turned out to be the most advantageous in this respect. An increase in the crumb hardness was observed after 7 days of storage in all analyzed samples. The increase in bread hardness results from its staleness. The main reason for this process is the transformation of starch from its amorphous to pseudocrystalline form (starch retrogradation). This form binds smaller amounts of water, which increases the brittleness and hardness of the crumb [49,50]. After 14 days of storage, an increase in the crumb hardness was observed only in the case of bread with the 20% addition of chestnut flour. Twenty-one days after baking, a significant increase in crumb hardness was observed in the control sample (18.87 N), while the breads with the 5, 10 and 15% addition of chestnut flour were characterized by lower hardness values compared to the crumb hardness on the 14th day of storage.

As in the case of the samples packed with air, an increase in the crumb hardness of the gluten-free bread packed in a barrier foil with a vacuum was observed after 7 days of storage in all analyzed variants (Table 4). After 14 days of storage, an increase in the crumb hardness was noticed only in the case of the breads with the 15 and 20% addition of chestnut flour. Twenty-one days after baking, the crumb hardness increased in the control sample and the breads with the 5% and 10% addition of chestnut flour. Gambuś [51] investigated the effect of an amaranth flour addition on the quality of gluten-free bread. The authors observed an increase in the crumb hardness with an increasing storage time. Similar results were also obtained by Kulczak [52], who assessed selected physical properties of gluten-free bread with the use of instant pea flour and buckwheat products. The crumb hardness of the tested samples was assessed 3 h, 24 h and 48 h after baking. An increase in the crumb hardness was observed with an increasing storage time. These results are partially consistent with the results presented in this study, because, both in the case of bread packed with air and in vacuum, an increase in the hardness of the crumb of the tested bread was observed after 7 days of storage.

Table 4. Hardness, elasticity, cohesiveness, gumminess and chewiness of gluten-free bread with the addition of chestnut flour in the amounts of 0, 5, 10, 15 and 20%.

	Sample	Time [day]	Hardness [N]	Elasticity	Cohesiveness	Gumminess	Chewiness [N]
Air	Control	0	7.32 c	2.37 a,b	0.41 a,b	2.83 b	7.89 b
		7	11.19 b$'$	4.81 a	0.51 a$'$	5.67 a$'$	27.32 a$'$
		14	10.82 E	1.37A	0.29 A	3.26 A	7.25 A
		21	18.87 D$'$	1.70 A$'$	0.58 B$'$	10.90 C$'$	17.32 A$'$,B$'$
	5%	0	1.25 a	0.98 a	0.17 a	0.25 a	1.21 a
		7	7.46 a$'$	4.94 a$'$	0.73 a$'$,b$'$	5.48 a$'$	27.06 a$'$
		14	4.14 A	3.67 C	0.72 B	2.96 A	11.12 A,B
		21	4.04 A$'$	1.63 A$'$	0.28 A$'$	1.11 A$'$	5.42 A$'$
	10%	0	3.21 b	2.13 a,b	0.46 a,b	1.59 a,b	6.24 b
		7	7.02 a$'$	4.59 a$'$	0.82 a$'$,b$'$	5.78 a$'$	26.53 a$'$
		14	5.86 B	1.70 A	0.41 A	2.43 A	7.74 A
		21	4.02 A$'$	4.88 B$'$	0.91 D$'$	3.67 A$'$,B$'$	17.92 A$'$,B$'$
	15%	0	2.43 b	3.39 a,b	0.64 a,b	1.53 a,b	6.97 b
		7	7.49 a$'$	4.31 a$'$	0.74 a$'$,b$'$	5.40 a$'$	23.29 a$'$
		14	6.55 C	2.52 B	0.59 A,B	3.92 A	13.75 A,B
		21	6.49 B$'$	4.70 B$'$	0.76 C$'$	4.88 A$'$,B$'$	23.06 B$'$
	20%	0	2.55 b	4.45 b	0.83 b	2.13 a,b	9.75 b
		7	9.46 b$'$	4.90 a$'$	0.92 b$'$	8.70 b$'$	42.63 b$'$
		14	9.63 D	3.21 C	0.81 B	7.82 B	26.06 B
		21	10.00 C$'$	1.73 A$'$	0.59 B$'$	5.89 B$'$	11.48 A$'$
Vacuum	Control	0	7.32 c	2.37 a,b	0.41 a,b	2.83 b	7.89 b
		7	7.63 c$'$	2.04 a$'$	0.61 a$'$,b$'$	4.61 b$'$	10.12 a$'$,b$'$
		14	3.92 A	3.60 C	0.72 B	2.80 A,B	9.46 A
		21	18.63 D$'$	1.00 A$'$	0.29 A$'$	5.34 A$'$,B$'$	8.31 A$'$
	5%	0	1.25 a	0.98 a	0.17 a	2.50 a	1.21 a
		7	4.42 a$'$	3.81 b$'$	0.77 b$'$	3.41 a$'$,b$'$	14.10 a,b
		14	4.31 A	1.66 A	0.27 A	1.12 A	5.57 A
		21	5.25 A$'$	3.02 B$'$	0.77 C$'$	4.07 A$'$	12.83 A$'$
	10%	0	3.21 b	2.13 a,b	0.46 a,b	1.59 a,b	6.24 b
		7	4.68 a$'$,b$'$	1.49 a$'$	0.34 a$'$	1.54 a$'$	4.96 a$'$
		14	3.55 A	2.61 B	0.68 B	2.39 A	6.53 A
		21	6.55 B$'$	4.10 C$'$	0.64 B$'$	4.18 A$'$	17.11 A$'$
	15%	0	2.43 b	3.39 a,b	0.64 a,b	1.53 a,b	6.97 b
		7	5.48 b$'$	4.02 b$'$	0.90 b$'$	4.88 b$'$	19.87 b$'$
		14	7.40 B	3.92 C	0.76 B	5.60 B	23.38 A,B
		21	5.32 A$'$	4.51 D$'$	0.77 C$'$	4.04 A$'$	18.28 A$'$
	20%	0	2.55 b	4.45 b	0.83 b	2.13 a,b	9.75 b
		7	10.00 d$'$	3.00 a$'$,b$'$	0.72 a$'$,b$'$	7.09 c$'$	21.23 b$'$
		14	11.76 C	4.43 D	0.76 B	8.99 C	40.08 B
		21	10.69 C$'$	2.16 A$'$B$'$	0.67 B$'$	7.12 B$'$	14.67 A$'$

Values in the same column marked with the same symbols mean no statistically significant differences (α = 0.05).

Comparing the method of storage of the tested variants, it was noticed that the samples packed in the vacuum (except for the bread with the 20% addition of chestnut flour) after 7 days were characterized by a lower crumb hardness compared to the samples packed with air. After 14 days from baking, the control sample and the vacuum-packed bread with the 10% chestnut flour had a lower crumb hardness compared to the samples packed with air, and 21 days after baking, the vacuum-packed breads with 5, 10 and 20% chestnut flour had a lower crumb hardness. harder crumb than air-packed bread.

3.6.2. Elasticity

Elasticity is defined as the quotient of the specimen deformation that occurs during the first and second compression cycles. It characterizes the degree of the shape recovery by the sample [48].

The lowest crumb elasticity of gluten-free bread packed in a barrier foil with air at the zero point was found in bread with the 5% addition of chestnut flour (0.98) (Table 4). As the amount of the addition of chestnut flour in the recipe increased, the elasticity of the bread crumb increased. The highest value of elasticity was characteristic for the bread with the 20% addition of chestnut flour (4.45). No statistically significant differences were observed between the control sample and the breads with the 10% and 15% addition of chestnut flour.

Seven days after baking, crumb elasticity increased significantly in all tested variants, and no statistically significant differences were observed between the samples. After 14 days of storage, a significant decrease in the crumb elasticity was observed compared to the values measured on the 7th day after baking. The lowest values of elasticity were found in the control sample (1.37) and the bread with the 10% addition of chestnut flour (1.70). There were no statistically significant differences between the breads with the 5% and 20% addition of chestnut flour. After 21 days of storage, a decrease in crumb elasticity was noted in the case of bread with the 5% and 20% addition of chestnut flour. No statistically significant differences were observed between the control sample, the breads with 5% and 20% addition of chestnut flour and between the bread with the 10% and 15% addition of chestnut flour. Similar relationships were observed in the studies by Marciniak-Lukasiak and Skrzypacz [52], where the addition of amaranth flour was used.

After 7 days of storage, a decrease in the crumb elasticity of the gluten-free breads packed in a barrier foil with a vacuum was observed in the case of the control and the breads with the 10% and 20% addition of chestnut flour (Table 4). In the case of the breads with the 5% and 15% addition of chestnut flour, an increase in crumb elasticity was noted. After 14 days from baking, the crumb elasticity of the control sample and the breads with the 10% and 20% addition of chestnut flour increased in comparison to the values measured after 7 days of storage. No statistically significant differences were observed for the control sample and for the bread with the 10% addition of chestnut flour. After 21 days of storage, a significant decrease in the crumb elasticity was noted in the control sample and in the bread with the 20% addition of chestnut flour.

Comparing the method of storage of the tested variants, it was noticed that the vacuum-packed samples after 7 days of storage were characterized by lower crumb elasticity values compared to the samples packed with air. However, such a tendency was not noticed 14 and 21 days after baking. After 14 days of storage, a lower value of the crumb elasticity was observed only in the case of bread with the 5% addition of chestnut flour, vacuum-packed, and after 21 days from baking, while the control sample and the breads with the 10% and 15% addition of chestnut flour showed lower values of crumb elasticity compared to the samples packed with air.

3.6.3. Cohesiveness

Cohesiveness (cohesiveness) characterizes the total strength of the internal bonds that hold the product together [53].

The values of the crumb cohesiveness parameter of the gluten-free bread packed with air at the zero point ranged from 0.17 to 0.83 (Table 4). On the basis of the obtained results, it was found that with the increase in the amount of the addition of chestnut flour in the recipe, the cohesiveness increases. However, the cohesiveness of the bread with the 5% addition of chestnut flour (0.17) was lower compared to the control sample (0.41). There were no statistically significant differences between the control sample and the breads with the 10% and 15% addition of chestnut flour.

After 7 days from baking, the value of the cohesiveness parameter increased. It was observed that, the larger the proportion of chestnut flour in the recipe, the smaller were the differences in the cohesiveness values. These results differ from the results obtained by Gambuś [51]. The authors observed in their research that the storage time causes a decrease in the cohesiveness. After 14 days from baking, there was a decrease in the cohesiveness in all the analyzed samples compared to their value after 7 days of storage. No statistically significant differences were observed between the control sample and the bread with the 10% addition of chestnut flour. Twenty-one days after baking, there was a decrease in cohesiveness in breads with the 5% and 20% addition of chestnut flour compared to their value after 14 days after baking.

In the case of the gluten-free breads packed in a barrier foil with a vacuum with the 10% and 20% addition of chestnut flour, lower cohesiveness values were obtained (0.34 for 10%; 0.72 for 20%) compared to the cohesiveness at the zero point (0.46 for 10%; 0.83 for 20%). The results obtained for these two variants confirm the observations made by Marciniak-Lukasiak and Skrzypacz [42], which, in their research, showed that with an increasing storage time, the cohesiveness decreases. However, it cannot be concluded in this study that the storage time caused a decrease in the cohesiveness. In the case of the control sample and the breads with the 5% and 15% addition of chestnut flour, an increase in the cohesiveness was observed 7 days after baking. After 14 days of storage, a decrease in the cohesion value for the breads with the 5 and 15% addition of chestnut flour was observed, while after 21 days of storage, a decrease was noted for the control sample and for the breads with the 10 and 20% addition of chestnut flour.

Comparing the method of storage of the tested variants, it was noticed that the vacuum-packed samples after 7 days of storage were characterized by lower crumb cohesiveness values compared to the samples packed with air. However, such a relationship was not observed with increasing the bread storage time. After 14 days from baking, the cohesiveness of the vacuum-packed variants was lower only in the case of the breads with the 5% and 20% addition of chestnut flour, and after 21 days of storage in the case of the control sample and the bread with the 10% addition of chestnut flour.

3.6.4. Gumminess

Gumminess is the quotient of the hardness and cohesion of the bread crumb [53].

The crumb chewiness of the gluten-free breads packed with air access at the zero point was from 0.25 to 2.83 (Table 4). Bread with the 5% addition of chestnut flour was characterized by the lowest gumminess, and the control sample was the highest. All variants with the addition of chestnut flour reduced the crumb gumminess compared to the control sample. No statistically significant differences were found between the gumminess of the bread crumb with the 10, 15 and 20% addition of chestnut flour.

After 7 days of storage, an increase in the gumminess was observed in all analyzed samples. No statistically significant differences were observed between the control sample and the breads with the 5, 10 and 15% addition of chestnut flour. Pajak [54] observed a decrease in the value of this parameter in the research on the impact of packaging on the quality of stored gluten-free bread. After 14 days from baking, there was a decrease in the gumminess in all analyzed samples compared to their values tested on the 7th day of storage. As in the samples tested after 7 days, no statistically significant differences were observed between the control sample and the breads with the 5, 10 and 15% addition of chestnut flour. On the 21st day after baking, a sharp increase in the gumminess was noted in the control sample (10.90). A decrease in the value of this parameter was observed in the case of the breads with the 5% and 20% addition of chestnut flour.

As with the air-wrapped bread, after 7 days of storage, the gumminess of the gluten-free bread wrapped in a barrier film with the vacuum increased, except for the bread with 10% chestnut flour. The zero point gumminess was 1.59, and after 7 days it was 1.54

(Table 4). Similar relationships were observed in the case of the cohesiveness value of this variant. After 14 days of storage, a decrease in gumminess was noted in the case of the control sample and the bread with the 5% addition of chestnut flour, while 21 days after baking, the gumminess of the control sample and the breads with the 5% and 10% addition of chestnut flour increased in comparison to the 14th day of storage.

Comparing the results obtained during the storage of the tested variants, it was noticed that the vacuum-packed samples, after 7 days of storage, were characterized by lower crumb gumminess values compared to the samples packed with air. After 14 days from baking, the gumminess of the vacuum-packed variants was lower in the control sample and the breads with the 5% and 20% addition of chestnut flour, and after 21 days of storage in the control sample and the bread with the 15% addition of chestnut flour.

3.6.5. Chewiness

Chewiness is the product of hardness, elasticity and cohesiveness. It characterizes the strength needed to chew a bite of food so that it is ready to be swallowed [53]

The chewiness of the crumb of gluten-free bread packed with air access at the zero point was from 1.21 N to 9.75 N (Table 4). On the basis of the obtained results, it was found that, with the increase in the amount of chestnut flour added to the recipe, the chewiness of the crumb grows. The lowest chewiness value was characteristic for the bread with the 5% addition of chestnut flour, and the highest value for the bread with the 20% addition of chestnut flour. The 5, 10 and 15% addition of chestnut flour in the gluten-free breads reduced the chewiness of their crumb. Pajak [54] observed in their research that the value of the chewing parameter of the crumb of gluten-free bread decreases during storage. After 14 days from baking, lower chewiness was noted in each of the tested variants. Both after 7 and 14 days of storage, the bread with the greatest addition of chestnut flour was characterized by the greatest chewiness of the crumb. After 21 days, a decrease in chewiness was observed only for the breads with the 5% and 20% addition of chestnut flour, as compared with its values on the 14th day after baking.

As with air-packed breads, after 7 days of storage, the chewiness of the gluten-free breads wrapped in a barrier film with a vacuum increased, with the exception of the bread with 10% chestnut flour (Table 4). The chewiness at the zero point was 6.24 N, and after 7 days, it was 4.96 N. Such relations were observed for the cohesiveness and gumminess values of this variant. After 14 days of storage, a decrease in the chewiness was noted for the control sample and the bread with the 5% addition of chestnut flour, while 21 days after baking, the chewiness of the breads with the 5% and 10% addition of chestnut flour increased compared to the 14th day of storage.

Comparing the method of storage of the tested variants, it was noticed that the vacuum-packed samples after 7 days of storage were characterized by lower crumb chewiness values compared to the samples packed with air. This relationship was not observed with an increasing bread storage time. After 14 days from baking, the chewiness of the vacuum-packed variants was lower in the case of the breads with the 5 and 10% addition of chestnut flour, and after 21 days of storage in the case of the control sample and the breads with the 10 and 15% addition of chestnut flour.

3.7. Sensoric Analysis

One of the discriminants subjected to the sensory assessment was taste (Figure 5). The assessed tastes are bready, salty, sour, yeast and foreign. A total of 100 UU was assumed as the highest intensity, and 0 for the lowest.

The bread with a 5% addition of chestnut flour (74.3 IU) was the most bready, i.e., the typical taste, and the second in terms of this feature was the bread with the 10% addition of chestnut flour (72.0 IU). The least bready taste (58.0), and at the same time having foreign aftertastes (13.8), which the evaluators described as slightly nutty, was the bread with the 20% addition of chestnut flour. According to the evaluators, all the breads with the addition of chestnut flour had a slightly more perceptible sour taste (4.0 IU) compared to the control sample (3.7 IU). The amount of the addition of chestnut flour used did not have a statistically significant effect on the taste perception of the added yeast.

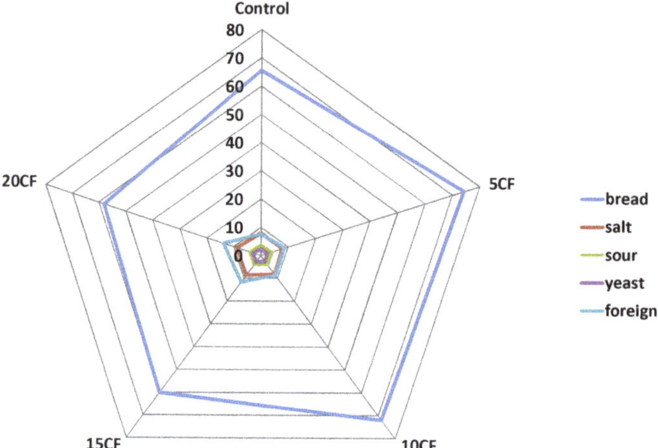

Figure 5. The taste of gluten-free bread with the addition of chestnut flour in the amount of 5% (5CF), 10% (10CF), 15% (15CF) and 20% (20CF).

Another distinguishing feature during the sensory evaluation was smell (Figure 6). The scents assessed were: bread-like, the smell of added yeast and the presence of foreign smells. In terms of the bread flavor, the best were the breads with the 5% and 10% addition of chestnut flour (76.0 u.u.). These breads were also rated the highest in terms of the typical bread flavor. Bread with a 20% addition of chestnut flour (58.3 u.u.) and the control sample (59.0 u.u.) had the least bread smell. According to the evaluators, these two variants were also characterized by an intense smell of added yeast (67.3 u.u. in the control sample and 63.7 u.u. in bread with 20% addition of chestnut flour). The presence of foreign smells was particularly significant in the case of the bread with the highest share of chestnut flour (45.3 u.u.), which the evaluators described as nutty, while in the control sample, the evaluators experienced slightly sour scent notes (16.0 u.u.).

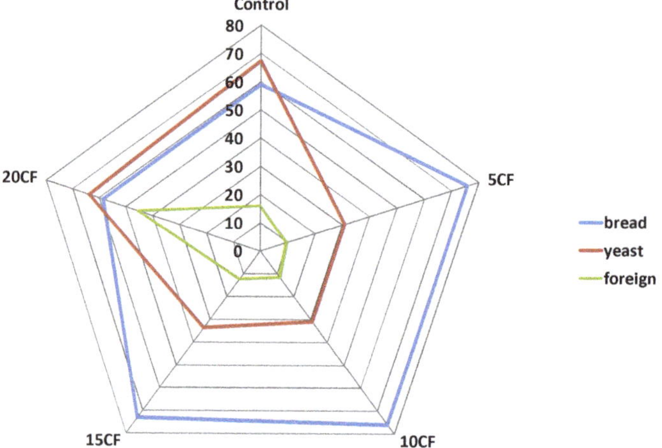

Figure 6. The smell of gluten-free bread with the addition of chestnut flour in the amount of 5% (5CF), 10% (10CF), 15% (15CF) and 20% (20CF).

The structure and texture of the crumb of the baked gluten-free bread were also assessed (Figure 7). The hardest (86.5 u.u.), and at the same time, the most fragile (81.3 u.u.) and the most compact (86.3 u.u.) crumb was characteristic of the control sample. The highest hardness of this bread was also confirmed in an instrumental TPA test. Similar values were obtained for the test with the 20% addition of chestnut flour—a hardness of 81.2 u.u., a brittleness of 73.2 u.u. and a compactness of 81.0 u.u. The lowest scores of all three features of the structure and texture were recorded for the bread with the 5% addition of chestnut flour (hardness 22.3 u.u.; brittleness 31.7 u.u.; compactness 23.3 u.u.), which was also confirmed in the TPA test. No significant differences were observed in the case of the bread with the 10% addition of chestnut flour, where the brittleness was assessed at the level of 33.3 u.u., the hardness of 23.0 u.u. and the compactness of 25.7 u.u.

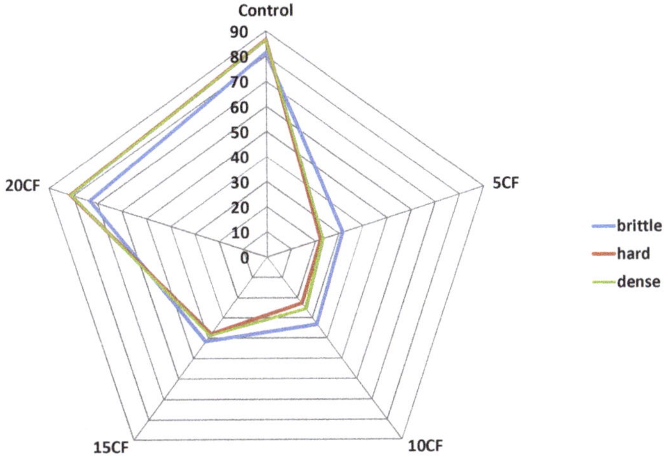

Figure 7. The structure and texture of the crumb of gluten-free breads with the addition of chestnut flour in the amount of 5% (5CF), 10% (10CF), 15% (15CF) and 20% (20CF).

The last of the assessed characteristics of baked gluten-free bread was the assessment of consumer desirability, which is presented in Figure 8.

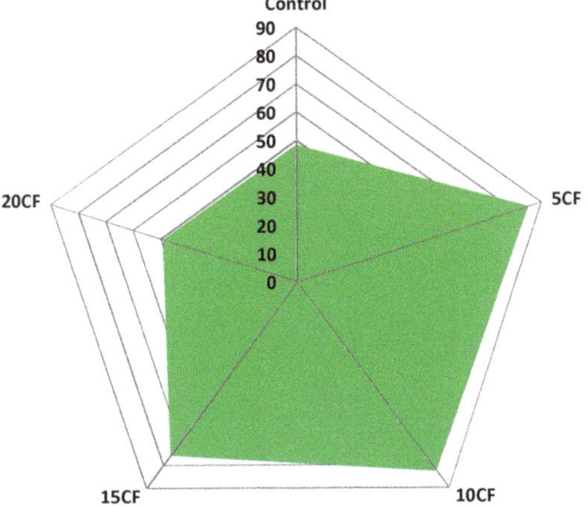

Figure 8. Consumer assessment for gluten-free breads with the addition of chestnut flour in the amount of 5% (5CF), 10% (10CF), 15% (15CF) and 20% (20CF).

Among the examined breads, the most desired by consumers was the bread with the 5% addition of chestnut flour (85.3 u.u.) and the bread with the 10% addition of chestnut flour (82.7 u.u.). The desirability of the bread with the 15% addition of chestnut flour was estimated at 75.7 u.u. In the evaluation of the gluten-free breads obtained, the control sample (48.3 u.u.) and the bread with the 20% addition of chestnut flour (49.2 u.u.) showed the lowest consumer desires.

3.8. Microbiological Quality

Microbiological quality of bread samples with the chestnut flour addition differed significantly depending on the method and the time of storage (Table 5). Immediately after baking, no microorganisms were detected in the bread samples; however, after 7, 14 and 21 days of storage, 3.47–5.87 log cfu/g of mesophilic aerobic microorganisms, as well as 2.11–5.47 log cfu/g of yeast and molds were denoted. Moreover, in the samples containing 20% of chestnut flour, *Bacillus* spp. was detected, but only after 7, 14 and 21 days of storage. Interestingly, the microbiological quality of the samples in vacuum storage was better than the aerobic conditions, which was seen in the TCV and Y&M counts.

Table 5. TVC, Y&M, LAB and *Bacillus* spp. of gluten-free bread with the addition of chestnut flour in the amount of 0, 5, 10, 15 and 20%.

	Chestnut Flour Addition	TVC		Y&M		LAB		*Bacillus* spp.	
		Air	Vacuum	Air	Vacuum	Air	Vacuum	Air	Vacuum
0	0%	<10	<10	<10	<10	<10	<10	Nd	nd
	5%	<10	<10	<10	<10	<10	<10	Nd	nd
	10%	<10	<10	<10	<10	<10	<10	Nd	nd
	15%	<10	<10	<10	<10	<10	<10	Nd	nd
	20%	<10	<10	<10	<10	<10	<10	Nd	nd
7	0%	3.53 ± 0.21 a	2.98 ± 0.15 b	2.11 ± 0.06 a	1.60 ± 0.09 a	<10	<10	Nd	nd
	5%	3.59 ± 0.16 a	2.47 ± 0.23 a	2.92 ± 0.02 b	2.17 ± 0.06 b	<10	<10	Nd	nd
	10%	3.48 ± 0.11 a	2.76 ± 0.20 ab	2.82 ± 0.08 b	2.60 ± 0.09 c	<10	<10	Nd	nd
	15%	3.78 ± 0.14 a	2.95 ± 0.14 b	3.48 ± 0.13 c	2.25 ± 0.15 b	<10	<10	Nd	nd
	20%	3.47 ± 0.09 a	3.25 ± 0.05 c	3.78 ± 0.11 c	2.92 ± 0.13 c	<10	<10	+	+
14	0%	4.58 ± 0.23 a'	3.14 ± 0.07 a'	5.33 ± 0.11 c'	3.60 ± 0.11 b'	<10	<10	Nd	nd
	5%	5.29 ± 0.12 c'	3.42 ± 0.20 b'	4.01 ± 0.08 a'	3.69 ± 0.13 b'	<10	<10	Nd	nd
	10%	4.60 ± 0.06 a'	3.10 ± 0.05 a'	4.95 ± 0.05 b'	3.30 ± 0.09 a'	<10	<10	Nd	nd
	15%	5.01 ± 0.08 b'	3.47 ± 0.09 b'	4.92 ± 0.08 b'	3.90 ± 0.04 c'	<10	<10	Nd	nd
	20%	4.60 ± 0.14 a'	3.60 ± 0.11 b'	5.47 ± 0.07 c'	3.98 ± 0.01 c'	<10	<10	+	+
21	0%	5.65 ± 0.14 C	3.11 ± 0.13 A	4.18 ± 0.12 A	3.91 ± 0.08 C	<10	<10	Nd	nd
	5%	5.87 ± 0.09 C	3.36 ± 0.07 A	4.05 ± 0.11 A	3.86 ± 0.09 C	<10	<10	Nd	nd
	10%	4.27 ± 0.21 A	3.94 ± 0.11 B	3.85 ± 0.09 A	2.89 ± 0.11 A	<10	<10	Nd	nd
	15%	4.81 ± 0.22 B	3.98 ± 0.12 B	3.90 ± 0.04 A	3.42 ± 0.14 B	<10	<10	Nd	nd
	20%	4.90 ± 0.09 B	4.33 ± 0.08 C	4.43 ± 0.12 B	4.32 ± 0.22 D	<10	<10	+	+

Explanatory: TVC—total viable counts, Y&M—total yeast and mold counts, BAC—*Bacillus* spp.; (+)—the presence of *Bacillus* spp. in 10 g of product. Values in the same column marked with the same symbols mean no statistically significant differences ($\alpha = 0.05$).

Many bakery products can be spoiled by different microorganisms, including fungi and bacteria. The fungal contamination is quite common in raw bakery materials; however, it is not considered as the most critical issue, as the life cells of microorganisms can be destroyed by the baking temperature. On the other hand, the postbaking contamination (air, product handling, equipment sanitized) seems to be an important issue in terms of the microbiological safety of bread [55]. For example, in the Morassi study [55], the raw materials of wheat flour and cornmeal exhibited the fungal counts of approx. 10^2 log cfu/g in 60% of the tested samples. However, in the other studies, the fungal counts have been reported as higher, at approx. 10^5 log cfu/g in similar samples [56,57]

Among bacteria contamination, *Bacillus* spp. is claimed as the main agent responsible for a spoilage process known as ropiness or rope. Rope spoilage occurring due to *Bacillus* bacteria may produce spores and therefore survive during the thermal processing of bakery products. Rope takes place when counts of *Bacillus* spores reach 10^3 spores/g. Once spores withstand baking, they may further germinate and grow [58]. A good solution for getting rid of *Bacillus* spp. can be the use of sourdough or natural substances like LAB-based bioingredients, which can lower the pH of dough and therefore decrease the thermal resistance of *Bacillus* cells and help in the inactivation of the microorganism through heat treatment [59]. In our study, we do not use sourdough in bread production, as the LAB count was < 10 log cfu/g in all samples during the whole storage period.

From a microbiological point of view, the storage conditions (vacuum or air) used as the association to control the microbiological risk and stability of the bread samples had the biggest influence, and therefore we recommend vacuum as the better solution.

3.9. PCA Analysis

A principal component analysis (Figure 9) of the results of the evaluated bread samples showed that the sample variation corresponds to the first main component (Factor 1), which accounted for 54.97% of the total variability and was related mainly to chewiness, elasticity, cohesiveness, crumb features, browning index, foreign and yeast smell as well as gumminess. The second component (Factor 2) constituted 37.87% of the general variable and was related mainly to tastes (yeast, sour, salt and foreign). Based on the obtained eigenvalues, the analysis can be limited to two factors explaining 92.84% of the total variability (Figure 9a).

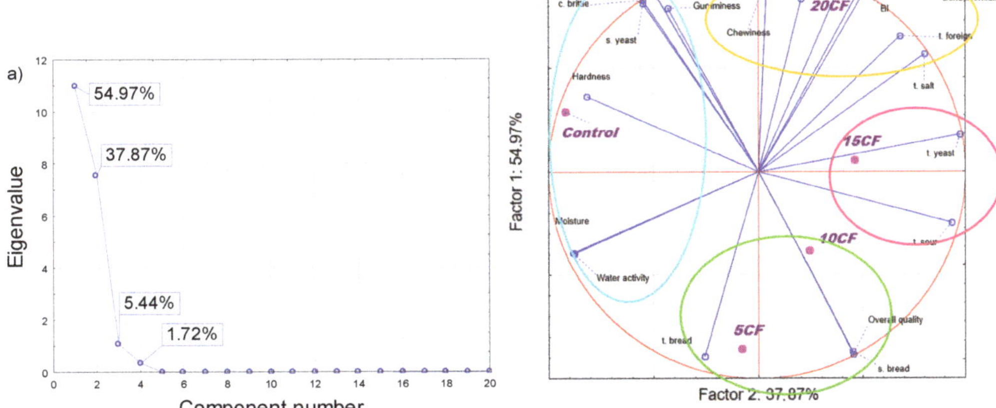

Figure 9. Principal component analysis (PCA) of the following samples: Control—control sample, gluten-free breads with the addition of chestnut flour in the amount of 5% (5CF), 10% (10CF), 15% (15CF) and 20% (20CF). BI—browning index, 'c.' stands for crumb, 's.' stands for smell, t. stands for taste. Eigenvalues for each individual principal component is presented in (**a**), but (**b**) presents importance of chosen factors.

The PCA results (Figure 9b) showed that the analyzed samples can be clustered into four distinctive groups. One of them consists of the control sample. The second cluster contains the breads with the 5% and 10% addition of chestnut flour. The next one contains the bread with addition of 15% of chestnut flour, and the last cluster consists of the bread sample with the 20% of chestnut flour addition.

The obtained clusters distinguish the samples with different amounts of chestnut flour additions. The control samples were characterized by the highest level of hardness, moisture and water activity. Additionally, the crumb was brittle, hard and a yeast smell was recognized. One can notice that control samples were accompanied by a higher water content in comparison with samples with the chestnut flour addition.

The addition of chestnut flour in amounts of 5% and 10% showed better organoleptic features than the other samples, such as smell and taste. The bread smell and taste were the most intense in such samples. As a result, both samples were recognized as the most demanding breads among the analyzed ones.

The increase of the addition of chestnut flour up to 15% caused more yeast and sour taste in the analyzed samples. The higher addition of chestnut flour resulted in a faster browning index, and a foreign taste and smell appeared. The bread samples were also characterized by higher level of chewiness, elasticity and cohesiveness.

Based on the provided analysis, one can conclude that the bread samples with different levels of the chestnut flour addition were characterized by different factors. The addition of the chestnut flour in small amounts decreased the level of the browning index, foreign taste and smell, and the level of the moisture. It also increased the feeling of the bread taste and smell in comparison to the control sample. Unfortunately, further increasing the amount of the chestnut flour addition did not improve the level of consumer demands. Samples with chestnut flour additions in the amount of 5 and 10% were better noticed by consumers than the other samples. The only problem is with the elasticity and crumb parameters, because the samples with the best quality are far from the expected values. One can conclude that the obtained results suggested that the addition of chestnut flour in amounts between 5 and 10% have promising organoleptic and sensory features for gluten-free breads.

4. Conclusions

Among the tested samples, the best physicochemical parameters were characteristic for breads with the 5% and 10% addition of chestnut flour. The least acceptable parameters, and at the same time the least acceptable quality, were noticed in the bread with the 20% addition of chestnut flour. The addition of chestnut flour in an amount up to 10% to baking gluten-free bread causes an increase in the volume of the tested breads.

The highest porosity of the crumb was characteristic for the bread with the 10% addition of chestnut flour, which also meets the requirements specified in the standards for traditional wheat bread. Moisture analysis showed that the tested gluten-free breads turn stale in a similar way. After 7 days of storage, the humidity of the crumb decreased for all analyzed variants packed with air and vacuum. The water activity for all tested loaves was characterized by the highest values at point zero. The TPA texture analysis showed that the addition of the chestnut flour to the recipe reduced the hardness of the gluten-free bread crumb. As the amount of the added chestnut flour increases, the elasticity and cohesiveness of the bread crumb increases. Bread with the 5% addition of chestnut flour was characterized by the lowest gumminess and chewiness.

As a main finding of the conducted research, we observed that the addition of chestnut flour to the recipe affects significantly ($p < 0.05$) the texture of the finished product, reducing the hardness and increasing the elasticity and cohesiveness of the bread crumb.

It is worth noticing that the use of chestnut flour in an amount of up to 10% increased significantly ($p < 0.05$) the volume of the resulting loaves.

Microbiological research has indicated vacuum packaging as a better way to protect and store gluten-free bread.

The sensory evaluation showed that the bread with the 5% addition of chestnut flour had the best taste and smell. This variant also obtained the highest marks in the consumer desirability survey.

For practical use in future production, it is recommended to replace corn starch in gluten-free breads by no more than 10% by chestnut flour.

Author Contributions: Conceptualization, K.M.-L.; methodology, K.M.-L. and D.Z.; validation, K.M.-L., A.Z., M.S., D.Z. and P.L. (Piotr Lukasiak).; formal analysis, P.L. (Patrycja Lesniewska) and D.Z.; resources, K.M.-L., M.S. and K.Z. data curation, K.M.-L., D.Z., P.L. (Piotr Lukasiak) and A.Z.; writing—original draft preparation, K.M.-L.; writing—review and editing, K.M-L., P.L. (Patrycja Lesniewska), M.S., K.Z., D.Z., P.L. (Piotr Lukasiak) and A.Z.; visualization, K.M.-L. and D.Z.; supervision, K.M-L.;

statistical analysis, P.L. (Piotr Lukasiak). All authors have read and agreed to the published version of the manuscript.

Funding: This research was funded by Warsaw University of Life Sciences.

Institutional Review Board Statement: Not applicable.

Informed Consent Statement: Not applicable.

Data Availability Statement: Not applicable.

Conflicts of Interest: The authors declare no conflict of interest.

References

1. Lohi, S.; Mustalahti, K.; Kaukinen, K.; Laurilla, K.; Collin, P.; Rissanen, H.; Lohi, O.; Bravi, E.; Gasparin, M.; Reunanen, A. Increasing prevalence of coeliac disease over time. *Aliment. Pharmacol. Ther.* **2007**, *26*, 1217–1225. [CrossRef] [PubMed]
2. Caio, G.; Volta, U.; Sapone, A.; Leffler, D.A.; Giorgio, R.; Catassi, C.; Fasano, A. Celiac disease: A comprehensive current review. *BMC Med.* **2019**, *17*, 142. [CrossRef] [PubMed]
3. Culetu, A.; Susman, I.E.; Duta, D.E.; Belc, N. Nutritional and functional properties of gluten-free flours. *Appl. Sci.* **2021**, *11*, 6283. [CrossRef]
4. Trynka, G.; Wijmenga, C.; Heel, D.A. A genetic perspective on coeliac disease. *Trends Mol. Med.* **2010**, *16*, 537–550. [CrossRef]
5. Bardella, M.T.; Velio, P.; Cesana, B.M.; Prampolini, L.; Casella, G.; Bella, C.; Lanzini, A.; Gambarotti, M.; Bassotti, G.; Villanacci, V. Coeliac disease: A histological follow-up study. *Histopathology* **2007**, *50*, 465–471. [CrossRef]
6. Catassi, C.; Verdu, E.F.; Bai, J.C.; Lionett, E. Coeliac disease. *Lancet*, **2022**; *in press*. [CrossRef]
7. Kaliciak, I.; Drogowski, K.; Garczyk, A.; Kopeć, S.; Horwat, P.; Bogdański, P.; Stelmach-Mdras, M.; Mardas, M. Influence of Gluten-Free Diet on Gut Microbiota Composition in Patients with Coeliac Disease: A Systematic Review. *Nutrients* **2022**, *14*, 2083. [CrossRef]
8. Ruszkowska, M.; Kropisz, P. Charakterystyka pieczywa bezglutenowego wytworzonego z mąk niekonwencjonalnych. *Inżynieria I Apar. Chem.* **2017**, *56*, 180–181.
9. Pastuszka, D.; Gambuś, H.; Sikora, M. Wartość odżywcza i dietetyczna pieczywa bezglutenowego z dodatkiem nasion lnu oleistego. *Żywność Nauka Technol. Jakość* **2012**, *3*, 155–167.
10. Kaim, U.; Harasym, J. Produkcja chelba bezglutenowego wyzwaniem dla współczesnego piekarnictwa. *Nauk. Inżynierskie I Technol.* **2017**, *4*, 41–54.
11. Ziemichód, A.; Różyło, R. Wpływ dodatku jagód goji na właściwości fizyczne chleba bezglutenowego. *Acta Ahrophysica* **2018**, *25*, 117–127. [CrossRef]
12. Codex Standard 118-1979. Available online: http://www.fao.org/fao-who-codexalimentarius/sh-proxy/en/?lnk=1&url=https%253A%252F%252Fworkspace.fao.org%252Fsites%252Fcodex%252FStandards%252FCXS%2B118-1979%252FCXS_118e_2015.pdf (accessed on 17 June 2022).
13. Alvarez-Jubete, L.; Arendt, E.K.; Gallagher, E. Nutritive value of pseudocereals and their increasing use as functional gluten-free ingredients. *Trends Food Sci. Technol.* **2010**, *21*, 106–113. [CrossRef]
14. Culetu, A.; Duta, D.E.; Papageorgiou, M.; Varzakas, T. The Role of Hydrocolloids in Gluten-Free Bread and Pasta; Rheology, Characteristics, Staling and Glycemic Index. *Foods* **2021**, *10*, 3121. [CrossRef]
15. Dłużewska, E.; Marciniak-Lukasiak, K.; Dojczew, D. Koncentraty chleba bezglutenowego z dodatkiem wybranych hydrokoloidów. *Żywność. Nauka. Technologia. Jakość* **2001**, *2*, 57–67.
16. Dłużewska, E.; Marciniak-Lukasiak, K.; Kurek, N. Effect of transglutaminase additive on the quality of gluten-free bread. *CyTA-J. Food* **2015**, *13*, 80–86. [CrossRef]
17. Moradi, M.; Bolandi, M.; Arabameri, M.; Karimi, M.; Baghaei, H.; Nahidi, F.; Eslami Kanafi, M. Semi-volume gluten-free bread: Effect of guar gum, sodium caseinate and transglutaminase enzyme on the quality parameters. *J. Food Meas. Charact.* **2021**, *15*, 2344–2351. [CrossRef]
18. Lynch, K.M.; Coffey, A.; Arendt, E.K. Exopolysaccharide producing lactic acid bacteria: Their techno-functional role and potential application in gluten-free bread products. *Food Res. Int.* **2018**, *110*, 52–61. [CrossRef]
19. Zys, A.; Garncarek, Z. Wykorzystanie transglutaminazy mikrobiologicznej do wytwarzania produktów bezglutenowych. *Pr. Nauk. Uniw. Ekon. We Wrocławiu* **2018**, *542*, 213–223. [CrossRef]
20. Cacak-Pietrzak, G.; Ceglińska, A.; Zaorska, A.; Rajkowska, A. Ocena jakości handlowego pieczywa bezglutenowego. *Przegląd Zbożowo-Młynarski* **2019**, *2*, 40–45.
21. Kupiec, M.; Zbikowska, A.; Marciniak-Lukasiak, K.; Zbikowska, K.; Kowalska, M.; Kowalska, H.; Rutkowska, J. Study on the introduction of solid fat with a high content of unsaturated fatty acids to gluten-free muffins as a basis for designing food with higher health value. *Int. J. Mol. Sci.* **2021**, *22*, 9220. [CrossRef]
22. Marciniak-Lukasiak, K.; Dłużewska, E. Pseudozboża w żywności bezglutenowej. *Przemysł Spożywczy* **2018**, *72*, 46–50. [CrossRef]
23. Paciulli, M.; Rinaldi, M.; Cavazza, A.; Ganino, T.; Rodolfi, M.; Chiancone, B.; Chiavaro, E. Effect of chestnut flour supplementation on physico-chemical properties and oxidative stability of gluten-free biscuits during storage. *LWT* **2018**, *98*, 451–457. [CrossRef]

24. Brochard, M.; Correia, P.; Barroca, M.J.; Guiné, R.P.F. Development of a New Pasta Product by the Incorporation of Chestnut Flour and Bee Pollen. *Appl. Sci.* **2021**, *11*, 6617. [CrossRef]
25. Montemurro, M.; Pontonio, E.; Rizzello, C.G. Design of a "clean-label" gluten-free bread to meet consumers demand. *Foods* **2021**, *10*, 462. [CrossRef]
26. Ribeiro, B.; Rangel, J.; Valentão, P.; Andrade, P.B.; Pereira, J.A.; Bölke, H.; Seabra, R.M. Organic acids in two Portuguese chestnut (Castanea sativa Miller) varieties. *Food Chem.* **2007**, *100*, 504–508. [CrossRef]
27. Sacchetti, G.; Pittia, P.; Mastrocola, D.; Pinnavaia, G.G. Stability and quality of traditional and innovative chestnut based products. *Acta Hortic.* **2005**, *693*, 63–70. [CrossRef]
28. Zhu, F. Properties and food uses of chestnut flour and starch. *Food Bioprocess Technol.* **2017**, *10*, 1173–1191. [CrossRef]
29. Masure, H.G.; Fierens, E.; Delcour, J.A. Current and forward looking experimental approaches in gluten-free bread making research. *J. Cereal Sci.* **2016**, *67*, 92–111. [CrossRef]
30. Grand View Research. Gluten-Free Products Market Size, Share & Trends Analysis Report by Product (Bakery Products, Dairy/Dairy Alternatives), By Distribution Channel (Grocery Stores, Mass Merchandiser), By Region, And Segment Forecasts, 2020–2027. 2020. Available online: https://www.grandviewresearch.com/industry-analysis/gluten-free-products-market (accessed on 17 July 2022).
31. AACC International. Methods 10-05 and 44-15A. In *Approved Methods of the American Association of Cereal Chemists*, 10th ed.; The Association: St. Paul, MN, USA, 2000.
32. Lazaridou, A.; Duta, D.; Papageorgiou, M.; Belc, N.; Biliaderis, C.G. Effects of hydrocolloids on dough rheology and bread quality parameters in gluten-free formulations. *J. Food Eng.* **2007**, *79*, 1033–1047. [CrossRef]
33. Papadakis, S.E.; Abdul-Malek, S.; Kamdem, R.E.; Yam, K.L. A versatile and inexpensive technique for measuring color foods. *Food Technol.* **2000**, *54*, 48–51.
34. *ISO 4833-2:2013*; Microbiology of the Food Chain—Horizontal Method for the Enumeration of Microorganisms—Part 2: Colony Count at 30 Degrees C by the Surface Plating Technique. International Organization for Standarization: Geneva, Switzerland, 2013.
35. *ISO 21527-1:2008*; Microbiology of Food and Animal Feeding Stuffs—Horizontal Method for the Enumeration of Yeasts and Moulds—Part 1: Colony Count Technique in Products with Water Activity Greater than 0.95. International Organization for Standarization: Geneva, Switzerland, 2008.
36. *ISO 21527-2:2008*; Microbiology of Food and Animal Feeding Stuffs—Horizontal Method for the Enumeration of Yeasts and Moulds—Part 2: Colony Count Technique in Products with Water Activity Less Than or Equal to 0.95. International Organization for Standarization: Geneva, Switzerland, 2008.
37. *ISO 21871:2006*; Microbiology of Food and Animal Feeding Stuffs—Horizontal Method for the Determination of Low Numbers of Presumptive Bacillus Cereus—Most Probable Number Technique and Detection Method. International Organization for Standarization: Geneva, Switzerland, 2006.
38. *PN-A-74123*; Diet Products. Gluten-Free Bread. Polish Standard: Warsaw, Poland, 1997.
39. Aguilar, N.; Albanell, E.; Miñarro, B.; Capellas, M. Chestnut flour sourdough for gluten-free bread making. *Eur. Food Res. Technol.* **2016**, *242*, 1795–1802. [CrossRef]
40. Demirkesen, I.; Mert, B.; Sumnu, G.; Sahin, S. Utilization of chestnut flour in gluten-free bread formulations. *J. Food Eng.* **2010**, *101*, 329–336. [CrossRef]
41. Ruszkowska, M. Pieczywo bezglutenowe—ocena trwałości z zastosowaniem metod sorpcyjnych. *Inżynieria I Apar. Chem.* **2014**, *53*, 110–112.
42. Marciniak-Lukasiak, K.; Skrzypacz, M. Koncentrat chleba bezglutenowego z dodatkiem mąki z szarłatu. *Żywność Nauka Technol. Jakość* **2008**, *4*, 131–140.
43. Rinaldi, M.; Paciulli, M.; Dall'Asta, C.; Cirlini, M.; Chiavaro, E. Short-term storage evaluation of quality and antioxidant capacity in chestnut–wheat bread. *J. Sci. Food Agric.* **2014**, *95*, 59–65. [CrossRef]
44. Maskan, M. Kinetics of colour change of kiwifruits during hot air and microwave drying. *J. Od Food Eng.* **2001**, *48*, 169–175. [CrossRef]
45. Fik, M. Czerstwienie pieczywa i sposoby przedłużania jego świeżości. *Żywność Nauka Technol. Jakość* **2004**, *2*, 5–22.
46. Demirkesen, I.; Campanella, O.H.; Sumnu, G.; Sahin, S.; Hamaker, B.R. A study on staling characteristics of gluten-free breads prepared with chestnut and rice flours. *Food Bioprocess Technol.* **2014**, *7*, 806–820. [CrossRef]
47. Pałacha, Z.; Makarewicz, M. Aktywność wody wybranych grup spożywczych. *Postępy Tech. I Przetwórstwa Spożywczego* **2011**, *2*, 24–29.
48. Giannone, V.; Lauroc, M.R.; Spinad, A.; Pasqualonee, A.; Auditoref, L.; Puglisig, I.; Puglisia, G. A novel α-amylase-lipase formulation as anti-staling agent in durum wheat bread. *LWT-Food Sci. Technol.* **2016**, *65*, 381–389. [CrossRef]
49. Michniewicz, J. Niektóre aspekty doboru opakowań jednostkowych do pakowania żywności. *Opakowanie* **2001**, *9*, 16–20.
50. Borowy, T.; Kubiak, M.S. Tekstura wyróżnikiem jakości produktów piekarskich i cukierniczych. *Cukiernictwo I Piekarstwo* **2010**, *7–8*, 33–36.
51. Gambuś, H.; Gambuś, F.; Sabat, R. Próby poprawy jakości chleba bezglutenowego przez dodatek mąki z szarłatu. *Żywność Nauka Technol. Jakość* **2002**, *31*, 99–112.

52. Kulczak, M.; Błasińska, J.; Słowik, E. Wybrane cechy fizyczne chleba bezglutenowego z udziałem preparowanej mąki grochowej i przetworów gryczanych. *Acta Agrophysica* **2014**, *21*, 445–455.
53. Szczesniak, A.S. Classification of textural characteristics a. *J. Food Sci.* **1963**, *28*, 385–389. [CrossRef]
54. Pająk, P.; Kuczera, D.; Fortuna, T. Wypływ opakowania na jakość przechowywanego pieczywa bezglutenowego. *Acta Agrophysica* **2013**, *20*, 633–649.
55. Morassi, L.L.; Bernardi, A.O.; Amaral, A.L.; Chaves, R.D.; Santos, J.L.; Copetti, M.V.; Sant'Ana, A.S. Fungi in cake production chain: Occurrence and evaluation of growth potential in different cake formulations during storage. *Food Res. Int.* **2018**, *106*, 141–148. [CrossRef]
56. Berghofer, L.K.; Hocking, A.D.; Miskelly, D.; Jansson, E. Microbiology of wheat and flour milling in Australia. *Int. J. Food Microbiol.* **2003**, *85*, 137–149. [CrossRef]
57. Dos Santos, J.L.P.; Bernardi, A.O.; Morassi, L.L.P.; Silva, B.S.; Copetti, M.V.; Sant'Ana, A.S. Incidence, populations and diversity of fungi from raw materials, final products and air of processing environment of multigrain whole meal bread. *Food Res. Int.* **2016**, *87*, 103–108. [CrossRef]
58. Pereira, A.P.M.; Stradiotto, G.C.; Freire, L.; Alvarenga, V.O.; Crucello, A.; Morassi, L.L.; Sant'Ana, A.S. Occurrence and enumeration of rope-producing spore forming bacteria in flour and their spoilage potential in different bread formulations. *LWT* **2020**, *133*, 110108. [CrossRef]
59. Park, H.W.; Yoon, W.B. A quantitative microbiological exposure assessment model for *Bacillus* cereus in pasteurized rice cakes using computational fluid dynamics and Monte Carlo simulation. *Food Res. Int.* **2019**, *125*, 108562. [CrossRef]

applied
sciences

MDPI

Article

Comparative Evaluation of the Antioxidative and Antimicrobial Nutritive Properties and Potential Bioaccessibility of Plant Seeds and Algae Rich in Protein and Polyphenolic Compounds

Joanna Miedzianka [1,*], Sabina Lachowicz-Wiśniewska [2,3], Agnieszka Nemś [1], Przemysław Łukasz Kowalczewski [4,*] and Agnieszka Kita [1]

[1] Department of Food Storage and Technology, Wroclaw University of Environmental and Life Sciences, 37 Chełmońskiego Street, 51-630 Wrocław, Poland
[2] Department of Health Sciences, Calisia University, 4 Nowy Świat Street, 62-800 Kalisz, Poland
[3] Department of Horticulture, West Pomeranian University of Technology Szczecin, 17 Słowackiego Street, 71-434 Szczecin, Poland
[4] Department of Food Technology of Plant Origin, Poznań University of Life Sciences, 31 Wojska Polskiego Street, 60-624 Poznań, Poland
[*] Correspondence: joanna.miedzianka@upwr.edu.pl (J.M.); przemyslaw.kowalczewski@up.poznan.pl (P.Ł.K.)

Citation: Miedzianka, J.; Lachowicz-Wiśniewska, S.; Nemś, A.; Kowalczewski, P.Ł.; Kita, A. Comparative Evaluation of the Antioxidative and Antimicrobial Nutritive Properties and Potential Bioaccessibility of Plant Seeds and Algae Rich in Protein and Polyphenolic Compounds. *Appl. Sci.* **2022**, *12*, 8136. https://doi.org/10.3390/app12168136

Academic Editor: Anna Lante

Received: 20 July 2022
Accepted: 12 August 2022
Published: 14 August 2022

Publisher's Note: MDPI stays neutral with regard to jurisdictional claims in published maps and institutional affiliations.

Abstract: Spice plants are not only a source of nutrition compounds but also supply secondary plant metabolites, such as polyphenols. Therefore, their bioaccessibility is an important issue. In order to understand the biological activity of polyphenols present in spice plants, it is necessary to broaden knowledge about the factors influencing their bioaccessibility, including nutritional factors. Therefore, the objective of this research was to determine the antioxidative and antimicrobial nutritive properties and potential bioaccessibility of plant seeds and microalgae rich in protein and polyphenolic compounds. Plant seeds rich in protein—i.e., black cumin, milk thistle, fenugreek, almonds, white sesame, white mustard, eggfruit and the two most popular algae, chlorella and spirulina—were analyzed for total polyphenolic compounds (TPC) and antioxidant properties (ABTS, FRAP), as well as their potential bioaccessibility, antimicrobial activity, basic chemical composition and amino acid profiles. With regard to the TPC, the highest levels were found in star anise, followed by milk thistle, white mustard and fenugreek, whereas the lowest were noted in white sesame, almonds, eggfruit, spirulina and chlorella. White mustard and milk thistle showed the highest antioxidant capacities and almonds, eggfruit, spirulina, and chlorella the lowest according to the ABTS and FRAP assays. The widest spectrum of microbial growth inhibition was detected for fenugreek extract, which showed antimicrobial activity against four analyzed microorganisms: *B. subtilis*, *P. mirabilis*, *V. harveyi* and *C. albicans*. The protein from seeds of black cumin, milk thistle, white mustard and eggfruit and chlorella was not limited by any essential amino acids. Among all analyzed plants, fenugreek seeds were judged to have potential for use in food formulation operations in view of their antioxidant activity and amino acid profile. Based on the results, intake of polyphenols together with protein in plant seeds does not have a major impact on the potential bioaccessibility of a range of polyphenols and phenolic metabolites.

Keywords: plant species; spirulina; chlorella; bioactive compounds; antiradical activity; ferric reducing antioxidative power; potential bioaccessibility; amino acid profile

1. Introduction

According to the International Organization for Standardization [1], spices are natural plant products used to improve the flavor, aroma, taste and color of food products that may also have additional properties; e.g., antioxidant or bacteriostatic properties. There are about 109 spices grown in different parts of the world, and India is known as the "Land of Spices". Spices are derived from different parts of plants: seeds (e.g., cardamom), leaves (e.g., bay leaf), flowers buds (e.g., clove), fruits (e.g., pepper), bark (e.g., cinnamon) and rhizome (e.g., ginger). A wide variety of spice plants provide edible seeds. Seeds are the dominant source of human calories and protein [2,3]. Seed protein is a composite of hundreds of different enzymes and structural proteins; however, its protein complement is dominated by a family of storage proteins [4]. The most important and popular seed food sources are cereals, followed by legumes and nuts. Spice plants are used in beverages, liquors and pharmaceutical, cosmetic and perfumery products. Some algae, such as spirulina and chlorella, are also used as color and flavor additives. They can be found in many applications around the world; e.g., powdered algae are used in a wide range of products, such as pasta, soups and sauces [5,6]. Nutritionists at the Food and Nutrition Institute have placed spice plants in a new nutrition pyramid along with updated guidance on the proper behavior and nutrition required to stay healthy. Spice plants, due to their aromatic and taste properties, can help diversify people's diets and, most importantly of all, limit salt consumption [7].

In terms of nutrition, spice plants contain variable amounts of protein, fat and carbohydrates and small quantities of vitamins and inorganic elements. Moreover, they supply secondary plant metabolites, including glucosides, saponins, tannins, alkaloids, essential oils, organic acids and others, that possess medicinal, antioxidant and antimicrobial properties [8,9]. The bioaccessibility of polyphenols—i.e., the extent to which they can be released and absorbed in the digestive tract and used by the body—is an important issue. The affinity and binding of proteins from different sources to phenolic compounds is a widely studied phenomenon [10]. Whether this interaction also impacts the bioaccessibility of polyphenols is still a matter of debate. The content and relative bioaccessibility of antioxidant components from plant raw materials and products (various types of nuts, cereal grains, groats, lupine seeds, etc.) depend on many factors, including the type of raw material used (species, variety, origin), the degree of purification and the type of technological or culinary processing, as well as the composition of the food product [3,11]. Compounds contained in the plant matrix may limit the absorption of antioxidant ingredients in the gastrointestinal tract; for instance, fiber (e.g., hemicellulose), bivalent elements and sticky and protein-rich meals may limit the bioaccessibility of polyphenols, while easily digestible carbohydrates, fats (especially for hydrophobic polyphenols; e.g., curcumin) and antioxidants may increase the availability of these compounds [12]. In order to understand the biological activity of polyphenols, it is necessary to broaden knowledge about the factors influencing their bioaccessibility, including nutritional factors.

The antimicrobial activity of spice plants has been an object of interest for scientists since before the 20th century. Since that time, numerous different plants have been tested for antibacterial and antifungal properties, and many of them have been found to have activities against microorganisms. However, the major antimicrobial components of spices are in their essential oils. The majority of the antimicrobial components of spices are phenol compounds with a hydroxyl group (-OH) [13,14]. The most active are carvacrol, eugenol and thymol [15,16]. Some of these compounds are also toxic to animals, but others may not be toxic. Indeed, many of these compounds have been used in the form of whole plants or plant extracts for food or medical applications for humans [14]. Additionally, plants and their extracts have important potential as manipulators of rumen fermentation, providing productivity and health benefits [17]. They have specific effects on members of the rumen microflora and fauna that can be beneficial for animal productivity and health [14].

Recently, there has been a shift in the food industry towards natural compounds. Consumers are increasingly looking for products that meet their requirements not only in terms of taste but also with regard to having a varied composition and beneficial nutritional value and functional. In response to the increase in consumer knowledge about the impact of food on health and well-being, producers are being forced to look for new solutions in food production. In highly developed countries, the market response to such a problem is the production of food with special health values, referred to as functional food. These are foods that, apart from nutrients, also contain compounds that have a beneficial effect on health, development and well-being [18]. These substances include dietary fiber, oligosaccharides, alcohol derivatives of sugars, amino acids, peptides, proteins, glycosides, alcohols, isoprenoids, vitamins, choline compounds, lactic acid bacteria, minerals and unsaturated fatty acids, as well as antioxidants and phytochemicals. Food referred to as functional must influence processes that significantly reduce the risk of developing certain diseases, the modulation of which enables improvements in health. Consumer expectations prompted our research aimed at determining the antimicrobial and antioxidative properties and amino acid profiles of proteins in selected spice plants and algae widely used in the kitchen. Since the ability to synthesize amino acids in human organisms is limited, it is necessary to analyze seed plants, and in the case of vegetarians and vegans, plants and plant products are the only sources of these valuable constituents indispensable for the normal functioning of the organism.

In the literature, much information on the antimicrobial and antioxidant activities of spice plants is available, but data on the potential bioaccessibility of their phenolic compounds and antioxidant properties are scarce. It is known that the activity of polyphenols may, however, be hampered when consumed together with protein-rich food products due to the interaction between polyphenols and proteins [19,20]. Therefore, the objective of this research was to determine the antiradical capacities, reducing power, antimicrobial properties and amino acid profiles of selected spice plants and microalgae with a view to investigating their potential as food supplements and food additives for use in various systems. Additionally, an experiment was performed to simulate gastrointestinal digestion in an in vitro model in order to determine the potential bioaccessibility of the polyphenols and antioxidant properties in the plants.

2. Materials and Methods

2.1. Plant Samples

Plant seeds rich in protein—namely, black cumin (*Nigella sativa*), milk thistle (*Silybum marianum*), fenugreek (*Trigonella foenum-graecum* L.), almond (*Prunus dulcis*), white sesame (*Sesamum indicum* L.), white mustard (*Sinapis alba*), star anise (*Illicium verum* Hook. f.), powdered biomass from eggfruit (*Pouteria lucuma* spp.)—as well as selected algae—spirulina (powdered cyanobacteria biomass *Arthospira plantensis* spp.) and chlorella (powdered biomass from *Chlorella vulgaris* spp.)—were purchased from the Market Hall in Wrocław (Lower Silesia, Poland). Before analysis, samples (except algae) were milled using a GM 200 mill (Retsch GmbH, Haan, Germany). All plant samples were stored in a freezer at $-18\,^\circ$C till further analysis.

2.2. Polyphenolic Compounds and Antioxidant Properties

2.2.1. Extraction Procedure

The material (~1 g) was mixed with 10 mL of 80% MeOH in deionized water with 1% HCl. Then, samples were sonicated twice (800 W, 40 Hz, Sonic 6D, Polsonic, Warsaw, Poland) for 20 min at 20 $^\circ$C and left for 24 h at 4 $^\circ$C. After this procedure, the extract was centrifuged for 10 min at $19,000\times g$, and the supernatant was filtered through a hydrophilic PTFE with a 0.20 µm membrane (Millex Samplicity Filter, Merck, Darmstadt, Germany) and used for tests.

2.2.2. Total Polyphenolic Compounds

Total polyphenols were measured using the Folin–Ciocalteu method [21]. Briefly, the extract (100 µL) was mixed with distilled water (2000 µL), Folin–Ciocalteu phenol reagent (200 µL) and 20% sodium carbonate solution in water (1000 µL). The sample was incubated for 1 h in the dark at 20 $^\circ$C. The absorbance was measured at 765 nm (UV-2401 PC, Shimadzu Corp., Kyoto, Japan). Total polyphenolics are presented in mg of gallic acid equivalents (GAE)/g dry matter (DM).

2.2.3. Antioxidant Activity

The procedures for the determination of antiradical activity, using 2,2′-azino-bis(3-ethylbenzothiazoline-6-sulfonic acid) diammonium salt (ABTS) and ferric reducing antioxidative power (FRAP) tests, have been described by Re et al. [22] and Benzie and Strain [23], respectively. Briefly, 30 µL of the sample was mixed with 3 mL of ABTS reagent, and 100 µL of the sample was mixed with 3 mL of FRAP reagent. After 6 and 10 min of reaction, respectively, the absorbance was measured at 734 nm for ABTS and 593 nm for FRAP (UV-2401 PC spectrophotometer; Shimadzu, Kyoto, Japan). The antioxidant potency is presented in µmol of Trolox/g DM.

2.3. Potential Bioaccessibility

In vitro digestion was carried out as described previously by Minekus et al. [24], with the modifications described by Lachowicz et al. [25]. For gastrointestinal digestion, the spice plants and algae were mixed with PBS buffer (pH 7.4) and simulated salivary fluid. After that, the material was shaken at 37 °C. The pH of the sample was changed to pH 3, and the mixture was shaken at 37 °C. Then, the pH of digests was changed to 7. The mixtures underwent intestinal digestion in vitro for 120 min at 37 °C. After in vitro digestion, the samples were centrifuged at 19,000× g per 10 min and used for the polyphenolic compound and antioxidant activity analyses.

2.4. Microorganism and Culture Conditions
2.4.1. Preparation

Evaluation of the antimicrobial properties of the analyzed seeds and algae was undertaken for the following strains of microorganisms: (1) the Gram-positive bacteria *Bacillus subtilis* (ATCC 6633), *Staphylococcus aureus* (ATCC 9538), *Enterococcus hirae* (ATCC 10542) and *Enterococcus faecalis* (ATCC 29212); (2) the Gram-negative bacteria *Pseudomonas aeruginosa* (ATCC 15442), *Escherichia coli* (ATCC 10536), *Proteus mirabilis* (ATCC 2011) and *Vibrio harveyi* (ATCC 12126); and (3) the yeast *Candida albicans* (ATCC 10231). All the microorganisms were obtained from the culture collection of the Department of Biotechnology and Food Microbiology (WUELS, Wrocław, Poland). Bacteria were grown in Luria broth (LB) (Sigma-Aldrich, Poznań, Poland; containing 5 g/L yeast extract 10 g/L tryptone and 5 g/L NaCl) at 37 °C; yeast was grown in yeast extract peptone dextrose (YPD) broth (Sigma-Aldrich, Poznań, Poland; containing 20 g/L bacteriological peptone, 10 g/L yeast extract and 20 g/L glucose) at 30 °C. Agar, at a concentration of 2%, was added to the medium when required. To prepare an inoculation culture for the agar diffusion assay, the microorganisms were grown in a 0.1 L flask containing 10 mL of proper medium on a rotary shaker at 37 °C or 30 °C and 180 rpm for 48 h. The cells was washed in sterile distilled water and adjusted to OD600 = 1.

2.4.2. Extraction Procedure

Extracts of plants were prepared by mixing 1 g of sample with 10 mL of ethanol and stirring with a vortex for 2 h at 25 °C. Then, the solution was centrifuged at 6000× g for 5 min and the supernatants were sterilized using the filtration method with a 0.22 µm syringe filter. The filtrates were kept in sterilized vials in refrigeration conditions.

2.4.3. Determination of Antimicrobial Properties with Agar Diffusion Assay

Antimicrobial activities of miscellaneous spice ethanol extracts were tested with the agar diffusion assay (the well diffusion assay). On each plate, which contained the proper (LB or YPD) medium, three wells were produced. Wells on vaccinated agar plates (200 µL of bacterial or yeast standardized overnight to OD600 = 1) were produced with a sterile Pasteur pipette (Ø 8.4 mm diameter). Two were filled with 100 µL of the tested extract, and the third contained 100 µL of ethanol as a control sample. Initially, the plates were incubated for 6 h at 4 °C to achieve full diffusion of the solution tested in the agar medium.

Then, the plates were incubated at 30 °C or 37 °C for 24 h and the diameters of the resulting inhibited growth around each well were measured. All tests were carried out in triplicate.

2.5. Chemical Composition

The dry matter (DM), total protein, fat and ash content were evaluated according to the methods described by the Association of Analytical Chemists [26]. The moisture content of the analyzed samples was determined on the basis of weight loss during thermal drying at 105 °C until a constant weight was achieved. Total nitrogen was determined with the Kjeldahl method using a Büchi Distillation Unit K-355 (Athens, Greece). A nitrogen-to-protein conversion factor of 6.25 was used as a standard [27]. Fat content was determined by means of the Soxhlet method in a Büchi B-811 apparatus (Flawil, Switzerland) with the use of diethyl ether after hydrolysis of the sample with 4N HCl. The total ash content was determined by adding 1 g of the protein preparation to a crucible, incinerating it in a muffle furnace at 550 °C and determining the weight of the residue. Total carbohydrates were calculated using a difference calculation (100 − the sum of protein, fat, ash and moisture). Data are reported as mean values ± standard deviation (SD) for two measurements.

2.6. Amino Acid Composition

The analyzed samples were previously degreased with the use of Soxhlet's apparatus. Then, the materials were acid hydrolyzed [28,29] and placed in an AAA400 automatic amino acid analyzer (INGOS s.r.o., Prague, Czech Republic). A two-wavelength photometer (440 and 570 nm) was employed as a detector. The length of the column packed with ion exchanger Ostion LG ANB (INGOS) was 350×3.7 mm, and the column temperature was maintained at 40–70 °C and the detector temperature at 121 °C. The amino acids were quantified using the ninhydrin method. Glutamine and asparagine was expressed as glutamic acid and aspartic acid, respectively. No analysis was carried out for tryptophan. Calculations were performed with the computer program Chromulan (Pikron s.r.o., Prague, Czech Republic). All amino acid profiles were analyzed in duplicate.

2.7. Scoring of Amino Acids

The amino acid score (AAS) was calculated for adults using the standard method recommended by the FAO/WHO [30]:

$$AAS = \frac{\text{essential amino acids contents}}{\text{recommended essential amino acids}}(\%)$$

2.8. Statistics

The data are presented as means ± standard deviation. The results were compared using the one-way Duncan's multiple range test (ANOVA). Statistica version 13.3 (Dell Software Inc., Round Rock, TX, USA) was used for the statistical analyses. Differences were considered significant when $p \leq 0.05$.

3. Results and Discussion

3.1. Polyphenolic Compounds, Antioxidant Activity and Potential Bioaccessibility

The results for total phenolics content (TPC) and for the potential bioaccessibility of polyphenolic compounds after in vitro digestion are presented in Figure 1A. The TPC was statistically significant for the chosen plant seeds and algae (Figure 1A). The TPC ranged from 29.44 to 1635 mg GAE/g DM. In order, TPC was highest in star anise, followed by milk thistle, white mustard and fenugreek, and the lowest content was noted in white sesame, almonds, eggfruit, spirulina and chlorella. The TPC analyzed in all plant seeds and algae was similar to data from previous studies [31–34]. The TPC in fenugreek was around give times higher, and that in white sesame seeds was around eight times lower, than the results obtained by Mashkor [35] and Lin et al. [36]. Generally, differences in the results for TPC were likely due to genotypic and environmental differences, such as temperature, climate, location, pest exposure, diseases within analyzed species and determination methods [37].

Moreover, each plant seed and algae could generally contain different mixtures of polyphenols, which influenced the TPC [38]. After the in vitro analysis in a simulated digestive system, the highest potential bioaccessibility for the TPC among the tested plant seeds and algae was noted in white sesame, almonds, spirulina and chlorella. The TPC determined in fenugreek, white mustard, black cumin and eggfruit was also potentially bioaccessible, as their values were above 1 [25]. Among all the analyzed materials, star anise and milk thistle presented the lowest potential bioaccessibility for TPC after in vitro digestion. Thus, the best plant seeds for the production of new functional food were found to be white sesame, almond, spirulina and chlorella. However, fenugreek, white mustard, black cumin and eggfruit can also be added to new nutraceutical food, but their potential bioaccessibility indexes will depend on various factors; e.g., the type of food, the enriched matrix food, the interaction of polyphenolic compounds with their chemical compounds and behavior of the ingredients in the digestive system [39]. The differences in the obtained values for the potential bioaccessibility of polyphenolic compounds may have resulted from the quality of the polyphenolic compounds. It has been previously found that the process of digestion itself can enhance the release of bioactive compounds, especially polyphenols [40–42], whereas the low relative bioaccessibility for the ingredients could suggest interactions with other components of the research material, such as proteins and/or enzymes in the digestive system [39]. The formation of protein–phenolic complexes can significantly affect protein hydrophobicity, structure and solubility, as well as the isoelectric point and thermal stability. Additionally, the combination of phenolic compounds with proteins may affect the blocking of various amino acid residues. Their complexation can also impact the bioaccessibility and antioxidant activity of phenolics [43].

The results for antioxidant activities (ABTS and FRAP assays), as well as the potential bioaccessibility of these activities after in vitro digestion, are presented in Figure 1B,C, respectively. The antioxidant capacity of the analyzed plant seeds and algae ranged from 0.01 (white sesame) to 1.79 µmol/g DM (star anise) in the ABTS assay (Figure 1B) and from 0.24 (fenugreek) to 71.75 µmol/g DM (star anise) in the FRAP assay (Figure 1C). Among the plants and algae, star anise, milk thistle and white mustard seeds showed the highest antioxidant capacity, whereas almond, spirulina and white sesame had the lowest in both tests. The high antioxidant activity in star anise, milk thistle and white mustard seeds was likely due to their having the highest TPC, as a positive linear correlation between TPC and antioxidant capacity in plant seeds and algae was observed. Additionally, these plants contained high amino acid and total proteins contents. The exception was eggfruit, which had the lowest concentrations of total proteins and low TPC. The antioxidant activities (ABTS and FRAP assays) measured in the milk thistle and fenugreek were similar to the results presented by Wojdyło et al. [34]. The antioxidant activity measured in the other samples was similar to the results from other authors [31–33]. According to the bioaccessibility index calculated for the analyzed materials, the antiradical ingredients, especially in fenugreek, white sesame and eggfruit, and the reducing compounds in white sesame, chlorella, almond and spirulina were highly bioaccessible. These plants with high relative accessibility indexes for polyphenols and antioxidant ingredients could be good functional additions for phenolics and protein-rich foods. The lowest potential bioaccessibility (below 1) for antioxidant activity was noted in the milk thistle and star anise. The low relative accessibility index for polyphenols could have resulted from complexes with proteins, limiting their absorption and also their ability to exert antioxidant activity [43]. Antioxidant properties mainly depend on the levels of bioactive compounds, including polyphenolic compounds, but also on the interactions with other compounds in the research material, such as proteins. The bioaccessibility of polyphenols depends on several factors, which include their release from the food matrix, their molecular size, their hydrophilic/lipophilic balance as related to their glycosylation and their different pH-dependent transformations (degradation, hydrolysis, epimerization and oxidation within the gastrointestinal tract), as well as their solubility and the interactions between polyphenols and food components [44]. In contrast, in research conducted on soy and dairy drinks, the bioaccessibility of polyphenols was not significantly affected when polyphenols were consumed in protein-rich products [19]. Therefore, the interactions of proteins with polyphenols and the formation of complexes are important factors influencing the functional quality of products and their pro-health properties and should be taken into account during the design of functional foods [43].

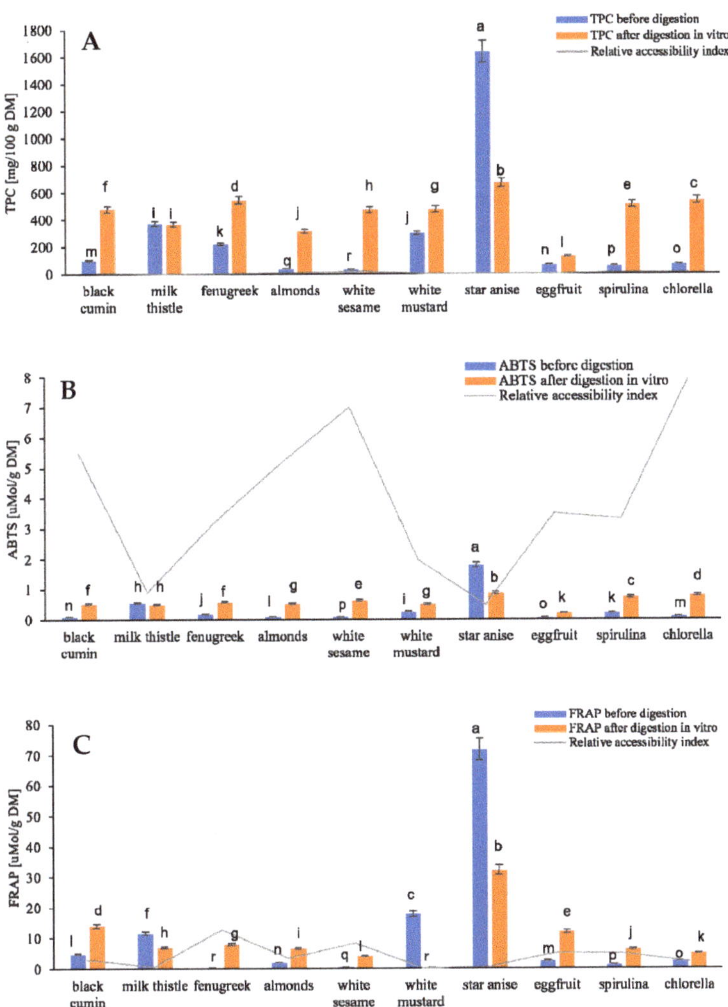

Figure 1. Total polyphenols content (**A**), antioxidant activity (ABTS+ (**B**); FRAP (**C**)) and their potential bioaccessibility in the analyzed plant materials.

Antioxidant activity tests are commonly used to assess the redox potential of compounds. Thus, the Person correlations between the antioxidant results (from the ABTS and FRAP assays) and the TPC were 0.992 and 0.986, respectively. As in the study by Wojdyło et al. [34] a strong correlation between antioxidant capacity and TPC in the analyzed plants was noted.

3.2. Antimicrobial Activity

The zones of inhibition of the growth of microorganisms in the analyzed plant materials are shown in Table 1. This study found that various bacteria and yeast were sensitive to several plant seed and algae ethanol extracts. Milk thistle extract was found to show activity against both Gram-positive (*B. subtilis*, *E. faecalis*) and Gram-negative bacteria (*P. mirabilis*). Growth of *E. faecalis* was also inhibited by cold-pressed milk thistle seed flours [45]. Food-borne *E. faecalis* is connected with common dental diseases and is able to colonize oral biofilm [46]. Accordingly, milk thistle extract may have beneficial properties as a food additive. Almond ethanol extract showed activity against *B. subtilis* and *V. harveyi*. The presented results expand the spectrum of plant seeds, with the ethanol extracts of fenugreek, white sesame, star anise and, as mentioned before, milk thistle, as well as almond, showing anti-*Bacillus* activity. Star anise extract achieved the highest ratings for inhibition of *B. subtilis* growth—approximately 9 mm. In 2009, there was a report about a novel toxin produced by *B. subtilis* strains isolated from foods that could be related to food poisoning outbreaks [47]. *V. harveyi* is the prevalent bacterial pathogen in farmed shrimp [48]. Activity against this Gram-negative bacteria was also shown by fenugreek, white sesame and eggfruit ethanol extracts. The World Health Organization identified *P. aeruginosa* as a quality indicator for drinking water [49]. Effective growth inhibition of *P. aeruginosa* was observed for white mustard seed extract. Previous studies have shown similar activity in extracts from basil, clove, cumin, dill, garlic and thyme [2]. In August 2008, *P. mirabilis* was identified as responsible for an outbreak of food poisoning in Beijing [50]. Eggfruit ethanol extracts may be useful to reduce the growth *P. mirabilis*—the extract revealed significant inhibition zones (4.18 mm). Activity against this Gram-negative bacteria was also shown by milk thistle, fenugreek, white mustard seed and star anise extracts. Previous research only pointed to anti-*Proteus* activity in garlic [2]. Fenugreek and white sesame extracts indicated minor antifungal activities against *C. albicans*. Likewise, *C. albicans* has been found to be sensitive to garlic and clove extracts [51]. Regarding the current and previous results, there is a need to evaluate the usefulness of spices in terms of their antimicrobial activities against bacteria and fungi. Among the selected plant seeds and algae, black cumin, spirulina and chlorella showed no inhibition of the chosen microorganisms. However, cumin oil has demonstrated antimicrobial and antifungal activities in laboratory tests against Gram-positive and Gram-negative bacteria species, proving to be effective against the genera *Clavibacter*, *Curtobacterium*, *Rhodococcus*, *Erwinia*, *Xanthomonas*, *Ralstonia* and *Agrobacterium* isolated from clinical isolates and foods [8], as well as against the fungi *Penicillium notatum*, *Aspergillus niger*, *A. fumigatus*, *Microsporum canis* [8]. The strong inhibition of the growth of microorganisms by the plant materials, especially milk thistle and fenugreek, resulted from the strong positive correlation between TPC and antimicrobial activity against bacteria and yeast (the Person correlation was higher than 0.6). This could be correlated with active constituents, such as isomeric flavonolignans—namely, silibinin (silybin), silychristin and silidianin, collectively known as silymarin, which is extracted from dried milk thistle seeds [52,53], and catechin, epicatechin, gallic acid, coumaric acid, cinnamic acid and vanillic acid, found in fenugreek ethanolic extract [35].

Table 1. Zones of inhibition (mm) of the growth of microorganisms in the analyzed plant materials.

Seeds and Algae	Gram-Positive Bacteria (G+)				Gram-Negative Bacteria (G−)				Yeast
	B. subtilis	*S. aureus*	*E. hirae*	*E. faecalis*	*P. aeruginosa*	*E. coli*	*P. mirabilis*	*V. harveyi*	*C. albicans*
Black cumin	-	-	-	-	-	-	-	-	-
Milk thistle	1.56 ± 0.02	-	-	6.83 ± 0.63	-	-	3.85 ± 0.76	-	-
Fenugreek	3.05 ± 0.64	-	-	-	-	-	3.75 ± 0.68	3.73 ± 0.53	0.28 ± 0.06
Almond	3.42 ± 0.77	-	-	-	-	-	-	2.89 ± 0.49	-
White sesame	2.98 ± 0.01	-	-	-	-	-	-	0.53 ± 0.01	1.25 ± 0.01
White mustard	-	-	-	-	6.95 ± 1.38	-	3.09 ± 0.01	-	-
Star anise	8.89 ± 0.03	-	-	-	-	-	2.71 ± 0.01	-	-
Eggfruit	-	-	-	-	-	-	4.18 ± 0.26	3.05 ± 0.50	-
Spirulina	-	-	-	-	-	-	-	-	-
Chlorella	-	-	-	-	-	-	-	-	-

-, Not identified (trace or % < 0.1).

3.3. Proximate Composition

Results regarding the chemical composition of the tested seeds and algae are presented in Table 2. The tested plant materials showed similar levels of humidity not exceeding 11%. Dry matter content in the seeds ranged from 91.17% (fenugreek) to 96.38% (white sesame). Across all analyzed seed samples, the protein content ranged from 16.33 g/100 g (milk thistle) to 24.58 g/100 g (almond). However, its content was almost three times lower compared to the studied algae, where it ranged from 55.40 g/100 g (chlorella) to 69.35 g/100 g (spirulina). The results of our study show the same range of protein content as in the report by Gonçalvesa et al. [51], and Barreca et al. [54] noted that almond presents similar overall nutrient profiles even when comparing different varieties, years of production and growing regions. Furthermore, the protein content in milk thistle seeds in this study was lower than the data obtained by El-Haak et al. [55]. This varied protein content may be related to seed maturity stage and environmental conditions. The key ingredients plants need for protein production are glucose and nitrates, which are taken up from the soil by the roots. When glucose and nitrates are joined, they produce amino acids. During protein synthesis, multiple amino acids are bound together to make proteins [56]. The fat content in the analyzed samples differed, ranging between 5.13 g/100 g (fenugreek) and 52.25 g/100 g (almond).

Table 2. Chemical compositions (g/100 g) of the analyzed plant materials.

Raw Material	Dry Matter	Total Protein	Fat	Ash	Carbohydrates
Black cumin	94.43 ± 0.08 [d]	20.39 ± 1.22 [d]	36.85 ± 1.14 [c]	4.54 ± 0.02 [e]	32.65 ± 3.11 [d]
Milk thistle	92.83 ± 0.02 [f]	16.33 ± 0.08 [e]	23.47 ± 0.87 [d]	6.51 ± 0.01 [c]	46.53 ± 4.52 [c]
Fenugreek	91.17 ± 0.03 [g]	23.70 ± 1.05 [c]	5.13 ± 0.22 [f]	3.47 ± 0.03 [f]	58.87 ± 3.85 [b]
Almonds	95.58 ± 1.02 [b]	24.58 ± 0.07 [c]	52.25 ± 4.27 [b]	4.81 ± 0.02 [e]	13.94 ± 1.75 [f]
White sesame	96.38 [a] ± 0.07 [a]	24.06 ± 1.01 [c]	51.02 ± 1.89 [b]	5.12 ± 0.02 [d]	16.18 ± 2.13 [e]
White mustard	95.08 ± 0.02 [c]	20.89 ± 1.52 [d]	36.24 ± 0.96 [b]	3.72 ± 0.01 [f]	34.23 ± 4.74 [d]
Star anise	89.19 ± 1.58 [h]	5.92 ± 0.18 [f]	68.75 ± 3.69 [a]	3.11 ± 0.03 [f]	11.41 ± 1.85 [f]
Eggfruit	92.75 ± 2.35 [f]	1.72 ± 0.98 [g]	2.23 ± 0.87 [g]	0.52 ± 0.01 [g]	88.28 ± 9.55 [a]
Spirulina	94.37 ± 3.12 [d]	69.35 ± 4.05 [a]	7.52 ± 1.45 [e]	11.47 ± 0.05 [b]	6.03 ± 0.15 [g]
Chlorella	94.07 ± 2.01 [e]	55.40 ± 3.21 [b]	8.15 ± 1.74 [e]	14.13 ± 0.07 [a]	16.39 ± 1.45 [e]

Values are means $\pm$ SD of two determinations; a–h, the same letters in columns mean homogenous groups.

The chosen plant seeds and algae were characterized by very diverse carbohydrate contents, ranging from 13.94 g/100 g (almond) to 88.28 g/100 g (eggfruit). Similar carbohydrate contents were observed by Kochhar et al. [56], who analyzed fenugreek seeds. Moreover, based on data presented by Lunn and Buttriss [57], presoluble carbohydrates constitute a major part of the total carbohydrate content, whereas, in higher plants, insoluble carbohydrates are a major constituent of total carbohydrates.

3.4. Amino Acid Composition

The basis for determining the nutritional value of products is the analysis of the amino acid composition. The amino acid scores (AASs) according to the FAO/WHO reference standard [30] are presented in Table 3. Among all the analyzed samples, the limiting amino acids were lysine and methionine with cysteine. Lysine was present in an insufficient amount in white sesame and star anise seeds and in spirulina (AAS = 35.33, 30.34 and 42.31%, respectively). However, it is worth mentioning that white sesame seeds were found to be a good source of aromatic amino acids (phenyloalanine and tyrosine), histidine, glutamic acid and arginine compared to the other analyzed plant seeds (Table S1). However, the opposite results were presented by Makinde and Akinoso [58], who found that the dominant amino acid in white sesame seeds was leucine. The star anise seeds' isoleucine, leucine, sulfur amino acids and valine values were lower than those recommended by the FAO/WHO/UNU [30]. This plant is also not a good source of non-essential amino acids (Table S2). Based on the present data, it can be concluded that protein in spirulina is also poor in histidine and sulfur amino acids. However, according to various authors [59–62], spirulina belongs to the group of products in which the protein contains sufficient amounts of essential amino acids except for methionine, cysteine and lysine.

Among the analyzed materials, fenugreek and almond were limited in sulfur amino acids (AAS = 19.87 and 16.67%, respectively). However, fenugreek seeds were found to have good quantities of isoleucine, lysine and aromatic amino acids as compared to the requirements for essential amino acids for schoolchildren. Similar data were also reported by Feyzi et al. [63], who analyzed fenugreek seeds. Therefore, this plant seed can be suggested for use in cereals and snack foods, including bread, biscuits and cakes, since it is rich in lysine and poor in histidine and methionine, in contrast to cereals. With regard to the almond seeds, the lysine value (38.33%) was lower than that recommended by the FAO/WHO/UNU [30]. In general, the protein in *Fabaceae* family crops is limited in sulfur amino acids [64]. On the other hand, these materials were found to be a good source of asparagine, glutamine, arginine and glycine compared to the other samples. Similar findings were presented by Barreca et al. [54], who analyzed almond proteins and found them to have good arginine content.

The rest of the analyzed plant seeds—i.e.: black cumin, milk thistle, white mustard, eggfruit and chlorella—demonstrated complete proteins (Table 3). Black cumin seeds were found to be a good source of aromatic amino acids (phenyloalanine and tyrosine) and glutamine. Furthermore, the amino acid composition of milk thistle protein was dominated by aromatic amino acids, but also threonine, asparagine, glutamine, arginine and serine, which was also confirmed by Sadowska et al. [65]. It is worth emphasizing that white mustard seeds were found to be good source of histidine, lysine and threonine. With regard to non-essential amino acids, there were significant amounts of aspartic acid in white mustard seeds. Powdered eggfruit was characterized by only low amounts of essential amino acids (i.e., leucine, lysine and arginine) compared to the other analyzed seeds. With regard to the non-essential amino acids, the analyzed material was found to be a good source of glutamic acid. For chlorella, the dominant essential amino acids were threonine, valine and leucine. Other authors [66] have also confirmed that chlorella's protein has the right amount of essential amino acids.

To sum up, the protein from the seeds of black cumin, milk thistle, white mustard, eggfruit and chlorella was not limited in any essential amino acids in comparison to the requirements for essential amino acids for schoolchildren (FAO/WHO/UNU [30]). Therefore, these plants can be considered a good sources of amino acids in the diet.

Table 3. Amino acid scores (AASs) for adults according to the FAO/WHO/UNU standards [30].

Amino Acid	Black Cumin	Milk Thistle	Fenugreek	Almond	White Sesame	White Mustard	Star Anise	Eggfruit	Spirulina	Chlorella	Standard
Essential											
His	31.70 ± 0.23 [d]	51.81 ± 0.13 [b]	31.35 ± 0.11 [d]	33.80 ± 0.12 [d]	35.73 ± 0.03 [d]	43.62 ± 0.11 [c]	26.12 ± 0.12 [e]	100.90 ± 1.98 [a]	15.20 ± 0.02 [f]	18.66 ± 0.08 [f]	16
Ile	39.64 ± 0.11 [e]	79.12 ± 0.14 [b]	53.37 ± 0.23 [c]	44.64 ± 0.15 [d]	47.45 ± 0.04 [d]	53.21 ± 0.15 [c]	28.60 ± 0.14 [f]	95.46 ± 1.65 [a]	44.94 ± 0.54 [d]	45.61 ± 0.07 [d]	30
Leu	65.68 ± 0.14 [f]	119.91 ± 0.15 [b]	75.08 ± 0.14 [e]	82.50 ± 0.08 [d]	84.17 ± 1.11 [d]	97.48 ± 1.17 [c]	45.05 ± 0.17 [g]	132.41 ± 1.77 [a]	73.28 ± 0.78 [e]	82.61 ± 0.15 [d]	61
Lys	48.64 ± 1.18 [d]	92.52 ± 0.09 [a]	75.67 ± 0.11 [c]	38.33 ± 0.03 [e]	35.33 ± 0.03 [e]	88.62 ± 0.18 [b]	30.34 ± 0.09 [f]	83.21 ± 0.88 [b]	42.31 ± 0.17 [e]	50.40 ± 0.26 [d]	48
Met + Cys	26.19 ± 2.01 [d]	42.80 ± 0.07 [b]	19.87 ± 0.05 [e]	16.67 ± 0.01 [f]	35.83 ± 0.03 [c]	38.39 ± 0.02 [b]	16.69 ± 0.08 [f]	65.15 ± 0.77 [a]	20.46 ± 0.08 [e]	75.43 ± 0.58 [f]	23
Phe + Tyr	82.30 ± 0.22 [e]	154.02 ± 1.12 [b]	79.91 ± 1.11 [e]	106.68 ± 2.04 [c]	102.77 ± 1.56 [c]	96.98 ± 0.09 [d]	59.85 ± 1.15 [g]	224.11 ± 2.12 [a]	69.62 ± 0.88 [f]	46.62 ± 0.23 [d]	41
Thr	46.99 ± 0.13 [d]	78.82 ± 1.13 [b]	41.88 ± 1.12 [d]	37.96 ± 1.04 [e]	49.76 ± 0.42 [d]	65.54 ± 1.02 [c]	25.77 ± 0.05 [f]	92.48 ± 0.88 [a]	43.45 ± 0.54 [d]	46.62 ± 0.23 [d]	25
Val	53.08 ± 1.15 [c]	100.78 ± 0.08 [a]	44.35 ± 0.08 [d]	53.92 ± 1.20 [c]	61.09 ± 0.91 [c]	71.37 ± 0.94 [b]	32.80 ± 0.19 [e]	112.81 ± 1.55 [a]	56.71 ± 0.74 [c]	58.74 ± 0.45 [c]	40
AAS (%)	101.3 ± 1.17 [c]	178.3 ± 0.05 [a]	82.8 ± 0.06 [d]	69.5 ± 1.17 [e]	73.6 ± 0.08 [e]	159.8 ± 2.01 [b]	63.3 ± 1.19 [f]	173.5 ± 1.44 [a]	88.1 ± 1.15 [d]	105.0 ± 1.25 [c]	100
Non-essential											
Asp	102.78 ± 2.22 [e]	190.59 ± 0.04 [b]	120.31 ± 2.06 [d]	145.02 ± 2.33 [c]	107.76 ± 1.17 [e]	96.57 ± 0.15 [f]	55.44 ± 1.23 [g]	296.71 ± 2.12 [a]	85.85 ± 1.19 [f]	88.98 ± 0.45 [f]	-
Glu	264.27 ± 0.11 [c]	417.83 ± 1.10 [a]	197.64 ± 2.11 [d]	371.79 ± 3.01 [b]	257.21 ± 2.21 [c]	260.87 ± 2.15 [c]	86.14 ± 1.78 [f]	204.03 ± 2.03 [d]	122.40 ± 1.42 [e]	126.93 ± 1.29 [e]	-
Ala	48.34 ± 0.08 [e]	85.58 ± 0.88 [b]	44.02 ± 1.14 [f]	56.80 ± 2.12 [d]	60.12 ± 1.15 [d]	62.27 ± 0.13 [d]	33.29 ± 0.33 [g]	150.37 ± 1.98 [a]	70.90 ± 0.58 [c]	70.20 ± 0.57 [c]	-
Arg	95.30 ± 1.16 [d]	208.01 ± 0.51 [a]	116.47 ± 0.14 [c]	161.82 ± 0.15 [b]	167.48 ± 2.03 [b]	85.90 ± 0.87 [e]	47.71 ± 0.77 [g]	86.75 ± 0.19 [e]	72.00 ± 0.95 [f]	66.05 ± 0.56 [f]	-
Gly	70.60 ± 0.09 [c]	109.81 ± 1.56 [a]	54.88 ± 0.13 [e]	84.67 ± 0.09 [b]	65.74 ± 0.08 [d]	71.99 ± 0.22 [c]	36.60 ± 0.55 [g]	109.68 ± 1.11 [a]	43.94 ± 0.49 [f]	49.51 ± 0.96 [f]	-
Pro	41.21 ± 1.13 [e]	79.94 ± 0.34 [c]	48.63 ± 0.01 [d]	54.96 ± 0.07 [d]	34.56 ± 0.17 [f]	86.85 ± 0.23 [b]	33.51 ± 0.21 [f]	118.33 ± 1.16 [a]	29.22 ± 0.09 [g]	30.07 ± 0.03 [f]	-
Ser	52.45 ± 2.02 [c]	102.55 ± 2.13 [a]	55.04 ± 0.14 [b]	50.75 ± 0.07 [c]	59.38 ± 0.26 [b]	59.50 ± 0.15 [b]	32.63 ± 0.11 [e]	109.29 ± 1.84 [a]	41.67 ± 0.23 [d]	41.93 ± 0.05 [d]	-

Data are means of two replicates. a–g, the same letters in columns mean homogenous groups; bold values indicate first limiting amino acids.

4. Conclusions

As shown by the present results, among the analyzed plants and algae, white mustard and milk thistle showed the highest antioxidant capacities in the ABTS and FRAP assays. The widest spectrum of microbial growth inhibition was indicated for fenugreek extract, which showed antimicrobial activity against four of the analyzed microorganisms, including *B. subtilis*, *P. mirabilis*, *V. harveyi* and *C. albicans*. Among all the analyzed plants, fenugreek seeds showed potential for use in food formulation operations due to their antioxidant activity and amino acid profile. Based on the results, the intake of polyphenols together with protein in plant seeds does not have a major impact on the potential bioaccessibility of a range of different polyphenols and phenolic metabolites.

This work has shown that fenugreek seeds have potential for use in food formulation operations due to their antioxidant activity and amino acid profile. The observed antibacterial activity in this plant suggests that it could play a dual role in food and in non-food systems, where it may also find uses. Bringing out its full potential for utilization in these systems is, however, dependent on the full characterization of biologically active components in the plant.

Supplementary Materials: The following supporting information can be downloaded at: https://www.mdpi.com/article/10.3390/app12168136/s1. Table S1. Concentrations of essential and relatively essential amino acids (g/100 g protein) in the analyzed raw materials; Table S2. Concentrations of non-essential and relatively non-essential amino acids (g/100 g protein) in the analyzed raw materials and accepted patterns.

Author Contributions: Conceptualization, J.M.; Investigation, J.M., S.L.-W., A.N. and P.Ł.K.; Methodology, J.M., S.L.-W. and A.K.; Validation, J.M.; Writing—original draft, J.M. and S.L.-W.; Writing—review and editing, J.M., S.L.-W., P.Ł.K. and A.K. All authors have read and agreed to the published version of the manuscript.

Funding: This research received no external funding.

Institutional Review Board Statement: Not applicable.

Informed Consent Statement: Not applicable.

Data Availability Statement: Not applicable.

Acknowledgments: The authors wish to thank Katarzyna Drzymała for her valuable technical support for this manuscript.

Conflicts of Interest: The authors declare no conflict of interest.

References

1. *ISO 927:2009*; Spices, Culinary Herbs and Condiments. International Organization for Standardization: Geneva, Switzerland, 2009.
2. Ceylan, E.; Fung, D.Y.C. Antimicrobial activity of spices. *J. Rapid Methods Autom. Microbiol.* **2004**, *12*, 1–55. [CrossRef]
3. Viuda-Martos, M.; Mohamady, M.A.; Fernández-López, J.; Abd ElRazik, K.A.; Omer, E.A.; Pérez-Alvarez, J.A.; Sendra, E. In vitro antioxidant and antibacterial activities of essentials oils obtained from Egyptian aromatic plants. *Food Control* **2011**, *22*, 1715–1722. [CrossRef]
4. Edelman, M.; Colt, M. Nutrient Value of Leaf vs. Seed. *Front. Chem.* **2016**, *4*, 32. [CrossRef] [PubMed]
5. Sánchez, M.; Bernal-Castillo, J.; Rozo, C.; Rodríguez, I. Spirulina (*Arthrospira*): An edible microorganism. A review. *Univ. Sci.* **2003**, *8*, 7–24.
6. Raczyk, M.; Polanowska, K.; Kruszewski, B.; Grygier, A.; Michałowska, D. Effect of Spirulina (*Arthrospira platensis*) Supplementation on Physical and Chemical Properties of Semolina (*Triticum durum*) Based Fresh Pasta. *Molecules* **2022**, *27*, 355. [CrossRef] [PubMed]
7. Całyniuk, B.; Grochowska-Niedworok, E.; Białek, A.; Czech, N.; Kukielczak, A. Food guide pyramid—its past and present. *Probl. Hig. I Epidemiol.* **2011**, *92*, 20–24.
8. Peter, K.V. (Ed.) *Handbook of Herbs and Spices*; Woodhead Publishing: Sawston, UK, 2012; ISBN 978-0-85709-039-3.
9. Ścieszka, S.; Klewicka, E. Algae in food: A general review. *Crit. Rev. Food Sci. Nutr.* **2019**, *59*, 3538–3547. [CrossRef] [PubMed]
10. Rashidinejad, A.; Birch, E.J.; Sun-Waterhouse, D.; Everett, D.W. Addition of milk to tea infusions: Helpful or harmful? Evidence from in vitro and in vivo studies on antioxidant properties. *Crit. Rev. Food Sci. Nutr.* **2017**, *57*, 3188–3196. [CrossRef]

11. Al-Jasass, F.M.; Al-Jasser, M.S. Chemical Composition and Fatty Acid Content of Some Spices and Herbs under Saudi Arabia Conditions. *Sci. World J.* **2012**, *2012*, 1–5. [CrossRef] [PubMed]

12. Li, H.; Tsao, R.; Deng, Z. Factors affecting the antioxidant potential and health benefits of plant foods. *Can. J. Plant Sci.* **2012**, *92*, 1101–1111. [CrossRef]

13. Ncube, N.; Afolayan, A.; Okoh, A. Assessment techniques of antimicrobial properties of natural compounds of plant origin: Current Methods and Future Trends. *AFRICAN J. Biotechnol.* **2008**, *7*, 1797–1806. [CrossRef]

14. Wallace, R.J. Antimicrobial properties of plant secondary metabolites. *Proc. Nutr. Soc.* **2004**, *63*, 621–629. [CrossRef] [PubMed]

15. Shimoda, K.; Kondo, Y.; Nishida, T.; Hamada, H.; Nakajima, N.; Hamada, H. Biotransformation of thymol, carvacrol, and eugenol by cultured cells of Eucalyptus perriniana. *Phytochemistry* **2006**, *67*, 2256–2261. [CrossRef]

16. Rajput, J.D.; Bagul, S.D.; Pete, U.D.; Zade, C.M.; Padhye, S.B.; Bendre, R.S. Perspectives on medicinal properties of natural phenolic monoterpenoids and their hybrids. *Mol. Divers.* **2018**, *22*, 225–245. [CrossRef] [PubMed]

17. Samtiya, M.; Aluko, R.E.; Dhewa, T.; Moreno-Rojas, J.M. Potential Health Benefits of Plant Food-Derived Bioactive Components: An Overview. *Foods* **2021**, *10*, 839. [CrossRef] [PubMed]

18. Damián, M.R.; Cortes-Perez, N.G.; Quintana, E.T.; Ortiz-Moreno, A.; Garfias Noguez, C.; Cruceño-Casarrubias, C.E.; Sánchez Pardo, M.E.; Bermúdez-Humarán, L.G. Functional Foods, Nutraceuticals and Probiotics: A Focus on Human Health. *Microorganisms* **2022**, *10*, 1065. [CrossRef]

19. Draijer, R.; van Dorsten, F.; Zebregs, Y.; Hollebrands, B.; Peters, S.; Duchateau, G.; Grün, C. Impact of Proteins on the Uptake, Distribution, and Excretion of Phenolics in the Human Body. *Nutrients* **2016**, *8*, 814. [CrossRef] [PubMed]

20. Cosme, P.; Rodríguez, A.B.; Espino, J.; Garrido, M. Plant Phenolics: Bioavailability as a Key Determinant of Their Potential Health-Promoting Applications. *Antioxidants* **2020**, *9*, 1263. [CrossRef] [PubMed]

21. Gao, X.; Ohlander, M.; Jeppsson, N.; Björk, L.; Trajkovski, V. Changes in Antioxidant Effects and Their Relationship to Phytonutrients in Fruits of Sea Buckthorn (*Hippophae rhamnoides* L.) during Maturation. *J. Agric. Food Chem.* **2000**, *48*, 1485–1490. [CrossRef]

22. Re, R.; Pellegrini, N.; Proteggente, A.; Pannala, A.; Yang, M.; Rice-Evans, C. Antioxidant activity applying an improved ABTS radical cation decolorization assay. *Free Radic. Biol. Med.* **1999**, *26*, 1231–1237. [CrossRef]

23. Benzie, I.F.F.; Strain, J.J. The Ferric Reducing Ability of Plasma (FRAP) as a Measure of "Antioxidant Power": The FRAP Assay. *Anal. Biochem.* **1996**, *239*, 70–76. [CrossRef] [PubMed]

24. Minekus, M.; Alminger, M.; Alvito, P.; Ballance, S.; Bohn, T.; Bourlieu, C.; Carrière, F.; Boutrou, R.; Corredig, M.; Dupont, D.; et al. A standardised static in vitro digestion method suitable for food—An international consensus. *Food Funct.* **2014**, *5*, 1113–1124. [CrossRef] [PubMed]

25. Lachowicz, S.; Świeca, M.; Pejcz, E. Biological activity, phytochemical parameters, and potential bioaccessibility of wheat bread enriched with powder and microcapsules made from Saskatoon berry. *Food Chem.* **2021**, *338*, 128026. [CrossRef]

26. AOAC. *AOAC Official Methods of Analysis*, 21st ed.; AOAC International: Rockville, MD, USA, 2019.

27. Mariotti, F.; Tomé, D.; Mirand, P.P. Converting Nitrogen into Protein—Beyond 6.25 and Jones' Factors. *Crit. Rev. Food Sci. Nutr.* **2008**, *48*, 177–184. [CrossRef] [PubMed]

28. Pęksa, A.; Miedzianka, J.; Nemś, A. Amino acid composition of flesh-coloured potatoes as affected by storage conditions. *Food Chem.* **2018**, *266*, 335–342. [CrossRef] [PubMed]

29. Spackman, D.H.; Stein, W.H.; Moore, S. Automatic Recording Apparatus for Use in Chromatography of Amino Acids. *Anal. Chem.* **1958**, *30*, 1190–1206. [CrossRef]

30. FAO. *Dietary Protein Quality Evaluation in Human Nutrition*; Report of an FAO Expert Consultation; FAO Food and Nutrition Paper 92; FAO: Rome, Italy, 2013; ISBN 978-92-5-107417-6.

31. Mariod, A.A.; Ibrahim, R.M.; Ismail, M.; Ismail, N. Antioxidant activity and phenolic content of phenolic rich fractions obtained from black cumin (*Nigella sativa*) seedcake. *Food Chem.* **2009**, *116*, 306–312. [CrossRef]

32. Summo, C.; Palasciano, M.; De Angelis, D.; Paradiso, V.M.; Caponio, F.; Pasqualone, A. Evaluation of the chemical and nutritional characteristics of almonds (*Prunus dulcis* (Mill). D.A. Webb) as influenced by harvest time and cultivar. *J. Sci. Food Agric.* **2018**, *98*, 5647–5655. [CrossRef] [PubMed]

33. Padmashree, A.; Roopa, N.; Semwal, A.D.; Sharma, G.K.; Agathian, G.; Bawa, A.S. Star-anise (*Illicium verum*) and black caraway (*Carum nigrum*) as natural antioxidants. *Food Chem.* **2007**, *104*, 59–66. [CrossRef]

34. Wojdylo, A.; Oszmianski, J.; Czemerys, R. Antioxidant activity and phenolic compounds in 32 selected herbs. *Food Chem.* **2007**, *105*, 940–949. [CrossRef]

35. Mashkor, I.M. Phenolic content and antioxidant activity of fenugreek seeds extract. *Int. J. Pharmacogn. Phytochem. Res.* **2014**, *6*, 841–844.

36. Lin, X.; Zhou, L.; Li, T.; Brennan, C.; Fu, X.; Liu, R.H. Phenolic content, antioxidant and antiproliferative activities of six varieties of white sesame seeds (*Sesamum indicum* L.). *RSC Adv.* **2017**, *7*, 5751–5758. [CrossRef]

37. Shan, B.; Cai, Y.Z.; Sun, M.; Corke, H. Antioxidant Capacity of 26 Spice Extracts and Characterization of Their Phenolic Constituents. *J. Agric. Food Chem.* **2005**, *53*, 7749–7759. [CrossRef]

38. Vitaglione, P.; Barone Lumaga, R.; Ferracane, R.; Radetsky, I.; Mennella, I.; Schettino, R.; Koder, S.; Shimoni, E.; Fogliano, V. Curcumin Bioavailability from Enriched Bread: The Effect of Microencapsulated Ingredients. *J. Agric. Food Chem.* **2012**, *60*, 3357–3366. [CrossRef] [PubMed]

39. Rodríguez-Roque, M.J.; Rojas-Graü, M.A.; Elez-Martínez, P.; Martín-Belloso, O. Soymilk phenolic compounds, isoflavones and antioxidant activity as affected by in vitro gastrointestinal digestion. *Food Chem.* **2013**, *136*, 206–212. [CrossRef] [PubMed]

40. Gawlik-Dziki, U.; Jeżyna, M.; Świeca, M.; Dziki, D.; Baraniak, B.; Czyż, J. Effect of bioaccessibility of phenolic compounds on in vitro anticancer activity of broccoli sprouts. *Food Res. Int.* **2012**, *49*, 469–476. [CrossRef]

41. Kowalczewski, P.Ł.; Olejnik, A.; Rybicka, I.; Zielińska-Dawidziak, M.; Białas, W.; Lewandowicz, G. Membrane Filtration-Assisted Enzymatic Hydrolysis Affects the Biological Activity of Potato Juice. *Molecules* **2021**, *26*, 852. [CrossRef]

42. Sęczyk, Ł.; Świeca, M.; Kapusta, I.; Gawlik-Dziki, U. Protein–Phenolic Interactions as a Factor Affecting the Physicochemical Properties of White Bean Proteins. *Molecules* **2019**, *24*, 408. [CrossRef] [PubMed]

43. Cantele, C.; Rojo-Poveda, O.; Bertolino, M.; Ghirardello, D.; Cardenia, V.; Barbosa-Pereira, L.; Zeppa, G. In Vitro Bioaccessibility and Functional Properties of Phenolic Compounds from Enriched Beverages Based on Cocoa Bean Shell. *Foods* **2020**, *9*, 715. [CrossRef] [PubMed]

44. Miedzianka, J.; Drzymała, K.; Nemś, A.; Kita, A. Comparative evaluation of the antioxidant, antimicrobial and nutritive properties of gluten-free flours. *Sci. Rep.* **2021**, *11*, 10385. [CrossRef]

45. Al-Ahmad, A.; Maier, J.; Follo, M.; Spitzmüller, B.; Wittmer, A.; Hellwig, E.; Hübner, J.; Jonas, D. Food-borne Enterococci Integrate Into Oral Biofilm: An In Vivo Study. *J. Endod.* **2010**, *36*, 1812–1819. [CrossRef] [PubMed]

46. Apetroaie-Constantin, C.; Mikkola, R.; Andersson, M.A.; Teplova, V.; Suominen, I.; Johansson, T.; Salkinoja-Salonen, M. *Bacillus subtilis* and *B. mojavensis* strains connected to food poisoning produce the heat stable toxin amylosin. *J. Appl. Microbiol.* **2009**, *106*, 1976–1985. [CrossRef] [PubMed]

47. Hervio-Heath, D.; Colwell, R.R.; Derrien, A.; Robert-Pillot, A.; Fournier, J.M.; Pommepuy, M. Occurrence of pathogenic vibrios in coastal areas of France. *J. Appl. Microbiol.* **2002**, *92*, 1123–1135. [CrossRef] [PubMed]

48. Tang, Y.; Ali, Z.; Zou, J.; Jin, G.; Zhu, J.; Yang, J.; Dai, J. Detection methods for Pseudomonas aeruginosa: History and future perspective. *RSC Adv.* **2017**, *7*, 51789–51800. [CrossRef]

49. Wang, Y.; Zhang, S.; Yu, J.; Zhang, H.; Yuan, Z.; Sun, Y.; Zhang, L.; Zhu, Y.; Song, H. An outbreak of *Proteus mirabilis* food poisoning associated with eating stewed pork balls in brown sauce, Beijing. *Food Control* **2010**, *21*, 302–305. [CrossRef]

50. Arora, D.S.; Kaur, J. Antimicrobial activity of spices. *Int. J. Antimicrob. Agents* **1999**, *12*, 257–262. [CrossRef]

51. de Oliveira Gonçalves, T.; Filbido, G.S.; de Oliveira Pinheiro, A.P.; Pinto Piereti, P.D.; Dalla Villa, R.; de Oliveira, A.P. In vitro bioaccessibility of the Cu, Fe, Mn and Zn in the baru almond and bocaiúva pulp and, macronutrients characterization. *J. Food Compos. Anal.* **2020**, *86*, 103356. [CrossRef]

52. Ezeagu, I.; Petzke, J.; Metges, C.; Akinsoyinu, A.; Ologhobo, A. Seed protein contents and nitrogen-to-protein conversion factors for some uncultivated tropical plant seeds. *Food Chem.* **2002**, *78*, 105–109. [CrossRef]

53. Chacón-Lee, T.L.; González-Mariño, G.E. Microalgae for "Healthy" Foods-Possibilities and Challenges. *Compr. Rev. Food Sci. Food Saf.* **2010**, *9*, 655–675. [CrossRef] [PubMed]

54. Barreca, D.; Nabavi, S.M.; Sureda, A.; Rasekhian, M.; Raciti, R.; Silva, A.S.; Annunziata, G.; Arnone, A.; Tenore, G.C.; Süntar, İ.; et al. Almonds (*Prunus dulcis* Mill. D. A. Webb): A Source of Nutrients and Health-Promoting Compounds. *Nutrients* **2020**, *12*, 672. [CrossRef]

55. El-haak, M.A.; Atta, B.M.; Abd Rabo, F.F. Seed yield and important seed constituents for naturally and cultivated milk thistle plants. *Egypt. J. Exp. Biol.* **2015**, *11*, 141–146.

56. Kochhar, A.; Nagi, M.; Sachdeva, R. Proximate Composition, Available Carbohydrates, Dietary Fibre and Anti Nutritional Factors of Selected Traditional Medicinal Plants. *J. Hum. Ecol.* **2006**, *19*, 195–199. [CrossRef]

57. Lunn, J.; Buttriss, J.L. Carbohydrates and dietary fibre. *Nutr. Bull.* **2007**, *32*, 21–64. [CrossRef]

58. Makinde, F.M.; Akinoso, R. Comparison between the nutritional quality of flour obtained from raw, roasted and fermented sesame (*Sesamum indicum* L.) seed grown in Nigeria. *Acta Sci. Pol. Technol. Aliment.* **2014**, *13*, 309–319. [CrossRef] [PubMed]

59. Spolaore, P.; Joannis-Cassan, C.; Duran, E.; Isambert, A. Commercial applications of microalgae. *J. Biosci. Bioeng.* **2006**, *101*, 87–96. [CrossRef]

60. Ahsan, M.; Habib, B.; Parvin, M.; Huntington, T.C.; Hasan, M.R. *A review on Culture, Production and Use of Spirulina as Food for Humans and Feeds for Domestic Animals*; FAO Fisheries and Aquaculture Circular No. 1034; FAO: Rome, Italy, 2008.

61. Koru, E. Earth Food Spirulina (*Arthrospira*): Production and Quality Standarts. In *Food Additive*; InTech: London, UK, 2012.

62. Holman, B.W.B.; Malau-Aduli, A.E.O. Spirulina as a livestock supplement and animal feed. *J. Anim. Physiol. Anim. Nutr.* **2013**, *97*, 615–623. [CrossRef]

63. Feyzi, S.; Varidi, M.; Zare, F.; Varidi, M.J. Fenugreek (*Trigonella foenum graecum*) seed protein isolate: Extraction optimization, amino acid composition, thermo and functional properties. *J. Sci. Food Agric.* **2015**, *95*, 3165–3176. [CrossRef]

64. Young, V.R.; Pellett, P.L. Plant proteins in relation to human protein and amino acid nutrition. *Am. J. Clin. Nutr.* **1994**, *59*, 1203S–1212S. [CrossRef]

65. Sadowska, K.; Andrzejewska, J.; Woropaj-Janczak, M. Others Effect of weather and agrotechnical conditions on the content of nutrients in the fruits of milk thistle (*Silybum marianum* L. Gaertn.). *Acta Sci. Pol. Hortorum Cultus* **2011**, *10*, 197–207.

66. Viegas, C.V.; Hachemi, I.; Mäki-Arvela, P.; Smeds, A.; Aho, A.; Freitas, S.P.; da Silva Gorgônio, C.M.; Carbonetti, G.; Peurla, M.; Paranko, J.; et al. Algal products beyond lipids: Comprehensive characterization of different products in direct saponification of green alga *Chlorella* sp. *Algal Res.* **2015**, *11*, 156–164. [CrossRef]

Article

Bird Cherry (*Prunus padus*) Fruit Extracts Inhibit Lipid Peroxidation in PC Liposomes: Spectroscopic, HPLC, and GC–MS Studies

Przemysław Siejak [1,*], Wojciech Smułek [2], Joanna Nowak-Karnowska [3], Anna Dembska [3], Grażyna Neunert [1] and Krzysztof Polewski [1]

[1] Department of Physics and Biophysics, Faculty of Food Science and Nutrition, Poznań University of Life Sciences, Wojska Polskiego 38/42, 60-637 Poznań, Poland; grazyna.neunert@up.poznan.pl (G.N.); krzysztof.polewski@up.poznan.pl (K.P.)
[2] Institute of Chemical Technology and Engineering, Poznań University of Technology, Berdychowo 4, 60-965 Poznań, Poland; wojciech.smulek@put.poznan.pl
[3] Department of Bioanalytical Chemistry, Faculty of Chemistry, Adam Mickiewicz University, Uniwersytetu Poznańskiego 8, 61-614 Poznań, Poland; j.nowak@amu.edu.pl (J.N.-K.); aniojka@amu.edu.pl (A.D.)
* Correspondence: przemyslaw.siejak@up.poznan.pl; Tel.: +48-61-848-7499

Abstract: The antioxidant potential of bird cherry fruit of water, methanol, ethanol, and acetone extracts and their antioxidant efficiency against oxidation of PC liposomes using spectroscopic and chromatographic methods were investigated. The chromatographic methods quantified and specified the presence of phenolic and flavonoid compounds in the investigated extracts. The characteristic peaks in the UV spectrum at 275 nm and 370 nm confirmed the presence of phenols and flavonoids and their derivatives. Their presence was also confirmed by FTIR spectra, which revealed the presence of its functional groups. The total luminescence spectra with maxima at 314–318 nm, 325–355 nm, and 428–435 nm were ascribed to the presence of phenolic acids and tocopherols. The antioxidant properties of extracts and its inhibition properties against lipid peroxidation in PC liposomes were determined by fluorogenic probes DCF-H and C11-BODIPY581/591. The measured antioxidant properties against generated free radicals in aqueous and lipogenic phases revealed differences between extracts depending on their physicochemical properties with the greatest potential for acetone extract and sirup. The presented quantitative analysis indicated that cherry bird extracts possess significant amounts of phenolics and flavonoids, thus having the opportunity to be used as a natural antioxidant agent source with a large potential for application in pharmaceutical and food industries.

Keywords: antioxidant; DCF; BOBIPY; phenolic compounds; flavonoids; liposomes; AAPH

Citation: Siejak, P.; Smułek, W.; Nowak-Karnowska, J.; Dembska, A.; Neunert, G.; Polewski, K. Bird Cherry (*Prunus padus*) Fruit Extracts Inhibit Lipid Peroxidation in PC Liposomes: Spectroscopic, HPLC, and GC–MS Studies. *Appl. Sci.* **2022**, *12*, 7820. https://doi.org/10.3390/app12157820

Academic Editor: Raffaella Boggia

Received: 18 July 2022
Accepted: 2 August 2022
Published: 3 August 2022

Publisher's Note: MDPI stays neutral with regard to jurisdictional claims in published maps and institutional affiliations.

1. Introduction

For centuries, plants were sources of medicines that were used to cure diseases in humans and animals. The presence of secondary metabolites with specific properties has been recognized and applied [1,2]. Especially phenols, polyphenols and flavonoids constitute major groups of compounds having natural antioxidant properties [3]. Phenolic compounds such as quercetin, rutin, catechin, genistein, caffeic acid, chlorogenic acid, and gallic acid are among those most popular [4,5]. The most powerful methods of detection of such compounds are chromatographic techniques such as LC–MS, GC–MS, HPLC, and NMR. The spectroscopic techniques such as UV–Vis or FTIR are less elaborate and faster to detect the presence of phytocomponents identifying functional groups [6].

The rapid oxidation of biologically important substances and agents, present in any living organism, as well as delivered to the human body in a variety of forms: in everyday diet, as a diet supplement lotion, cream, drugs, etc., is a problem of great importance. Therefore, many methods have been developed to prevent this effect, especially when it comes to medical products of human health importance. Among those, the main one is the encapsulation of biologically active agents (especially unsaturated fatty acids) with the use of saccharide shells [7,8] or yeast cells [9–11]. The aim of the capsule is to produce a mechanical barrier—wall, preventing invasive agents, including oxidizing species, from access to the active substance and, thus, maintaining the designed therapeutical action of the drug. The presence of natural antioxidants in the medical products, diet supplements, or edible products may reduce the need for encapsulation. Moreover, the presence of easily available natural antioxidants in the human body may also prevent from oxidation unsaturated fatty acids both building cell membranes and delivered with the diet.

Recently, much attention has been given to the influence of natural compounds present in everyday diet on human health. Much emphasis has been put on nonmedical or semi-medical cosmetics, food, and drink products, such as herbs (fresh and dried), teas, and infusions, as well as juices and extracts, and their antioxidant performance is one of the crucial factors [12,13]. One of the possible antioxidant sources are fruits of European bird cherry (*Prunus padus*), which is a European and Asian native tree of the *Rosaceae* family. The bark, leaves, and fruits have been known in the field of folk medicine, considering their antibacterial, diuretic, antirheumatic, styptic, and other performances. Nevertheless, properties of any part of the tree, including fruits and fruit extracts, are poorly known, and only a few reports on the topic are available [14,15]. The results presented in those publications showed that bird cherry fruits contain a number of compounds including polyphenols and bioactive compounds, especially vitamins, and many of the above are known antioxidant agents. The studies showed that 3-rutinoside and 3-glucoside of cyanidin are the main flavonoids detected in bird cherry fruits [16]. The strong antioxidant properties of cyanidin 3-glucoside was reported by Smyk et al. [17].

A review regarding phytopharmacological properties and its bioactive compounds was presented by Telichowsaka et al. [18]. The authors showed that different parts of the plant, including fruit, contain many active compounds such as polyphenols, flavonoids, tocopherols, vitamins, terpenes, and cyanogenic glycosides. Most of them possess a beneficial influence on health, having antioxidant, antimicrobial, and anti-inflammatory activities.

A very thorough examination of the wild prunus fruit composition was carried out by Mikulic-Petkovsek at al. using HPLC-DAD-MSN analysis [15]. They confirmed the presence of sugars, organic acids, carotenoids, tocopherols, chlorophylls, and phenolic compounds. The antibacterial action of extracts of bird cherry was also reported [19]. In addition, shrouds of bird cherry are very invasive and its presence is almost everywhere including forestry where it may become a problem [20]. Thus, any practical utilization or application of any part of this plant will be of great importance.

The above presented consideration indicates that there is some knowledge regarding the chemical composition of bird cherry extracts; however, there is scarce knowledge regarding its possible action and role during its antioxidant action in biological membranes. As lipid peroxidation causes oxidative damage to cell membranes and all other systems that contain lipids, in investigation of the total antioxidative activity of plant extracts, it is necessary to investigate their effects on lipid peroxidation.

To understand and explain the putative antioxidant properties of the reported presence of phenolics in bird cherry extracts, its composition and scavenging properties against generated peroxyl radicals were determined. The antioxidant properties of sirup and solvent extracts from bird cherry fruits against free radical oxidation as well as their inhibition potential of lipid peroxidation in soybean L-α-phosphatidylcholine (PC) liposomes used as a model of oxidation of one of the important membrane components were investigated.

2. Materials and Methods

2.1. Chemicals

Methanol, ethanol, chloroform, acetone, and DL-α-tocopherol (α-T) were purchased from Merck (Darmstadt, Germany). L-α-phosphatidylcholine (PC) from egg yolk, the generator of free radicals 2,2'-azobis(2-amidinopropane) dihydrochloride (AAPH), 2',7'-

dichlorofluorescein diacetate (DCFH-DA), and 1,1-diphenyl-2-picrylhydrazyl (DPPH), was obtained from Sigma-Aldrich (Steinheim, Germany). Fluorescent probe C11-BODIPY581/591 [4,4-difluoro-5-(4-phenyl-1,3-butadienyl)-4-bora-3a,4a-diaza-s-indacene-3-undecanoic acid] was purchased from Invitrogen (Carlsbad, CA, USA). The used water was purified by a MicroPure Water System (TKA, Niederelbert, Germany).

2.2. Preparation of Extracts

First, the bird cherry fruits were lyophilized, and then they were crushed with a pestle in a ceramic mortar. The obtained powder was suspended in a proportion of 10 g per 60 mL in different solvents: acetone, methanol, ethanol, and water. Next, the suspensions were incubated during 1 h at a 50 °C temperature in an ultrasound bath. After that, the obtained supernatant was filtered to remove solids and the remaining solid material in the sample was extracted again. Then, both extracts were added and, using a vacuum evaporator, the dry mass of each extract was obtained. Only pure native juice was directly obtained from squeezed fruits. Stock samples were diluted directly before measurements. The dilution was adjusted as required for absorption and fluorescence measurements.

2.3. Spectroscopic Measurements

To estimate the composition of obtained extracts for each sample, we measured the absorption spectra in the UV–Vis range and the fluorescence spectra at different excitation wavelengths. The steady-state absorption and fluorescence spectra of extracts were measured for each extract dissolved in a water concentration of 10 mg/mL, and additionally diluted directly before measurements to a concentration of 0.48 mg/mL (0.1 mL of stock solution was added to 2 mL of water). The absorption spectra were collected with the use of a Shimadzu UV-1201 spectrophotometer (Kyoto, Japan), and the fluorescence spectra—Shimadzu RF 5001PC fluorimeter (Kyoto, Japan). For fluorescence measurements, 90° (L-shaped) geometry was applied.

2.4. HPLC Measurements

High-performance liquid chromatography (HPLC) was performed with a Waters system with a binary gradient-forming module Waters 1525, diode-array UV–Vis detector Waters 2998, and fluorescence detector Waters 2475. HPLC analyses were performed on an Inertsil ODS-3 Column (5 μm, 4.6 × 250 mm) at 25 °C, eluted with 1% HCOOH aq., using a linear gradient of 0–30% of acetonitrile over 15 min, then 40% of acetonitrile over 10 min at a flow rate of 1 mL/min. Then, 10 μL of solution was injected. ESI–MS spectra were recorded using a ZQ Waters-Micromass Spectrometer.

2.5. GC-TOFMS Measurements

In order to identify extracts' components and to indicate differences in composition between them, GC-TOFMS analyses were conducted. At the beginning, 10 mg of the freeze-dried extract was mixed with 0.1 mL of derivatization reagent BSTFA (N,O-bis(trimethylsilyl)trifluoroacetamide, 99%) and incubated for 60 min at 75 °C. Then, 0.3 mL of hexane was added. The samples were filtered and then injected (1 μL) to the chromatograph inlet. The analysis was conducted using the PEGASUS 4D GCxGC-TOFMS gas chromatograph (LECO Corp., St. Joseph, MO, USA) equipped with a BPX5 (5% phenyl equivalent, 28 m × 0.25 mm; 0.25 μm) capillary column (SGE Int., Melbourne, Australia).

Helium was used as a carrier gas (1.0 mL/min), and the ion source and transfer line temperature was set at 250 °C. The oven temperature program was as follows: 30 °C maintained for 2 min, followed by a heating ramp to 300 °C at a rate of 10 °C min^{-1}, and the final temperature maintained for 15 min. The ionization source was operated in the positive ion mode (-70 V), and the acquisition rate was set to 10 spectra/s.

2.6. FTIR Measurements

The FTIR spectra were obtained using a Spectrum Two FT-IR spectrometer equipped with a Universal ATR with a diamond crystal (PerkinElmer, Waltham, MA, USA). The data were collected over a spectral range of 4000–500 cm^{-1}. The measurements were repeated three times for each sample.

2.7. Oxidation Tests

2.7.1. Determination of Total Antioxidant Capacity by Inhibition of DCF-H Oxidation

The antioxidant properties of extracts in isotropic medium were also evaluated by fluorometric methods—we tested the influence of the extract and reference sample (gallic acid at final concentration of 12 μM) on free radicals produced by AAPH with the DCF-H fluorescence probe (obtained from the diacetate form—DCFH-DA). For that purpose, to 2 mL of phosphate buffer (pH = 7.4) stored at 40 °C of water, a 15 μL extract solution in water (0.01 g/mL; final content in the sample was 0.07 μg/mL) was added followed by the addition of a 20 μL DCF-H solution (1 mM) and 50 μL AAPH water solution (0.2 M) under constant stirring at 40 °C. The efficiency of free radical inhibition was assessed by relative changes in the fluorescence intensity of DCF (520 nm) created as a result of peroxidation. Measurements were conducted for 10 min with a 1 min interval. The gallic acid at a concentration range of 0–20 μM was chosen as reference samples.

2.7.2. DPPH Method

Scavenging of DPPH free radicals is one of most commonly used tests for antioxidant properties. To determine the scavenging efficiency of DPPH, we applied slight modifications to the traditional method (which includes usually only the measurement of absorbance after 20 min of incubation). Briefly, we prepared the following samples to perform the test. A 15 μL volume of each extract examined (0.01 g/mL in water) was added to a 2.2 mL methanol solution of DPPH, and its absorbance changes at 517 nm in specified time periods was evaluated, starting with the moment of addition. The 12 μM water solution of gallic acid in the final sample was chosen as a reference (aliquot volume of stock solution was added to 2.2 mL of DPPH methanol solution).. Measurements were conducted until the stabilization of each curve individually to establish the maximum scavenging ability of each extract examined. For scavenging performance assessment, the following formula was used:

$$\text{MSE [\%]} = (A_0 - A_{min}) \times 100/A_0, \tag{1}$$

where A_0 and A_{min} are absorbance values of DPPH at 517 nm before antioxidant addition and for the plateau (the lowest absorbance value detected) as a result of antioxidant agent addition, respectively.

2.7.3. Thermal Oxidation of PC Liposomes

The liposomes were prepared by the extrusion method according to Neunert at al. [13]. For liposomes preparation, dry PC was dissolved in chloroform without any compounds (blind probe) or mixed in required proportions with DL-α-tocopherol dissolved in methanol, for final concentrations of tocopherol 12 μM ("tocopherol in membrane"). Then, the solvents were removed under vacuum at 30 °C by a rotary evaporator (Büchi Labortechnik AG, Flawil, Switzerland). The formed PC dry film was hydrated with 0.1 M phosphate buffer (pH 7.4) and vortexed for 10 min at 22 °C until the solution became clear. Next, the liposome suspension was extruded repeatedly 11 times through a 100 nm pore polycarbonate filter using the LiposoFast Basic LF-1 extruder (Avestin, Mannheim, Germany). The final concentration of lipids was 0.08 mg/mL with a mean liposome size of 100 nm as confirmed by dynamic light scattering (DLS) measurements using a Zetasizer Nano (Malvern Instruments, Worcestershire, UK) at 20 °C under an angle of 90°. The mean values of the liposome size of 100 nm was determined from analysis of the number of peaks from a few different samples (from different series). The mean results of DLS measurements are presented in Supplementary Material (Figure S1). The stability of the liposomes was confirmed by zeta potential measurements. The average zeta potential of the liposomes was −17.6 mV. Moreover, the low average values observed for the polydispersity index (Pdl) equal to 0.15 revealed a good degree of homogeneity among the investigated samples.

The thermal oxidation of PC liposomes was realized as follows. The 15 μL of water solution of each extract (or aliquot volume of gallic acid water solution and tocopherol methanol solution—to obtain the final concentration of this compound of 12 μM) was added to 2 mL of PC liposomes in phosphate buffer (0.08 mg/mL) and left at ambient conditions for 30 min under stirring (magnetic stirrer at the rate of 150 RPM). After that time, the blind probe and the samples with extracts were placed in 40 °C under constant stirring and 16 μL of methanol solution of C11-BODIPY581/591 (0.5 mg/mL) was added. Fluorescence measurements of each sample (λ_{ex} = 505 nm, λ_{obs} = 520 nm) were repeated in a 10 min time interval starting with C11-BODIPY581/591 addition up to stabilization or a noticeable drop in fluorescence intensity.

2.8. Data Evaluation and Statistical Analysys

Each sample was measured at least three times and average values with standard deviations were calculated (where necessary).

3. Results

The chemical composition of extracts was estimated with absorbance and fluorescence spectroscopy, HPLC, and GS–MS methods. The antioxidant properties of extracts in aqueous phase against 2,2′-azobis(2-amidinopropane) dihydrochloride (AAPH)-generated radicals were determined using a fluorescent probe of 2′,7′-dichlorofluorescein (DCF-H). The inhibition of lipid peroxidation in PC liposomes of the extracts was estimated using a fluorescent sensor C11-BODIPY581/591 [4,4-difluoro-5-(4-phenyl-1,3-butadienyl)-4-bora-3a,4a-diaza-s-indacene-3-undecanoic acid]. To approach a more realistic model of biological membranes, or at least a model of one of the important components of real biological membranes, the unilamellar- and homogenous-size-distribution PC liposomes in liquid-crystalline phase were used. As the applied PC liposomes are not a direct model of the cell membrane, it is worth noticing that we did not aim to create a model of a biological membrane; thus, such important components as cholesterol or proteins were not included. The use of PC was intentional as it possesses a low-main-phase transition temperature; thus, at room temperature, it exists in the fluid phase. It is a phase where biological processes occur, including the fastest permeation and inclusion of different compounds into the membrane interior. In addition, in PC from egg yolk, there is a number of unsaturated fatty acids prone to fast oxidation.

3.1. Absorption Spectra

The fact that phenolic compounds are organic compounds containing s-bonds, p-bonds, lone pairs of electrons, and chromophores where most of them in their structure composes the aromatic ring with many attached moieties, they exhibit strong absorption in the UV region originated from electronic transitions in molecules. Its presence is a source of vibrational molecular motions, thus giving the possibility of using IR spectroscopy to detect the presence of characteristic chemical groups, allowing the identification of polyphenols, flavonoids, and alkaloids.

The absorption spectrum of acetone extract shows one clearly visible peak at 270 nm and two wide ranges: at about 350 nm and the second at 475 nm. This indicates that this solvent is more specific than the other as the spectra of the other extracts and sirup (juice) are less structured with only one clearly visible peak at 280 nm. However, in the visible part of spectra, we may detect some specific peaks, as shown in the inset in Figure 1. Therefore, local absorption maxima are observed at 400 nm in water extract, at 527 nm in methanol, and at 479 and 516 nm in acetone. The smoothest spectrum with barely visible peaks is observed for sirup, however, with absorption values in the whole range of the investigated spectrum area. This means that the recorded juice spectrum is an overlapped sum of absorption of all compounds present in natural juice, which is partially detected with other extracts. The distinct absorption bands at 270–280 nm are characteristic of tocopherols, gallic acid, or other phenolic acids and flavonoids. The range of 350–400 nm can be assigned to ascorbic acid and indicate the presence of flavonoids (II absorption band). The range between 410 and 550 nm is characteristic for flavonols and alkaloids, and the presented assignment is similar to that reported [4,21–24].

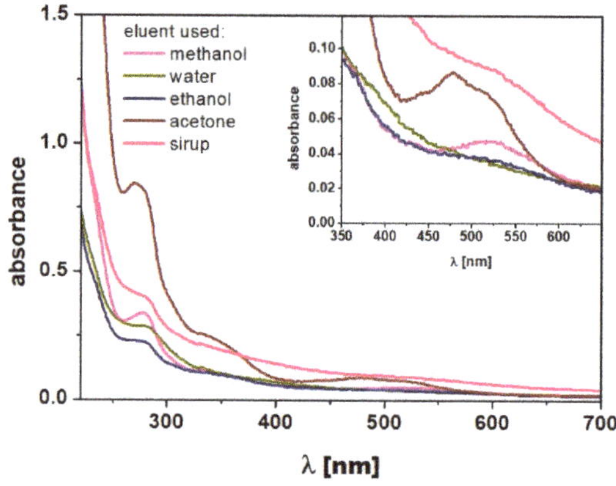

Figure 1. Absorption spectra of extracts and sirup.

3.2. Fluorescence Emission and Total Fluorescence Spectra

Fluorescence spectra were collected at the excitation range of 200 to 400 nm with 10 nm steps. This led to maps of total fluorescence presented in Figure 2.

Analysis of the obtained fluorescence spectra revealed that three basic bands of fluorescence spectra shapes can be distinguished upon the excitation wavelength range. Representative fluorescence spectra are presented also in Figure 2 (each set of fluorescence spectra is placed above the corresponding fluorescence map). From total fluorescence maps (Figure 2), it follows that for all investigated samples, at least two fluorophores are present in each extract and in sirup. The longer the excitation wavelength is, the longer the fluorescence wavelength that takes place. The presence of two different fluorophores excited at wavelengths of 280 nm and 310 nm, respectively, is clearly seen for water and acetone extract, as well as for fresh sirup. Moreover, detailed insight into the set of fluorescence spectra of each extract reveals the presence of a third fluorophore (less manifested due to the much lower fluorescence intensity of the compound). The third component is characterized with a broad fluorescence band with a maximum close to 316 nm, which can be considered as the main fluorescence band, but slightly shifted to longer wavelengths (the shift depends on the extract) and substantially different shape.

Finally, we conclude that in the investigated samples, at least three fluorophores are present with fluorescence bands placed at 314–318 nm, 325–355 nm, and 428–435 nm, respectively (dependent on type of eluent used), which confirms the presence of phenolic acids in each sample [25] and tocopherols [13]. The most diversified in terms of the fluorescence (especially fluorescence intensity) of each fluorescent compound is acetone and methanol extract. For water extract, variations in the fluorescence intensity of each compound are much less visible. This leads to the conclusion that each solvent used elutes fluorescent compounds present in fruits and sirup with different efficiencies. The most efficient globally is the use of acetone as an eluent as the fluorescence of each compound is the most intensive among all investigated samples.

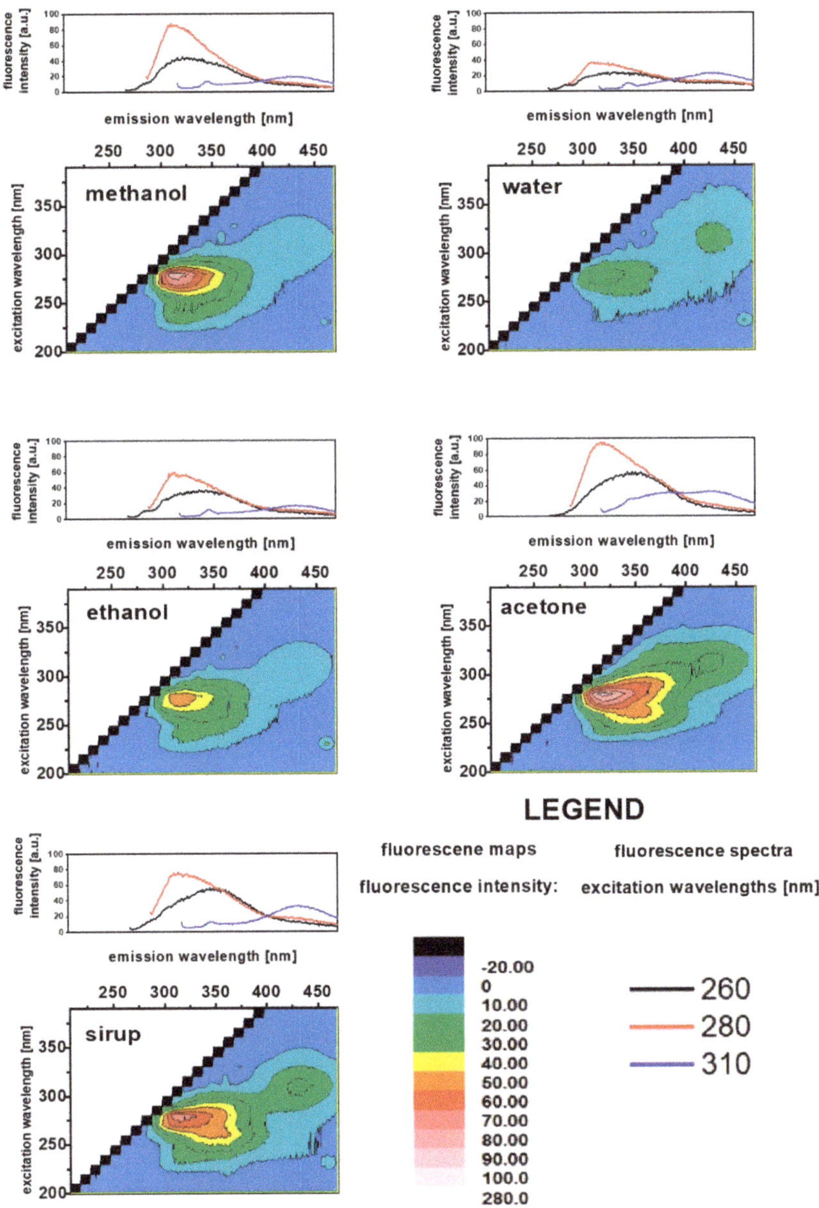

Figure 2. Fluorescence spectra registered for λ_{ex} = 260, 280, and 310 nm and total fluorescence maps of investigated extracts and sirup.

3.3. HPLC Results

The chromatograms obtained for the investigated extracts and sirup are presented in Figures 3 and S2. The four main peaks were clearly observed in nonaqueous extracts. The corresponding UV–Vis spectra of compounds **1–4** are presented in Figure S3. ESI–MS spectra were recorded for isolated compounds. The ESI–MS spectrum of compound **1** (r.t. 12.0 min) in positive mode (Supplementary Figure S4) shows the three main ions at m/z

287, 449, and 595, which correspond to the cyanidin chromophore itself, and glucoside and rutinoside of cyanidin, respectively. The compounds with retention times of 17.5 min (marked **3**) and 19.0 min (marked **4**) are glycosides of quercetin, as, in both cases, ESI–MS spectra contain the same ion signals at m/z 303 in positive mode. In the case of compound **3**, it is difficult to distinguish between glucoside or galactoside of quercetin as the obtained m/z signal is 463 (negative mode) (Supplementary Figure S5). The ESI–MS spectrum of compound **4** contains a signal at m/z 433 (negative mode), which can be arabinoside of quercetin (Supplementary Figure S6). In the case of compound **2** with r.t. 13.8, its UV–Vis spectrum and retention time were identical to the chlorogenic acid standard sample (Supplementary Figures S2 and S3).

Table 1. The list of compounds found in nonaqueous extracts with possible identification.

Peak Mark	Retention Time [min]	Found Mass [m/z]	Compound
		287 [M]$^+$	cyanidin
1	12.0	449 [M]$^+$ 447 [M−H]$^-$	3-glucoside of cyanidin
		595 [M]$^+$ 593 [M−H]$^-$	3-rutinoside of cyanidin
2	13.8	-	chlorogenic acid
3	17.5	303 [M+H]$^+$	quercetin
		463 [M−H]$^-$	quercetin glycoside
4	19.0	303 [M+H]$^+$	quercetin
		433 [M−H]$^-$ 457 [M+Na]$^+$	quercetin glycoside

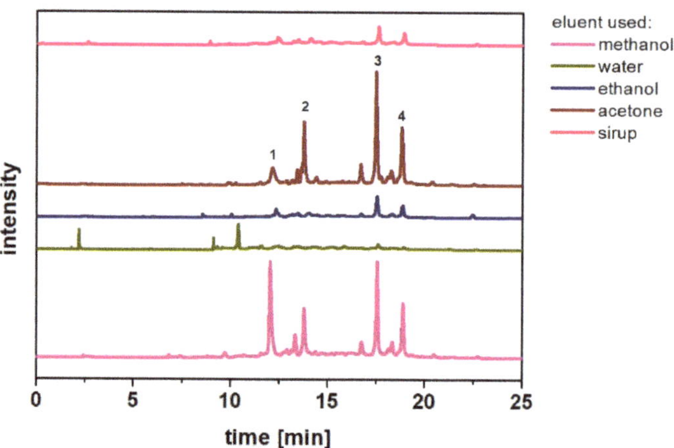

Figure 3. HPLC elution profiles (330 nm) of extracts and sirup. The proposed identification of main components (marks with numbers) was collected in Table 1.

Therefore, the HPLC chromatographic method was used to quantify and specify the presence of phenolic and flavonoid compounds in the investigated extracts, which was supported by UV–Vis and ESI–MS spectra of isolated compounds (Table 1). The obtained results show the highest amount of cyanidin in methanol extract; however, in acetone and ethanol extracts, it was also detected. The chlorogenic acid and quercetin glycosides were clearly visible in sirup and all extracts except water extract in which no presence of any of compounds **1–4** were detected.

3.4. GC–MS Results

Gas chromatography analysis is mainly dedicated to volatile and/or hydrophobic compound analysis and identification. However, the derivatization procedure gives the possibility to increase its applicability to more polar compounds. The chromatograms of the analyzed extract are presented in Figure 4. The list of compounds with possible identification (based on MS spectra of standard compounds) is presented in Table 2. The results obtained show the variation in the composition of the different extracts. The acetone and methanol extract samples proved to be the most complex (after derivatization), and the water extract sample the least complex. In general, these observations are in good agreement with the HPLC results. However, it should be taken into account that silanized samples were analyzed by GC–MS, which will significantly change the selectivity of the analytical method; similarly, the identification of compounds should be considered as approximate, due to the complexity of the matrix, different susceptibility of the components to derivatization, and limitations in the availability of spectra of standards. However, it can be noted that the acetone extract sample contained relatively high amounts of carboxylic acids (such as citric and malic acids). In contrast, the methanol extract sample contained the most glucose and its derivatives (compared to other samples).

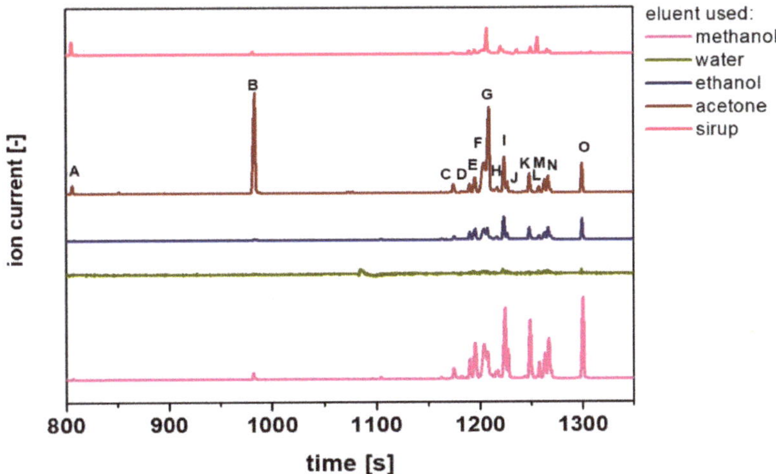

Figure 4. The chromatograms (ion current for m/z 73) of derivatized samples of tested extracts. The proposed identification of main components (marks with capital letters) was collected in Table 2.

Table 2. The list of compounds with possible identification.

Compound Mark	Retention Time [s]	Compound Name
A	806	glycerol
B	981	malic acid
C	1174	ribitol
D	1190	fructose
E	1195	erythrofuranose
F	1204	gluconic acid
G	1207	citric acid
H	1209	ribose
I	1224	glucose derivatives (with oxidized certain hydroxyl groups)
J	1227	glucose derivatives (with oxidized certain hydroxyl groups)
K	1249	glucose derivatives (with oxidized certain hydroxyl groups)
L	1256	mannonic acid

Table 2. *Cont.*

Compound Mark	Retention Time [s]	Compound Name
M	1263	levoglucosan
N	1267	sorbitol
O	1300	glucose

3.5. FTIR Results

FTIR spectra were used to determine the presence of active compounds identifying its functional groups. The results indicated the presence of various types of compounds including phenolic, flavonoids, and glycosides. The peaks' positions are presented in Figure 5.

The strong and broad band at 3285 cm^{-1} observed in extracts belongs to stretching vibrations of phenolic hydroxyl groups conforming to the presence of phenolic compounds in the investigated extracts. The peak at 2922 cm^{-1}, 2890, and 2855 cm^{-1} belong to C-H stretching in the methyl and methylene group, indicating the presence of aromatic compounds. The band at 1632 cm^{-1} is related to the presence of the stretching vibration of C=O in aromatic ring deformation, related to the presence of flavonoids and amino acids. The band at 1717 cm^{-1} is related to stretching vibrations of the carboxyl group and indicates the presence of carboxylic acids. The band at 1410 cm^{-1} arises from C=C ring stretching. The band at 1344 cm^{-1} could be related to C-O stretching of acid groups or bending vibrations of CH_3 or CH_2 groups in carboxylic acid, indicating the presence of compounds related to alkenes. The presence of the cyanide group is particularly seen in sirup, at the range of 2260 cm^{-1}. The lack of this peak in the investigated extracts shows its nontoxic nature, probably due to applied procedures. The band at 1252 cm^{-1} indicates the presence of aliphatic esters and would be assigned to C-O vibrations, indicating the presence of hydroxyflavonoids. The peak at 1024 cm^{-1} is connected with the C-O stretching band of alcohols, whereas the band at 1089 cm^{-1} would be related to secondary alcohols.

The presented FTIR spectra showed that fruit of bird cherry contains phenols, amino acids, carboxylic acids, alkanes, secondary alcohols, glycosides, and flavonoids and the assignment of groups and compounds are similar to those previously reported [3,11,26].

3.6. Oxidation Tests

Considering the presence of antioxidant compounds, we performed additional tests to estimate the antioxidant potential of extracts of bird cherry fruits. The antioxidant properties of the obtained extract were determined with standard methods. For this purpose, we used 1,1-diphenyl-2-picrylhydrazyl (DPPH) free radicals and the oxidation of DCF-H, and the obtained results were compared with gallic acid, considered as standard. Moreover, to determine the possible role of the investigated extracts in membranes, we carried out our studies in a model membrane system consisting of PC liposomes in a solid-crystalline phase with the used C11-BODIPY581/591 probe.

3.6.1. Determination of Total Antioxidant Capacity by Inhibition of DCF-H Oxidation

The total antioxidant capacity of a biological material is defined as the resultant capacity of a test material to resist a specific oxidation reaction. As foods usually contain different antioxidants with often different mechanisms of action and different activities, the total antioxidant capacity is a useful concept for determining and comparing the antioxidant potential of different foods, body fluids, and other biological fluids [27,28].

On the basis of data presented in Figure 6, the antioxidant efficiency of each extract was estimated as an equivalent of gallic acid concentrations. The results are presented in Table 3.

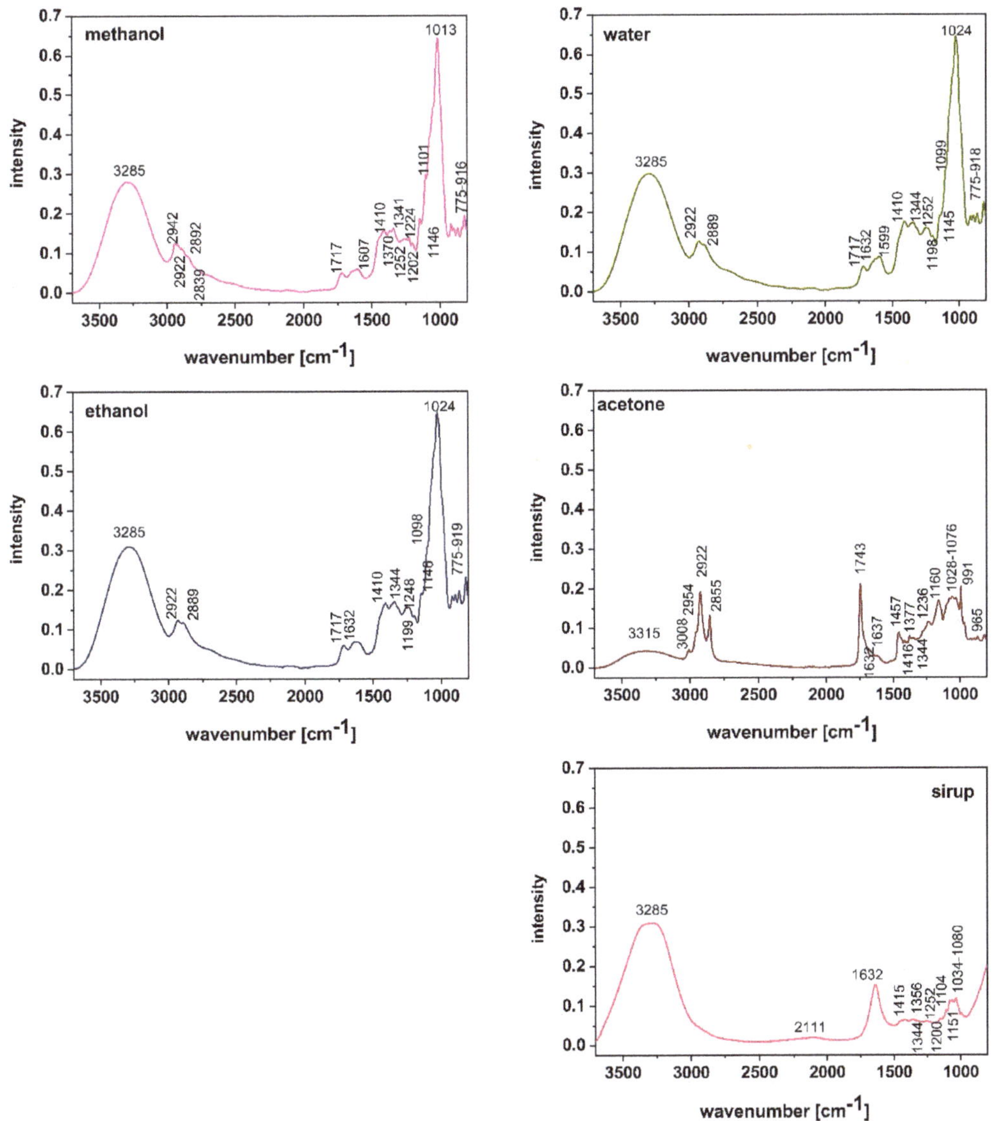

Figure 5. FTIR spectra of bird cherry extracts using hexane, methanol, ethanol, and water and measured directly from fruit.

Table 3. Antioxidant efficiency of examined extracts with standard deviation.

Eluent	MeOH	Water	EtOH	Acetone	Sirup (with no Treatment)
Antioxidant efficiency in equivalent of gallic acid (μM)	12.59 ± 0.44	2.51 ± 0.31	6.26 ± 0.36	40.4 ± 1.8	19.6 ± 0.7

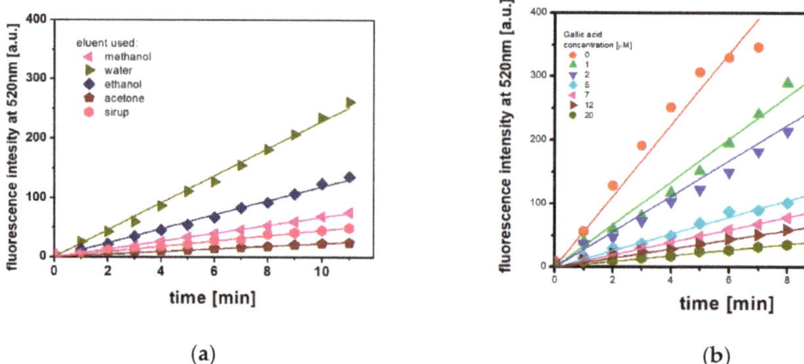

(a)   (b)

Figure 6. Time changes of DCF fluorescence maximum (λ_{ex} = 490 nm; λ_{obs} = 520 nm); (**a**) for water extract solutions; (**b**) for gallic acid water solutions.

3.6.2. The Free Radical DPPH Scavenging Activity Assay

Determination of the antiradical activity of biological fluids with the use of DPPH is often included among the methods for determining the total antioxidant capacity based on the reduction in the indicator substance. This method is commonly used to measure the anti-radical activity of products such as fruit, juices, and plant extracts [29–32]. The kinetics of DPPH scavenging in methanol solutions by the examined extracts and sirup with reference to gallic acid (12 µM) is presented in Figure 7.

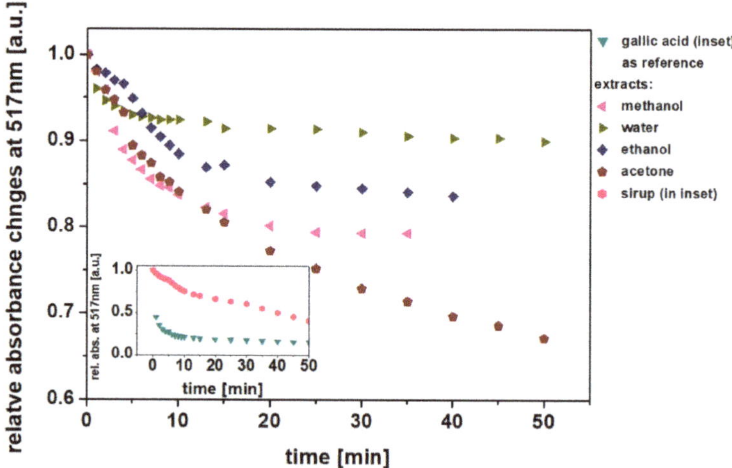

Figure 7. Scavenging kinetics of DPPH free radical in methanol by examined extracts. Inset shows the kinetics of gallic acid and sirup, as they are very effective scavengers.

On the basis of the kinetics presented in Figure 7, we calculated the maximum scavenging efficiency according to Formula (1), where A_{min} is the lowest value of DPPH absorbance (at 517 nm) observed in the plateau. The plateau was obtained in different times after extract addition, depending on the eluent used. The results are presented in Figure 8.

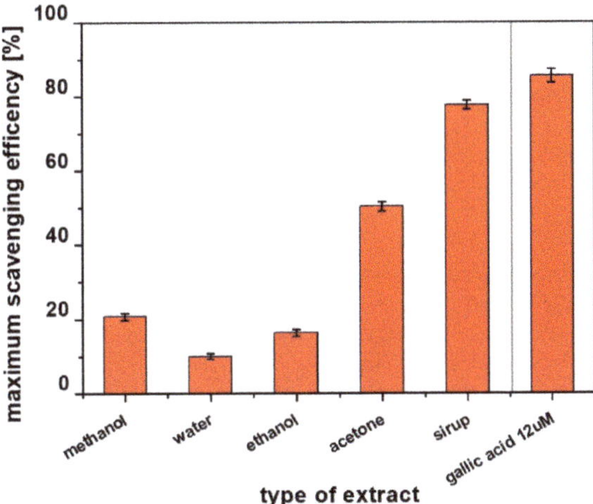

Figure 8. Maximum scavenging of DPPH free radical by extract examined.

3.6.3. Oxidation in Liposomes

The oxidation process of the lipids was measured with the C11-BODIPY581/591 probe used as a fluorescent indicator. C11-BODIPY581/591 has been commonly used to determine lipid peroxidation in liposomes and mammalian cells [13,33]. The time changes of maximum fluorescence intensity of the C11-BODIPY581/591 probe embedded in the PC membrane is presented in Figure 9. The figure reveals the thermal autooxidation of phospholipid-building PC liposomes. The blind probe is a PC membrane without any additional compounds. The "tocopherol in membrane" is the sample of PC with tocopherol embedded in the membrane in the PC liposomes preparation process (see Materials and Methods).

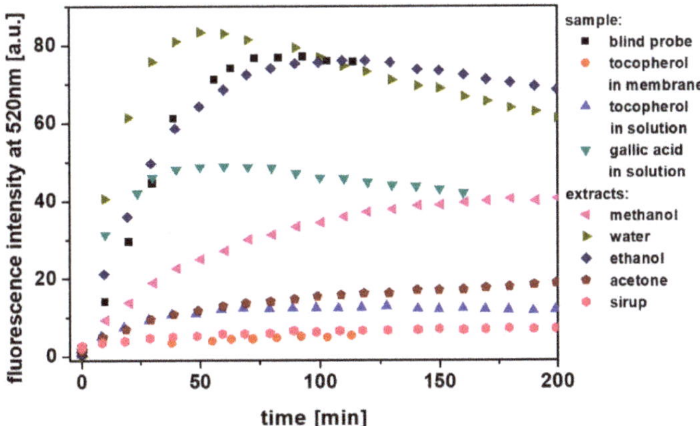

Figure 9. C11-BODIPY581/591 fluorescence changes as a function of time (excitation 505 nm; emission 520 nm).

Thermal autooxidation leads to the significant oxidation of phosphatidylcholine observed as increasing C11-BODIPY581/591 fluorescence. The fastest oxidation occurs with extracts from water and ethanol. When compounds with antioxidant potential were added, the rate of oxidation was suppressed. The presence of gallic acid decreased oxidation by about 40%. The presence of tocopherol in solution very efficiently protects lipid oxidation, where, when embedded in the PC membrane, almost completely suppressed oxidation, and this value is comparable with the antioxidant power of raw sirup (juice). Interestingly, acetone extracts exhibit a similar antioxidant power as tocopherol. We may also notice that extracts from water and ethanol show biphasic kinetics, whereas the other extracts show a monotonic behavior of action. These results indicate the efficient ability of tocopherol to scavenge formed radicals, protecting the membrane lipids from autooxidation.

4. Discussion

To obtain the full spectrum of compounds present in bird cherry fruits, we applied solvents with different protic and hydrophobic properties. During these studies, we identified in different extracts a number of polyphenols and flavonoids and some minor components, including tocopherols, alkaloids, and sugars, and this composition was similarly reported in [14,15,18,34,35]. Donno et al. [14] detected some polyphenolic compounds including cinnamic acids (relatively high content of caffeic, coumaric, and ferulic), flavonols (especially quercetin and quercitrin), benzoic acids, catechins, and tannins, as well as monoterpenes, other organic acids (especially citric and quinic), and vitamin C in fresh fruits of bird cherry. Mikulic-Petrovsek et al. [15] presented very detailed results of bird cherry (*Prunus padus*), as well as other similar species such as mahaleb cherry (*Prunus mahaleb*), wild cherry (*Prunus avium*), and blackthorn (*Prunus spinosa*) fruit content, including not only compounds presented by [14] with much more detailed insight into each group, but also other natural compounds such as chlorophylls, carotens and carotenoids, xanthophylls, and different anthocyanins including cyanidin derivatives. Similar findings were recently presented by Brozdowski et al. [34] for fruits of another *Prunnus* species—black cherry (*Prunus serotina*). A high content of quercetin and cyanidins in bird fruit was also reported by Lenchyk et al. [35].

The focus of our studies is to estimate the antioxidant potential of the aqueous solution of the obtained extracts and raw sirup because the detailed phytochemical characterization of fruit was already described by others [14,15,18,34,35]. The presence of bioactive compounds confirmed in this work, including antioxidants (polyphenols, cyanidins, and quercetins), as well as sugars, organic acids, etc., in extracts indicates the great potential of extracts to be used as antioxidant agents in everyday diet. Lipid oxidation, especially of unsaturated lipids, is one of the main concerns in the food industry as this process leads to food deterioration. To prevent or inhibit oxidation processes, the most effective way is to include antioxidants into food. Knowing that the investigated extracts contain antioxidant compounds, we apply DCF-H and BODIPY fluorescent probes, very sensitive markers of oxidation processes, to determine the antioxidant capacity of extracts against lipid peroxidation.

Investigation of lipid peroxidation in PC liposomes allow one to directly estimate the influence of free radicals in pure systems in the absence of the other interfering reactions. Additionally, it is possible to control the structural properties of phospholipids by choosing an appropriate phase; in our case, it was the fluid phase of the PC liposome that gives a better opportunity of the oxidation reaction carried out with unsaturated fatty acids chains present in PC liposomes. In addition, it is much easier to evaluate the antioxidant activity of the investigated extracts by monitoring the lipid peroxidation using the fast oxidation test based on the physical method of fluorescence of the specialized fluorescent dyes, DCF-H and BODIPY, sensitive to the presence of free radicals and oxidized species, as has been used in this study.

The antioxidant activity can be traditionally correlated with the number of phenolic compounds in plants. Phenolic compounds and flavonoids can be considered their secondary metabolites. Phenolic compounds would be the ones directly correlated with the antioxidant activity, but flavonoids could also be responsible for this effect. Their mechanism of action is related to their ability to donate hydrogen and scavenge free radicals. The ability to scavenge the free radicals allows these compounds to interact with reactive oxygen species (ROS), which can lead to oxidative stress and damage to tissues. Phenolic compounds and flavonoids, by reacting with ROS, avoid oxidative stress that could delay damaging processes, thus contributing to their antioxidant activity. Our HPLC, GC–MS and FTIR studies, shown in Figures 3–5 and Tables 1 and 2, confirmed the presence of phenolics and flavonoids in the investigated extracts, and the test carried out confirmed their antioxidant properties against the generated free radicals. The use of different extracts showed that the antioxidant properties of the eluted compounds are related to their chemical composition/structures shown in Tables 1 and 2. The lowest antioxidant efficiency is observed for water and ethanol extracts (Table 3), which mainly include sugars and phenolic acids, suggesting that this type of chemical compound exhibits a low inhibition efficiency against peroxidation, as shown in Figure 9. The highest antioxidant index was obtained for acetone extract, containing tocopherols and flavonoids, as shown by fluorescence and HPLC data, where both classes of compounds are well-documented inhibitors of free radical oxidation. Obtained for raw sirup from fruits, a high value of antioxidant index confirmed its potent antioxidant properties. Despite the measured concentration of tocopherols being relatively small due to the presence of many compounds with antioxidant properties, we may assume the existence of synergy between antioxidants, explaining such a high value of inhibition against the observed lipid peroxidation in PC membranes.

5. Conclusions

Chromatographic and spectroscopic analysis of extracts of *Prunus padus* allowed for the identification and characterization of phenolic and flavonoids. The results show that fruit of bird cherry is full of secondary metabolites, which constitute a mixture with strong antioxidant activity. Our studies identified compounds that have been reported previously; however, the presented studies with membrane model systems using PC liposomes in the liquid-crystalline state have shown that the investigated extracts are able to delay oxidation processes not only in homogenic but also in heterogenous environments, including model biological membranes. The reported high antioxidant properties of raw sirup, squeezed directly from bird cherry fruits, may also arise from the synergistic effect that has been observed between compounds detected in other systems. Finally, the extracts and sirup could be considered as food supplements of naturally occurring antioxidants with large application potential in the food industry.

Supplementary Materials: The following supporting information can be downloaded at: https://www.mdpi.com/article/10.3390/app12157820/s1, Figure S1: Size distribution of liposomes used. Figure S2: Absorption spectra of compounds **1–4** (HPLC) and standard samples. Figure S3: HPLC elution profiles (330 nm) of methanol extract of bird cherry fruit and standard sample of chlorogenic acid. Figure S4: ESI–MS spectrum of compound **1**. Figure S5: ESI–MS spectrum of compound **3**. Figure S6: ESI–MS spectrum of compound **4**.

Author Contributions: Conceptualization, P.S., K.P., and G.N.; methodology, P.S., K.P., and G.N.; software, K.P., G.N., and P.S.; validation, K.P., G.N., and P.S.; formal analysis, P.S., K.P., and G.N.; investigation, P.S., W.S., G.N., J.N.-K., and A.D.; resources, K.P.; data curation, P.S., K.P., and G.N.; writing—original draft preparation, P.S., K.P., and G.N.; writing—review and editing, P.S., W.S., J.N.-K., A.D., K.P., and G.N.; visualization, P.S., W.S., J.N.-K., K.P., and G.N.; supervision, K.P. All authors have read and agreed to the published version of the manuscript.

Funding: This work was partially supported by grant 508.782.00 from the Poznan University of Life Sciences.

Institutional Review Board Statement: Not applicable.

Informed Consent Statement: Not applicable.

Data Availability Statement: Data are available from the corresponding author on reasonable request.

Conflicts of Interest: The authors declare no conflict of interest.

References

1. Hasana, H.; Desalegn, E. Characterization and quantification of phenolic compounds from leaf of *Agarista salicifolia*. *Herb. Med. Open Access* **2017**, *3*, 1. [CrossRef]
2. Kowalczewski, P.Ł.; Olejnik, A.; Świtek, S.; Bzducha-Wróbel, A.; Kubiak, P.; Kujawska, M.; Lewadowicz, G. Bioactive compounds of potato (*Solanum tuberosum* L.) juice: From industry waste to food and medical applications. *Crit. Rev. Plant Sci.* **2022**, *41*, 52–89. [CrossRef]
3. Oliveira, R.N.; Mancini, M.C.; Oliveira, F.C.S.D.; Passos, T.M.; Quilty, B.; Thiré, R.M.D.S.M.; McGuinness, G.B. FTIR analysis and quantification of phenols and flavonoids of five commercially available plants extracts used in wound healing. *Matéria* **2016**, *21*, 767–779. [CrossRef]
4. Merken, H.M.; Beecher, G.R. Measurement of food flavonoids by high-performance liquid chromatography: A review. *J. Agric. Food Chem.* **2000**, *48*, 577–599. [CrossRef]
5. Singh, R.; Mendhulkar, V.D. FTIR studies and spectrophotometric analysis of natural antioxidants, polyphenols and flavonoids in Abutilon indicum (Linn) Sweet leaf extract. *J. Chem. Pharm. Res.* **2015**, *7*, 205–211.
6. Ashokkumar, R.; Ramaswamy, M. Phytochemical screening by FTIR spectroscopic analysis of leaf extracts of selected Indian Medicinal plants. *Int. J. Curr. Microbiol. Appl. Sci.* **2014**, *3*, 395–406.
7. Karaca, A.C.; Nicerson, M.; Low, N.H. Microcapsule production employing chickpea or lentil protein isolates and maltodextrin: Physicochemical properties and oxidative protection of encapsulated flaxseed oil. *Food Chem.* **2013**, *139*, 448–457. [CrossRef]
8. Karrar, E.; Mahdi, A.A.; Sheth, S.; Ahmed, I.A.M.; Manzoor, M.F.; Wei, W.; Wang, X. Effect of maltodextrin combination with gum arabic and whey protein isolate on the microencapsulation of gurum seed oil using a spray-drying method. *Int. J. Biol. Macromol.* **2021**, *171*, 208–216. [CrossRef]
9. Paramera, E.I.; Konteles, S.; Karathanos, V.T. Microencapsulation of curcumin in cells of *Saccharomyces cerevisiae*. *Food Chem.* **2011**, *125*, 892–902. [CrossRef]
10. Cichocki, W.; Czerniak, A.; Smarzyński, K.; Jeżowski, P.; Kmiecik, D.; Baranowska, H.M.; Walkowiak, K.; Ostrowska-ligęza, E.; Różańska, M.B.; Lesiecki, M.; et al. Physicochemical and Morphological Study of the *Saccharomyces cerevisiae* Cell-Based Microcapsules with Novel Cold-Pressed Oil Blends. *Appl. Sci.* **2022**, *12*, 6577. [CrossRef]
11. Kavosi, M.; Mohammadi, A.; Shojaee-Alibadi, A.; Khaksar, R.; Hosseini, S.M. Characterization and oxidative stability of purslane seed oil microencapsulated in yeast cells biocapsules. *J. Sci. Food. Aric.* **2017**, *98*, 2490–2497. [CrossRef]
12. Niculae, G.; Lacatusu, I.; Badea, N.; Stan, R.; Vasile, B.S.; Meghea, A. Rice bran and raspberry seed oil-based nanocarriers with self-antioxidative properties as safe photoprotective formulations. *Photochem. Photobiol. Sci.* **2014**, *13*, 703–716. [CrossRef]
13. Neunert, G.; Górnaś, P.; Dwiecki, K.; Siger, A.; Polewski, K. Synergistic and antagonistic effects between alpha-tocopherol and phenolic acids in liposome system: Spectroscopic study. *Eur. Food Res. Technol.* **2015**, *241*, 749–757. [CrossRef]
14. Donno, D.; Mellano, M.G.; De Biaggi, M.; Riondato, I.; Rakotoniaina, E.N.; Beccaro, G.L. New findings in *Prunus padus* L. fruits as a source of natural compounds: Characterization of metabolite profiles and preliminary evaluation of antioxidant activity. *Molecules* **2018**, *23*, 725. [CrossRef]
15. Mikulic-Petkovsek, M.; Stampar, F.; Veberic, R.; Sircelj, H. Wild *Prunus* fruit species as a rich source of bioactive compounds. *J. Food Sci.* **2016**, *81*, C1928–C1937. [CrossRef]
16. Kucharska, A.Z.; Oszmianand'ski, J. Anthocyanins in fruits of *Prunus padus* (bird cherry). *J. Sci. Food Agric.* **2002**, *82*, 1483–1486. [CrossRef]
17. Smyk, B.; Pliszka, B.; Drabent, R. Interaction between Cyanidin 3-glucoside and Cu(II) ions. *Food Chem.* **2008**, *107*, 1616–1622. [CrossRef]
18. Telichowska, A.; Kobus-cisowska, J.; Szulc, P. Phytopharmacological possibilities of bird cherry *Prunus padus* L. and *Prunus serotina* L. Species and their bioactive phytochemicals. *Nutrients* **2020**, *12*, 1966. [CrossRef]
19. Kumarasamy, Y.; Cox, P.J.; Jaspars, M.; Nahar, L.; Sarker, S.D. Comparative studies on biological activities of *Prunus padus* and P. *spinosa*. *Fitoterapia* **2004**, *75*, 77–80. [CrossRef]
20. Halarewicz, A.; Gabryś, B. Probing behavior of bird cherry-oat aphid *Rhopalosiphum padi* (L.) on native bird cherry *Prunus padus* L. and alien invasive black cherry *Prunus serotina* Erhr. in Europe and the role of cyanogenic glycosides. *Arthropod. Plant. Interact.* **2012**, *6*, 497–505. [CrossRef]
21. Michalak, A. Phenolic compounds and their antioxidant activity in plants growing under heavy metal stress. *Pol. J. Environ. Stud.* **2006**, *15*, 523–530.
22. Sathish, S.S. Phytochemical analysis of *Vitex altissima* L. using UV-VIS, FTIR and GC-MS. *Int. J. Pharm. Sci. Drug Res.* **2012**, *4*, 56–62.
23. Komal Kumar, J.; Devi Prasad, A.G. Identification and comparison of biomolecules in medicinal plants of *Tephrosia tinctoria* and *Atylosia albicans* by using FTIR. *Rom. J. Biophys.* **2011**, *21*, 63–71.
24. Saxena, M.; Saxena, J. Evalution of phytoconstituents of *Acorus Calamus* by FTIR and UV-VIS spectroscopic analysis. *Int. J. Biol. Pharm. Res. Int. J. Biol. Pharm. Res.* **2012**, *3*, 498–501.
25. Carvalho, D.G.; Ranzan, L.; Jacques, R.A.; Trierweiler, L.F.; Trierweiler, J.O. Analysis of total phenolic compounds and caffeine in teas using variable selection approach with two-dimensional fluorescence and infrared spectroscopy. *Microchem. J.* **2021**, *169*, 106570. [CrossRef]

26. Heneczkowski, M.; Kopacz, M.; Nowak, D.; Kuźniar, A. Infrared spectrum analysis of some flavonoids. *Acta Pol. Pharm.* **2001**, *58*, 415–420.
27. Górnaś, P.; Dwiecki, K.; Siger, A.; Tomaszewska-Gras, J.; Michalak, M.; Polewski, K. Contribution of phenolic acids isolated from green and roasted boiled-type coffee brews to total coffee antioxidant capacity. *Eur. Food Res. Technol.* **2016**, *242*, 641–653. [CrossRef]
28. Nazeri, S.; Yaghmaie, B.; Hedayati, M. Simple and sensitive method of fluorometry for determination of total antioxidant capacity. *Mod. Med. Lab. J.* **2017**, *1*, 29–35. [CrossRef]
29. Rubalya Valantina, S.; Neelamegam, P. Selective ABTS and DPPH- radical scavenging activity of peroxide from vegetable oils. *Int. Food Res. J.* **2015**, *22*, 289–294.
30. Troup, G.J.; Navarini, L.; Liverani, F.S.; Drew, S.C. Stable radical content and anti-radical activity of roasted arabica coffee: From in-tact bean to coffee brew. *PLoS ONE* **2015**, *10*, e0122834. [CrossRef]
31. Rijai, H.R.; Fakhrudin, N.; Wahyuono, S. Isolation and identification of DPPH radical (2,2-diphenyl-1-pikrylhidrazyl) scavenging active compound in ethyl acetat fraction of Piper acre Blume. *Maj. Obat Tradis.* **2019**, *24*, 204. [CrossRef]
32. Adejoh, I.; Barnabas, A.; Chiadikaobi, O. Comparative anti-radical activity of five indigenous herbal plants and their polyherbal extract. *Int. J. Biochem. Res. Rev.* **2016**, *11*, 1–10. [CrossRef]
33. Pap, E.H.W.; Drummen, G.P.C.; Winter, V.J.; Kooij, T.W.A.; Rijken, P.; Wirtz, K.W.A.; Op Den Kamp, J.A.F.; Hage, W.J.; Post, J.A. Ratio-fluorescence microscopy of lipid oxidation in living cells using C11-BODIPY(581/591). *FEBS Lett.* **1999**, *453*, 278–282. [CrossRef]
34. Brozdowski, J.; Waliszewska, B.; Loffler, J.; Hudina, M.; Veberic, R.; Mikulic-Petkovsek, M. Composition of phenolic compounds, cyanogenic glycosides, organic acids and sugars in fruits of Black Cherry (*Prunus serotina* Ehrh.). *Forests* **2021**, *12*, 762. [CrossRef]
35. Lenchyk, L.V.; Upyr, D.V.; Ovezgeldiyev, D. Phytochemical investigation of bird cherry fruits. *Der Pharm. Lett.* **2016**, *8*, 73–76.

Article

Effect of Co-Encapsulated Natural Antioxidants with Modified Starch on the Oxidative Stability of β-Carotene Loaded within Nanoemulsions

Ahmad Ali [1], Abdur Rehman [1], Seid Mahdi Jafari [2], Muhammad Modassar Ali Nawaz Ranjha [3,*], Qayyum Shehzad [4], Hafiz Muhammad Shahbaz [5], Sohail Khan [1], Muhammad Usman [4,6], Przemysław Łukasz Kowalczewski [7], Maciej Jarzębski [8] and Wenshui Xia [1,*]

[1] State Key Laboratory of Food Science and Technology, School of Food Science and Technology, Collaborative Innovation Center of Food Safety and Quality Control in Jiangsu Province, Jiangnan University, Wuxi 214122, China; ahmadaliju@yahoo.com (A.A.); hafizje145@gmail.com (A.R.); sohailkhan6566@yahoo.com (S.K.)

[2] Department of Food Materials and Process Design Engineering, Gorgan University of Agricultural Science and Natural Resources, Gorgan 49165, Iran; jafari01@gmail.com

[3] Institute of Food Science and Nutrition, University of Sargodha, Sargodha 40100, Pakistan

[4] Beijing Advanced Innovation Center for Food Nutrition and Human Health, School of Food and Health, Beijing Technology and Business University, Beijing 100048, China; qayyum.shehzad.5@gmail.com (Q.S.); ch.usman1733@gmail.com (M.U.)

[5] Department of Food Science and Human Nutrition, University of Veterinary and Animal Sciences, Lahore 54000, Pakistan; muhammad.shahbaz@uvas.edu.pk

[6] Department of Food Science and Technology, Riphah International University, Faisalabad 38000, Pakistan

[7] Department of Food Technology of Plant Origin, Poznań University of Life Sciences, 31 Wojska Polskiego St., 60-624 Poznań, Poland; przemyslaw.kowalczewski@up.poznan.pl

[8] Department of Physics and Biophysics, Poznań University of Life Sciences, Wojska Polskiego 38/42, 60-637 Poznań, Poland; maciej.jarzebski@up.poznan.pl

* Correspondence: modassarranjha@gmail.com (M.M.A.N.R.); xiaws@jiangnan.edu.cn (W.X.)

Citation: Ali, A.; Rehman, A.; Jafari, S.M.; Ranjha, M.M.A.N.; Shehzad, Q.; Shahbaz, H.M.; Khan, S.; Usman, M.; Kowalczewski, P.Ł.; Jarzębski, M.; et al. Effect of Co-Encapsulated Natural Antioxidants with Modified Starch on the Oxidative Stability of β-Carotene Loaded within Nanoemulsions. *Appl. Sci.* **2022**, *12*, 1070. https://doi.org/10.3390/app12031070

Academic Editor: Anna Lante

Received: 15 December 2021
Accepted: 17 January 2022
Published: 20 January 2022

Publisher's Note: MDPI stays neutral with regard to jurisdictional claims in published maps and institutional affiliations.

Abstract: β-Carotene (vitamin A precursor) and α-tocopherol, the utmost energetic form of vitamin E (VE), are known to be fat-soluble vitamins (FSVs) and essential nutrients needed to enhance the growth and metabolic functions of the human body. Their deficiencies are linked to numerous chronic disorders. Loading of FSVs within nanoemulsions could increase their oxidative stability and solubility. In this research, VE and β-Carotene (BC) were successfully co-entrapped within oil-in-water nanoemulsions of carrier oils, including tuna fish oil (TFO) and medium-chain triglycerides (MCTs), stabilized by modified starch and Tween-80. These nanoemulsions and free carrier oils loaded with vitamins were stored for over one month to investigate the impact of storage circumstances on their physiochemical characteristics. Entrapped bioactive compounds inside the nanoemulsions and bare oil systems showed a diverse behavior in terms of oxidation. A more deficiency of FSVs was found at higher temperatures that were more noticeable in the case of BC. VE behaved like an antioxidant to protect BC in MCT-based nanoemulsions, whereas it could not protect BC perfectly inside the TFO-loaded nanoemulsions. However, cinnamaldehyde (CIN) loading significantly enhanced the oxidative stability and FSVs retention in each nanoemulsion. Purity gum ultra (PGU)-based nanoemulsions comprising FSVs and CIN presented a greater BC retention (42.3%) and VE retention (90.1%) over one-month storage at 40 °C than Twee 80. The superior stability of PGU is accredited to the OSA-MS capabilities to produce denser interfacial coatings that can protect the entrapped compounds from the aqueous phase. This study delivers valuable evidence about the simultaneous loading of lipophilic bioactive compounds to enrich functional foods.

Keywords: fat-soluble vitamins; cinnamaldehyde; nanoemulsions; purity gum ultra; physicochemical stability

1. Introduction

Fat-soluble vitamins (FSVs) are biologically active compounds with numerous therapeutic biological roles and are known to be essential nutrients that can enhance the growth and metabolic functions of the human body. Among FSVs, α-tocopherol (the utmost energetic form of vitamin E) and β-carotene (vitamin A precursor) have been exploited as potent antioxidant agents. In previous studies, it has been well clarified that chronic ailments, including aging, cancers, diabetes, and cardiovascular disorders, can be reduced by the consumption of FSVs [1–3]. Thus, the food industries are showing their keen attention in the production and enrichment of diverse functional foods such as supplements and beverages with lipophilic b, e.g., ω-3 fatty acids, curcuminoids, and FSVs [4,5]. Though, the addition of lipophilic bioactives into the aqueous-based food matrices is restricted because of their structure break-down and poor solubility nature, as well as high sensitivity to environmental temperature, oxygen, and light [6].

Various lipid-based and biopolymeric nanocarriers have been established, including nanoemulsions, nanoliposomes, nanohydrogels, and solid-lipid nanoparticles for the entrapment of vitamins [6–8]. Among these, oil-in-water (O/W) nanoemulsions have been considered to be efficient delivery systems because of their superior stability in contradiction of gravitational separation, flocculation, and coalescence [9]. However, the nanometric diameter of nanoemulsions provides a higher surface area that facilitates interaction between the aqueous phase and unsaturated lipids present in the dispersed phase droplets, which are known to be extremely susceptible to oxidation [10,11]. On the other hand, the structural composition of nanoemulsions, particularly the type of antioxidants, carrier oils OSA-modified starch, can leave negative impacts on the oxidation of loaded lipophilic bioactive compounds [12,13].

It should be noted that FSVs first need to be dissolved into an appropriate carrier oil before loading into the nanoemulsions. Previous literature has revealed that since carrier oils have a higher unsaturation degree, they exhibit poor stability in the direction of oxidation rather than those oils that have lower unsaturation levels [14,15]. For this purpose, antioxidants have been recognized as promising candidates, which are being used to avoid oxidation issues of lipid-based matrices; though, the behavior of each antioxidant agent could be different for each system; because oxidation is a complicated phenomenon [16,17]. For instance, Chaiyasit et al. [18] reported that α-tocopherol was more effective as compared to δ-tocopherol inside the menhaden oil; but they found a different trend in terms of antioxidant effect inside the menhaden oil-loaded emulsions. In addition, butylated hydroxyl toluene (BHT) displayed an identical efficiency in each formulation of emulsions. In another research, Serfert et al. [19] found dissimilar properties of lecithin when combined with ascorbyl palmitate used for O/W nanoemulsions rather than spray-dried nanoemulsions.

Usage of innovative techniques has become common for production and processing [20–23]. Recently, there is a growing interest in the use of natural antioxidants instead of synthetic ones due to consumer preference for natural ingredients [24–26]. Cinnamon is recommended as an ancient spice that encompasses cinnamaldehyde, has been exploited as an antioxidant, anticancer, anti-inflammatory, antimicrobial, and insecticidal agent [27,28]. Additionally, it has been gained safe status by the Joint WHO/FAO Expert Committee on Food Additives [29]; thus, being used as an antioxidant in the nano-emulsified systems [30].

Emulsifiers can decrease the interfacial tension among the aqueous and lipid phase as well as their outstanding characteristics, such as layer thickness, binding capacity, and surface charge, which could contribute to improving the bioactivity and stability of entrapped bioactive compounds [31–33]. Among biopolymer-based emulsifiers, modified starches have been successfully used owing to their superior stability against ionic strength, a diverse range of pH values, and higher temperatures. Octenyl succinic anhydride-modified starches (OSA-MS) have also been used in several investigations as emulsifiers and stabilizers for entrapment of various lipophilic bioactive compounds, such as FSVs, curcumin, flavoring oils, and resveratrol [9,34,35].

Herein, the aim of this study was to fabricate FSVs-loaded nanoemulsions stabilized by OSA-MS and Tween 80. In addition, the effect of co-entrapped antioxidants (cinnamaldehyde) and carrier oils was also investigated.

2. Materials and Methods

2.1. Materials

Tuna fish oil as a long-chain triacylglycerol was kindly gifted by Novosana (Taicanag) Ltd., (Suzhou, China). Medium-chain triacylglycerols (MCTs) with capric acid ($\sim$44%) and caprylic acid ($\sim$56%) were supplied by TA Foods Ltd. (Yorkton, SK, Canada) and Stepan (Maywood, NJ, USA). OSA-modified starch (Purity Gum Ultra, PGU) was bought from Ingredion China Limited (Shanghai, China). Tween-80 (TW-80) was provided by Sinopharm Chemical Reagents Co., Ltd. (Shanghai, China). β-Carotene (BC) and α-tocopherol (VE) with ≥97.0% purity were obtained from Sigma-Aldrich (St. Louis, MO, USA). Cinnamaldehyde (CIN) (>98%) was purchased from Jian Ju Peng Natural Flavor Oil Co LTD (Jilin, China). Double distilled water was used for all aqueous solutions, and all other reagents and chemicals used in this study were of analytical grade.

2.2. Preparation of Nanoemulsion

The aqueous suspension was prepared by dissolving modified starch (PGU, 1.5% w/w) into double distilled water and kept stirring for the whole night at 30 °C in order to confirm perfect hydration. Aqueous suspension of Tween-80 (TW) was also dispersed (2% w/w) into double distilled water while stirring for 1 h at 25 °C. In the next stage, diverse concentrations of different bioactives (0.1% (w/w) of BC, 2% and 3% (w/w) of VE, and 1% (w/w) of CIN) were loaded into TFO and MCT oils for fabrication of lipid phases (10% w/w) as illustrated in Figure 1. In brief, 0.1% (w/w) of BC was dispersed into oil TFO and MCT and stirred for 10 min at 55 °C, followed by further stirring for 1 h at 25 °C in order to confirm perfect dissolvation. After 45 min, 2% and 3% (w/w) of VE was poured into the BC-loaded oil phase with additional stirring for 60 min. Next, 1% (w/w) of CIN was added inside the vitamin-loaded oil phase while stirring over 10 min. Then, nanoemulsions were formulated by adopting the way proposed formerly by Rehman et al. [9], with minor amendments. Firstly, the oil phase was added into aqueous at a ratio of 10:90 and then homogenized by Ultra-Turrax for 3–4 min at 18,000 rpm to obtain coarse emulsions. Secondly, 40 mL of coarse emulsions were proceeded to ultrasonication (40% of the maximum power at 13 min) to achieve nanoemulsions. In this regard, an ultrasonic processor (JY98-IIIDN, Ningbo Scientz Biotechnology Co., Ningbo, China) with a 20 mm probe diameter, with a maximum power of 1200 W and a frequency of 20 kHz, was applied. Sonication processing time and resting time were put at 5 and 7 s, respectively. Throughout the ultrasonication process, the temperature of all nanoemulsions was maintained through ice jackets and observed by a thermometer.

2.3. Measurement of Mean Particle Size (MPS), Zeta-Potential (ZP), and Polydispersity Index (PDI)

MPS, PDI, and ZP of nanoemulsions were examined through Zetasizer (Nano-ZS90, Malvern Instruments, Malvern, UK) according to the dynamic light scattering (DLS) technique at room temperature by adopting the procedure of Rehman et al. [9]. All formulations were diluted 200 times with distilled water to minimize scattering prior to the measurements and stirred to make sure consistency among samples. Every analysis was carried out three times. The surface charge on all nanoemulsion particles was obtained by 100-times dilution.

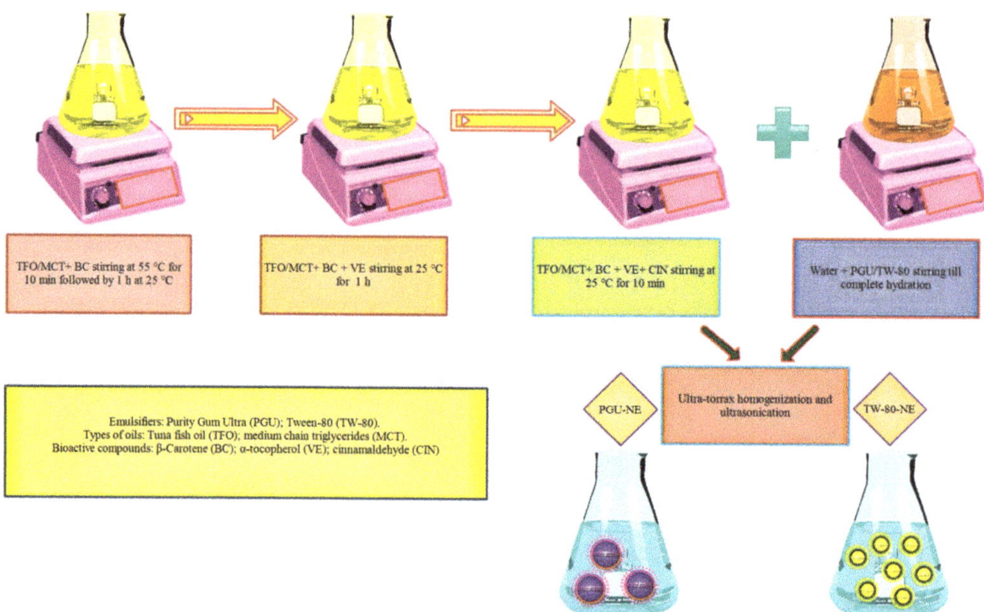

Figure 1. Schematic illustration of nanoemulsion formulations in this study.

2.4. Storage Stability of Vitamin-Loaded Nanoemulsions

All nanoemulsions were packed into tightly capped glass bottles directly after fabrication and kept at 25 and 40 °C for 30 days in the shady apartment. Each collected nanoemulsions was analyzed occasionally in terms of MPS, oxidative stability, and entrapped contents of vitamins (VE and BC).

2.4.1. Measurement of Peroxide Value (POV)

Lipid hydroperoxides of freshly prepared nanoemulsions and one-month stored nanoemulsions at 25 °C were analyzed based on oxidative stability according to the procedure proposed by Rashidinejad et al. [36], with minor modifications. Briefly, about 1 mL of each nanoemulsion sample was combined with isooctane: 2-propanol (2:1 v/v) and vortexed 3-times (10 s each time). Next, centrifuged at $1300 \times g$ for 3 min, and the solvent phase was collected, which was then added to 3.2 mL of 2:1 v/v methanol/1-butanol mixture with the addition of 3.94 M ferrous iron reagent/ammonium thiocyanate/(1:1 v/v; 30 mL). After 20 min in the dark, it was analyzed at 510 nm wavelength in a UV-visible spectrophotometer (Bckman Coulter Inc., Fullerton, CA, USA). Each sample was analyzed in triplicate.

2.4.2. Measurement of Thiobarbituric Acid Reactive Substances (TBARS)

TBARS, secondary oxidative products were analyzed by following the procedure of Qiu et al. [37], with small variations. First, a solution of thiobarbituric acid (TBA) was made by mixing 0.375 g of TBA, 15 g of trichloroacetic acid, 1.76 mL of 12 M HCl, and 82.9 mL of deionized water. For the analysis, 0.5 mL sample of oil or nanoemulsion was mixed with 2.5 mL of TBA solution in test tubes and heated for 15 min in a hot water bath at 25 °C. After this step, tubes were cooled down with tap water for 10 min and then centrifuged at $11,000 \times g$ for 15 min. The absorbance of all samples was observed at 532 nm using a UV-VIS spectrophotometer (Synergy HT, BioTek Instruments Inc., Winooski, VT, USA). The TBARS concentration was accessed by means of a 1,1,3,3-tetraethoxypropane standard curve (standard curve equation Y = 0.7499 X + 0.0366 and coefficient correlation (R2) = 0.9968). Each sample was analyzed in triplicate.

2.4.3. Measurement of BC by HPLC

To extract BC contents, 0.4 mL of each emulsion was packed in a glass tube. In brief, 1 mL of dimethyl sulfoxide was added to the emulsion sample inside the glass tube and vortexed for 3 min in order to de-emulsify the nanoemulsion. For extraction of BC contents, a 4 mL combination of dichloromethane/n-hexane (1:4 v/v) was poured into the de-emulsified sample. The obtained extracts were poured using a nylon syringe (having filter diameter of 0.22 μm) into a 2 mL amber container. After that, BC contents were measured using HPLC by adopting the procedure proposed by Yi et al. [38] in triplicate. The standard curve among BC contents and absorption peak area was designed and fitted using a linear function. All values were presented as the BC retentions, which are well-defined employing C_t/C_o, where C_t is the BC contents at storage time, and C_o is the preliminary BC contents accessed directly after fabrication of all nanoemulsions.

2.4.4. Measurement of VE by HPLC

To extract VE contents, 0.05 mL of the emulsion was de-emulsified by using 0.2 mL of dimethyl sulfoxide followed by incorporation of 5 mL composition of acetonitrile/methanol (3:97 v/v). Briefly, a 0.22 μm nylon syringe filter was used to suck the aforementioned extract, which was shifted into a 1.5 mL amber container. Next, HPLC was used to measure the VE concentration by adopting the method of Hategekimana et al. [39], with slight amendments in triplicate. The standard curve among VE contents and absorption peak area was designed and fitted using a linear function. All results were stated as the VE retentions, which are well-defined by means of C_t/C_o, where C_t is the VE contents at storage time, and C_o is the primary VE contents accessed directly after construction of all nanoemulsions.

2.5. Statistical Analysis

The experimental data were evaluated using Minitab 17 (Minitab Inc., State College, PA, USA) and one-way ANOVA; the significant differences among samples were analyzed with Duncan's multiple range test on a 95% confidence level ($p \leq 0.05$).

3. Results and Discussion

3.1. Physiochemical Stability of Vitamin-Loaded Nanoemulsions

3.1.1. MPS, ZP, and PDI of BC- and VE-Loaded Nanoemulsions

In this study, three bioactive compounds, i.e., BC, CIN, and VE, were loaded within the emulsion-based matrices by using two kinds of carrier oils, which were emulsified by PGU and TW-80. In the primary section, BC alone and with diverse levels of VE were loaded into TFO and MCT-based lipid phases, stabilized by PGU. Nanoemulsions comprising 0.1% w/w BC-loaded TSO (NE1) and MCT (NE2) exhibited a droplet size < 250 nm, as illustrated in Table 1. On the other side, mixtures of BC with diverse levels (2% and 3% w/w) of VE were equipped in both kinds of carrier oils, which were loaded into nanoemulsions emulsified by PGU and TW-80. An increment in MPS was observed for NE3 and NE5, both encompassing 2% VE, and nanoemulsions (NE4 and NE5) enriched with 4% VE showed NPS above 250 nm [40].

It has been concluded in previous studies that, incorporation of the maximum quantity of VE in any nano-emulsified system reduces the droplet size owing to the interfacial tension lowering effect of CIN [39,41]. PDI, a significant factor, has known to evaluate the distribution of particle size of any nano-emulsified system; thereby, smaller PDI results elaborate fine and smooth distribution of fabricated particles [9]. Table 1 clearly depicts that VE loading could not exhibit any significant influence on PDI as well as there was no direct relationship observed among the kinds of carrier oils and PDI variations. ZP is a key parameter that is used for evaluating the stability of emulsion-based matrices; thus, greater ZP values (>±30) are well thought-out to offer superior stability to the emulsified systems [42]. Nanoemulsions coded by NE1, NE3, and NE5 holding TFO displayed greater

ZP rather than MCT-loaded nanoemulsions (NE2, NE4, and NE6). In addition, the VE loading resulted in a decrease in the ZP values except for NE4. A negative charge on the exterior layers of OSA-MS is accredited to the succinylation used for modification of native starches. A very minor disparity in ZP findings of all nanoemulsions might be owing to the existence of diverse nature of oil combinations as well as entrapped lipophilic bioactive compounds [43].

Table 1. Effect of carrier oils, emulsifiers, and entrapped bioactive compounds on MPS, ZP, and PDI of produced nanoemulsions.

Nanoemulsions Sample Codes	Emulsifier Type	Oil Type	Bioactive Type	MPS (nm)	ZP (mV)	PDI
NE1	PGU	TFO	BC (0.2%)	235 ± 2.23 [c]	−38.2 ± 1.34	0.162 ± 0.023
NE2	PGU	MCT	BC (0.2%)	233 ± 1.42 [c]	−33.6 ± 0.45	0.134 ± 0.009
NE3	PGU	TFO	BC (0.2%) + VE (2%)	246 ± 3.12 [b]	−38.4 ± 0.76	0.129 ± 0.049
NE4	PGU	MCT	BC (0.2%) + VE (2%)	243 ± 1.45 [b]	−35.6 ± 0.54	0.112 ± 0.045
NE5	PGU	TFO	BC (0.2%) + VE (3%)	257 ± 4.67 [a]	−33.9 ± 0.32	0.147 ± 0.078
NE6	PGU	MCT	BC (0.2%) + VE (3%)	252 ± 3.22 [a]	−33.1 ± 0.92	0.103 ± 0.035
NE7	PGU	TFO	BC (0.2%) + CIN (1%)	222 ± 4.34 [d]	−35.1 ± 0.07	0.132 ± 0.003
NE8	PGU	MCT	BC (0.2%) + CIN (1%)	224 ± 5.87 [d]	−33.7 ± 0.65	0.092 ± 0.075
NE9	PGU	TFO	BC (0.2%) + VE (2%) + CIN (1%)	244 ± 3.12 [b]	−31.3 ± 0.34	0.083 ± 0.062
NE10	PGU	MCT	BC (0.2%) + VE (2%) + CIN (1%)	232 ± 4.89 [c]	−29.7 ± 0.89	0.010 ± 0.013
NE11	TW-80	TFO	BC (0.2%) + VE (2%) + CIN (1%)	186 ± 3.41 [e]	N/D	0.012 ± 0.062
NE12	TW-80	MCT	BC (0.2%) + VE (2%) + CIN (1%)	181 ±1.38 [e]	N/D	0.014 ± 0.061

Values with different letters are significantly different ($p < 0.05$).

3.1.2. Impact of Storage Conditions on Vitamin-Loaded Nanoemulsions

Storage time and temperature are those parameters that can be used to assess the stability of nanoemulsions. In order to observe the coalescence, Ostwald ripening and flocculation phenomena, all formulations were stored for 30 days at 25 and 40 °C, as well as to access the effect of storage time and temperature on mean droplet size with the aim of determining the stability of all emulsions. In order to explore the influence of temperature and time on MPS, nanoemulsions having codes (NE1-NE6) comprising BC alone or combined with VE were stored at 25 and 40 °C over 30 days. After storage of 30 days, all nanoemulsions that were kept at 25 °C showed more stability, and an extreme increment in MPS around 21 nm was observed in the case of NE6 nanoemulsion, as portrayed in Figure 2A. NE3 nanoemulsion presented the smallest increase in MPS, about 9 nm over 30 days' storage. At 40 °C, all nanoemulsions displayed a significant increase in MPS and MCT-loaded nanoemulsions encapsulating BC and VE represented as NE4 and NE6 showed an extreme rise in MPS (∼29 nm), as expressed in Figure 2B. Many researchers have covered the increasing phenomenon of MPS caused by storage circumstances in their investigations [9,44,45]. As an example, Zhao and co-workers found superior stability of nanoemulsions enriched with FSO as compared to MCT because of greater interfacial tension among aqueous suspension of nanoemulsion and FSO [45].

3.1.3. Oxidative Stability of Vitamin-Loaded Nanoemulsions

Nanoemulsions having codes (NE1-NE6) were kept at 25 °C in order to access the primary and secondary oxidation stability. Figure 3A,B shows that TFO-loaded nanoemulsions produced greater POV and TBARS oxidative products rather than MCT entrapped nanoemulsions. Additionally, NE5 nanoemulsions equipped by TFO, BC, 3% *w/w* VE had presented more instability in terms of oxidation. The MCT-loaded nanoemulsions showed higher stability toward oxidation that might be owing to containing medium-chain triglycerides, which are known for being less vulnerable toward oxidation rather than long-chain triglycerides [46]. In our earlier research, the oxidative instability of borage seed oil-loaded nanoemulsions was observed significantly [9]. In this study, loading of

BC and VE could not deliver safety; therefore, nanoemulsions encompassing 3% w/w VE presented higher oxidative instability. For instance, Sánchez et al. [47] observed more oxidative instability in emulsified systems encapsulating polyunsaturated oil and VE during storage. This contradicts with renowned phenomenon "polar paradox" that elaborated that hydrophobic and hydrophilic natural antioxidants could perform suitable results inside the emulsion-based matrices as well as in free oils, respectively. Though, numerous researches have successfully proved wrong to this hypothesis [48,49]. In order to investigate the oxidative behavior (primary and secondary oxidative products), free carrier oils, i.e., TFO, MCT, or both enriched with BC and 2% w/w VE, were kept at 25 °C. For this purpose, Figure 3C clearly depicts that both oils (TFO and MCT) showed diverse attitudes for POV. TFO alone or in combination with any studied bioactive compounds was found to be more unstable owing to more production of POV, but MCT alone or in combination with any studied bioactive compounds revealed more stability toward oxidation. In the case of secondary oxidation, TFO combined with 2% (w/w) VE had presented higher TBARS values, as illustrated in Figure 3D. MCT alone or in combination with any studied bioactive compounds proposed a minor rise in TBARS as compared to POV findings but results were significantly less rather than TFO. Our results about primary and secondary oxidative products were in accordance with findings of Jacobsen and his group [50]. Again, the greater stability of bare MCT alone in combination with bioactive compounds in terms of oxidation might be due to medium-chain triglycerides including capric acid and caprylic acid. The oxidation values of TFO seem very interesting and unique. The loading of 2% w/w VE resulted in a significant rise in the case of POV; on the other side, there was no significant influence observed on TBARS results for TFO alone after storage of 30 days, as explained in Figure 3C,D. Nevertheless, due to the loading of BC, a reduction in oxidation phenomenon was reported instead of TFO alone. In conclusion, the oxidation findings of the current study exposed that loading of VE in highly unsaturated free oil did not perform well being an antioxidant owing to the production of hydroperoxides through the accretion of tocopherols radicals. Our results were supported by the theory of the "polar paradox" and found to be in accordance with the discoveries of Chaiyasit et al. [18] and Serfert et al. [19]. The amount of an antioxidant could play a crucial part in governing oxidation. It has been investigated in previous studies that loading of higher amounts of VE in order to stabilize the bare oil rich in PUFAs is not a competent tactic, and as compared to antioxidants, it behaves as prooxidant [18,19,51]. In prior researches, the influence of VE as a prooxidant as well as an antioxidant in an emulsified system has been investigated; but this is interlinked with its entrapped amounts [52,53]. In the current study, superior oxidation level of nanoemulsions might be owing to use of greater amounts of 3% w/w VE and superior unsaturation degree of TFO.

3.1.4. Storage Stability of Vitamin-Loaded Nanoemulsions

All BC and VE enriched nanoemulsions were kept at 25 and 40 °C over 30 days of storage in order to explore the influence of storing temperatures on the retention of BC as well as of VE. A smooth declining trend in the concentration of BC was reported for all nanoemulsions except NE2 that disclosed a sharp declining trend on storage at 25 °C, and only 2.34% BC was retained over 30 days storage as depicted in Figure 4A. Loading of VE in NE4 and NE6 based on MCT provided greater BC retention around 35.17% and 33.89%, respectively, in comparison with NE2 nanoemulsion (Table 1). NEI nanoemulsions encapsulating (TFO and BC) showed the greater BC retention around 42.98% throughout storage, and further loading of VE in NE3 and NE5 nanoemulsions encapsulating (TFO and BC) reduced the BC retention. It seems that reduction in BC was reported contrarywise relational to VE concentrations. Higher amounts of 3% w/w VE resulted in a greater reduction in each oil phase, and this behavior is in accordance with our findings of oxidation as highlighted in Section 3.1.3. Although MCT showed superior oxidative stability, further assessment of BC retention explored that oxidation study is not so reliable standard to observe the stability of any system. However, further experiments are needed to explore the real providence of bioactive. As TFO has a larger molecular structure, allowing it to hold up a higher quantity of BC, but at a similar moment, it displayed very poor solubility in the direction of oxidation that might be owing to its higher unsaturation degree. Though, in earlier research, Sharif and co-workers had reported a great BC retention loaded inside MCT and stabilized via OSA-MS. The most suitable reason for higher BC retention could be due to the usage of a higher quantity of OSA starches (>30%) to stabilize emulsions rather than the present study (2%). At 40 °C, the results of BC retention could be due to the sensitivity of BC at higher temperatures. The greater declining trend in BC loss was observed even in the occurrence of VE that might be owing to the sensitivity of TFO with respect to oxidation. As our oxidation findings of nanoemulsions also showed greater oxidation for those nanoemulsions that comprise TFO alone or combined with VE, as presented in Figure 4A,B. However, NE4 and NE6 nanoemulsions framed with MCT had displayed 34.98% and 31.22% BC retention after 30 days of storage, respectively. This behavior of MCT-based nanoemulsions indicated that the existence of VE had successfully provided protection to the BC contents, but this trend was not observed for TFO-based nanoemulsions. The VE-loaded nanoemulsions were further stored at 25 and 40 °C in order to access the VE retention. Nanoemulsions equipped with 2% *w/w* of VE concentrations exhibited greater VE retention at 25 °C. After 30 days of storage at 25 and 40 °C, NE4 nanoemulsion encompassing MCT showed supreme retention behavior of VE about 94.3% and 90.1%, respectively, as highlighted in Figure 4C,D. On the other hand, NE5 nanoemulsion enclosing TFO showed 88% and 79.8% at 25 and 40 °C, correspondingly. The minimum retention of VE loaded inside the TFO-based nanoemulsions could be due to the sensitivity of TFO with respect to oxidation, as covered in Section 3.1.3. Though, this study indicated superior retention for VE ~80% throughout storage at 40 °C, exhibiting that VE is less sensitive toward oxidation than BC. Our findings of greater VE retention inside the MCT-based nanoemulsions were in accordance with an earlier study [54].

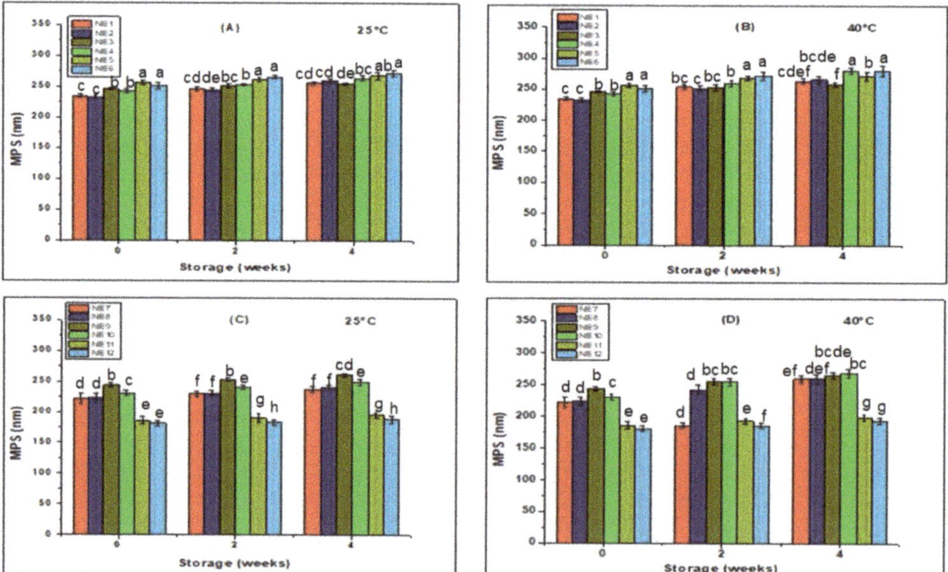

Figure 2. Effect of storage conditions on MPS of vitamin-loaded nanoemulsions; (**A,C**) 25 °C and (**B,D**) 40 °C. NE1–NE12 are given codes to the produced nanoemulsions, as illustrated in Table 1. Values with different letters are significantly different (*p* < 0.05).

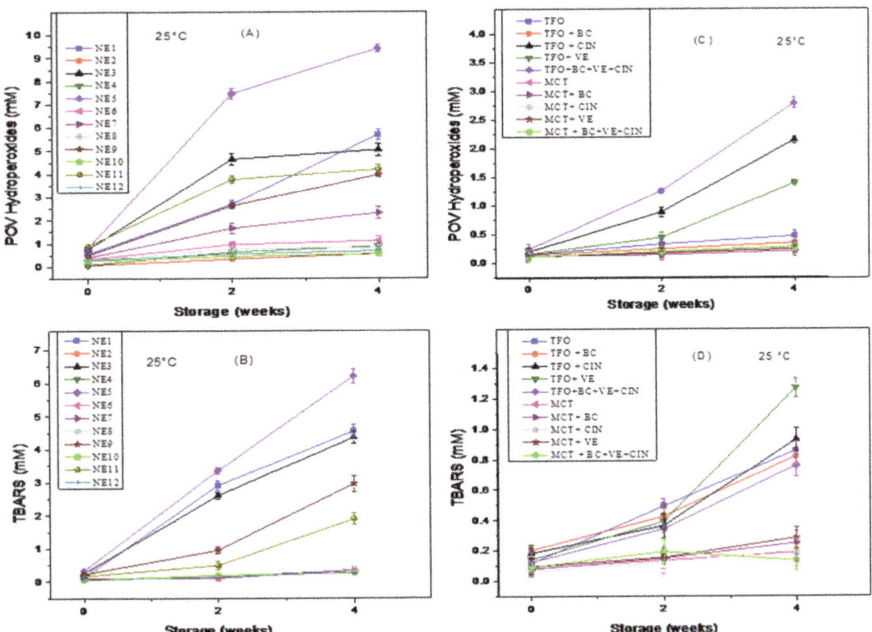

Figure 3. Oxidative stability of vitamin-loaded nanoemulsions over one month of storage at 25 °C. (**A**,**C**) peroxide value. (**B**,**D**) thiobarbituric acid reactive substances NE1–NE12 are given codes to the produced nanoemulsions, as illustrated in Table 1.

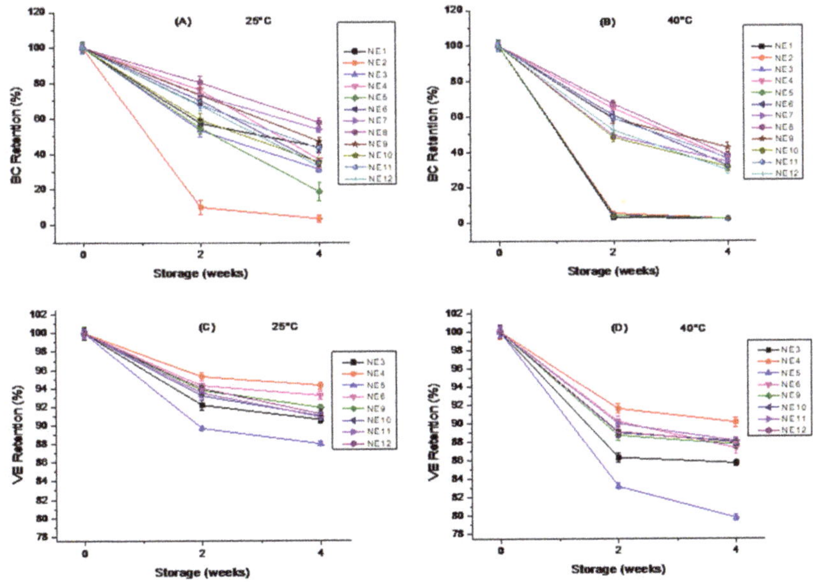

Figure 4. Changes of BC and VE encapsulated within nanoemulsions over one-month storage; (**A**,**C**) 25 °C and (**B**,**D**) 40 °C. NE1–NE12 are given codes to the produced nanoemulsions, as demonstrated in Table 1.

3.2. Effect of Cinnamaldehyde (CIN) on the Physiochemical Stability of Vitamin-Loaded Nanoemulsions

In the first section, PGU-based nanoemulsions (NE1-NE6) encapsulating TFO and MCT oil phases loaded with BC alone or in combinations with diverse levels of VE were fabricated. The physicochemical stability of fabricated nanoemulsions was evaluated, and findings of MPS and oxidation showed that loading of 3% *w/w* VE increased the size of particles as well as the system exhibited instability in response to oxidation. It has been investigated in the above section that loading of VE decreased and increased retention of BC in TFO-based nanoemulsions and TFO-based nanoemulsions, respectively. In the second section, 1% *w/w* cinnamaldehyde was used as an antioxidant in order to investigate its influence on the physicochemical stability of 0.1% *w/w* BC and 2% *w/w* VE-loaded nanoemulsions. Additionally, TW-80 was also used for the fabrication of nanoemulsions in order to shed light on a significant difference with PGU.

3.2.1. Impact of Loading Cinnamaldehyde on the Size and Z-Potential of VE-Loaded Nanoemulsions

Loading of 1% *w/w* CIN resulted in a decrease in MPS for PGU-based nanoemulsions covering TFO and MCT enriched with BC (NE7 and NE8), as expressed in Table 1. A declining trend in MPS by loading of CIN was observed that might be due to the lowering interfacial tension effect of CIN. After loading of CIN in PGU-based nanoemulsions (coded by NE9 and NE10) equipped with 0.1% BC and 2% VE showed intermediate MPS results in comparison with their individual encapsulation findings. However, a significant loss in MPS was noted for NE11 and NE12 (Table 1). The smallest particle size of TW-80-based nanoemulsions was previously examined instead of sodium caseinate emulsions [55]. As TW-80 is known to be the smallest molecular emulsifier that has the potent capability to lower interfacial tension among aqueous and oil phases in comparison with larger molecular emulsifiers [56]. In short, loading of CIN could not make any significant difference in terms of PDI results of TW-80 and PGU-based nanoemulsions, and values were quite lower than 0.15, as stated in Table 1. A significant loss in ZP results was observed for PGU-based nanoemulsions by loading of CIN with respect to those nanoemulsions that have no CIN. However, the findings of the current study revealed ZP values above -29, indicating potent electrostatic repulsion force among the dispersed oil particles within the aqueous suspension. TW-80-based nanoemulsions (NE11 and NE12) exhibited neutral behavior, demonstrating non-ionic characteristics of TW-80 [56].

3.2.2. Impact of Storage Conditions on MPS of BC-, VE-, and CIN-Loaded Nanoemulsions

Loading of CIN offered physical stability to the system, and even after the storage period, nanoemulsions NE7 and NE8 presented 15.2 and 36.2 nm rises in MPS at 25 and 40 °C, respectively as presented in Figure 2C,D. Overall, each nanoemulsion exhibited a superior increase in MPS that was occurred because of greater storage temperature (40 °C), and NE10 demonstrated the extreme rise in MPS around 38.3 nm. However, NE11 and NE12 nanoemulsions (Table 1) based on TW-80 were observed to be more stable rather than PGU-based nanoemulsions. TW-80 provided great stability over PGU because smaller molecular emulsifiers can adsorb more proficiently at the O/W boundary [56].

3.2.3. Impact of Loading of CIN on Oxidative Stability of BC- and VE-Loaded Nanoemulsions and Bare Carrier Oils

Numerous researches are available that oppose the "polar paradox" and recommend that the efficacy of antioxidants is based on several aspects besides the polarity, such as surface accessibility, the structure of antioxidant, splitting, relationship between emulsifier and antioxidant, and concentration [18,52,57]. Thus, we considered exploring the impact of CIN loading on oxidative stability of BC and VE-loaded nanoemulsions equipped with PGU and TW-80. Figure 3A depicts that, loading of CIN significantly reduced the POV production inside the NE7 nanoemulsions as compared to NE1, NE3, and NE5 nanoemulsions emulsified by PGU. Further, by loading CIN into NE9 and NE11, a reduction trend was observed in the case of POV production, and NE9 nanoemulsion emulsified by PGU showed the least production of POV over NE11 emulsified by TW-80. On the other side, loading of CIN into MCT-based nanoemulsions did not exhibit any significant difference; however, MCT-loaded nanoemulsions presented the very least production of POV. Secondary oxidation findings were in accordance with primary oxidation results in this study. However, differing from POV findings, NE9 nanoemulsion embedded with PGU presented a little bit greater TBARS results rather than NE11 emulsified with TW-80 (Figure 3B). However, an opposite trend has been observed in previous studies regarding the delivery of oxidative stability to the nanoemulsions emulsified by different emulsifiers, such as the smaller and greater molecular structure of emulsifiers. As an example, Berton and his colleagues observed greater oxidative stability for Tween-20-based nanoemulsions over proteins [58], while Kargar el. al. [59] investigated different trends regarding the stability of nanoemulsions. Overall, MCT-loaded nanoemulsions showed greater oxidative stability, which was found to be more stable. Among TFO-based nanoemulsions, NE7, NE9, and NE11 nanoemulsions revealed comparatively greater stability. Though, loading of CIN alone or in mixing with BC and VE offered superior oxidative stability to the produced nanoemulsions, demonstrating its great potential as an antioxidant. Figure 3C,D openly demonstrates the impact of entrapped CIN on the oxidative stability of free oils where greater values of POV and TBARS were observed for TFO over MCT. Loading of CIN increased the oxidation level for both kinds of free oils and could not act as an antioxidant. This contradicts the behavior of CIN in free oils, and nanoemulsions appealed that lipophilic antioxidant performs well in lipid-based nanoemulsions matrices as compared to free oils that endorse the "polar paradox" mechanism.

3.2.4. Impact of CIN Loading on Retention of BC and VE

Effect of loaded CIN was accessed on the BC retention throughout storage for 30 days at 25 and 40 °C. After 30 days of storage at 25 °C, NE7 and NE8 nanoemulsions (Table 1) revealed the highest BC retention around ~52.23% and ~57.02%, respectively, as elaborated in Figure 4A. Furthermore, loading of CIN within NE9 to NE12-based nanoemulsions presented loss in BC; but TW-80-based nanoemulsions enriched with MCT oil showed more BC reduction. NE7 and NE8 nanoemulsions reserved a great amount of BC at 25 °C. An identical trend regarding BC retention was observed in NE9, NE10, NE11, and NE12 nanoemulsions at 25 °C. Though, NE9 nanoemulsion presented higher BC retention, around 42.3% over a storage period of one month at 40 °C, as stated in Figure 4B. Loading of CIN within NE10 and NE12 nanoemulsions reduced VE contents, while an opposite trend was observed regarding reduction in VE contents in NE9 and NE11 nanoemulsions at 25 °C for 30 days of storage (Figure 4C). However, there was no significant impact observed on the VE retention by any kind of emulsifier. Even at 40 °C, nanoemulsions revealed an identical trend in the case of VE retention, and NE5 nanoemulsion presented the least VE retention ~79.8%, as seen in Figure 4D. Overall, each nanoemulsion exhibited VE retention above than ~80%, and NE4 nanoemulsion presented ~90.1% VE retention rather than others over a storage period of one month at 40 °C. Antioxidants perform differently within diverse emulsified systems; however, they deliver different oxidative stability when they are used together. For instance, Qian and his team used aqueous and lipid-soluble antioxidants in single form as well as in combined form to enhance the BC stability. They reported that EDTA performed excellently in individual form as compared to VE, and their combined form offered greater stability rather than VE acetate but not higher than EDTA, signifying antagonistic potential among VE acetate and EDTA [14]. In another previous research, better VE retention was observed in whey protein isolate (WPI)-based nanoemulsions, while the defensive potential of ascorbic acid was based on molar percentages of WPI and ascorbic acid [60]. In current work, CIN significantly enhanced BC retention in all nanoemulsions covering BC and CIN, whereas it displayed intermediate consequences when entrapped within the nanoemulsions encapsulating BC, VE, and EU together (Figure 4A). Additionally, throughout storage at 40 °C, nanoemulsions encompassing a mixture of BC with VE and CIN presented to some extent greater BC retention rather than nanoemulsions holding of

BC + VE as well as BC + CIN as illustrated in Figure 4B. It is an understood phenomenon that emulsifiers could plan a crucial role in the stability of nanoemulsions. As reported by Qian et al. [61] that emulsifiers such as proteins having greater molecular weight exposed a smaller amount of degradation rather than Tween-20-based nanoemulsions because of their smaller molecular weight. Mao et al. [62] fabricated BC-loaded nanoemulsions stabilized by larger and smaller molecular-based emulsifiers where they reported BC retention in flowing order protein > Tween-20 > OSA starch during the storage period. Similarly, a great BC retention of OSA-MS stabilized nanoemulsions was observed over the storage of 30 days by Liang et al. [44]. However, in this work, TW-80-based nanoemulsions showed slightly lesser BC retention that could be owing to their smaller particle size instead of PGU-based nanoemulsions. A reduction in MPS causes a rise in interfacial surface area that further upsurges the oxidation rate, resulting in a supreme loss in BC [63]. The most suitable reason for the healthier character of PGU might be owing to the OSA-MS capabilities to produce denser interfacial coatings that can protect the entrapped lipids from the aqueous phase [40].

4. Conclusions

Nanoemulsions comprising FSVs and CIN emulsified with PGU and TW-80 provided excellent physical stability over a storage period of one month at diverse temperatures. Different kinds of emulsifiers (PGU and TW-80), carrier oils (TFO and MCT), and CIN loading all directed toward prominent changes in the properties of bioactive compounds toward oxidation inside the nanoemulsions as well as within free oils. Herein, we investigated BC to be more sensitive as well as superior deprivation of VE, and BC was detected from nanoemulsions kept at 40 °C. Loading of CIN enhanced the oxidative stability and improved FSVs retention of produced nanoemulsions throughout a prolonged storage period. PGU-based nanoemulsions offered a little bit more or equivalent stability to the entrapped bioactive rather than TW-80-based nanoemulsions. Additionally, loading of CIN and VE performed dissimilar in each nanoemulsions holding dissimilar carrier oils. CIN entrapment performed being an antioxidant in each lipid phase that framed with TFO and MCT, but VE acted as an antioxidant and prooxidant in MCT and TFO, correspondingly. The changes in BC retention and oxidative stability proposed that the oxidative investigation is not enough to observe the stability of any emulsified system; however, still more studies on entrapped bioactive compounds are needed to find out the real fortune of bioactive compounds. This study could be more useful for designing oxidative stability of nano-emulsified systems in order to ensure the simultaneous incorporation of diverse lipophilic bioactive compounds as well as their application inside the food and beverage matrices.

Author Contributions: Investigation, A.A., A.R., S.M.J., M.M.A.N.R., Q.S., H.M.S., S.K., M.U. and W.X.; Methodology, A.A., A.R., S.M.J., M.M.A.N.R. and W.X.; Supervision, M.M.A.N.R. and W.X.; Writing—original draft, A.A., A.R., S.M.J., M.M.A.N.R., Q.S., H.M.S., S.K., M.U., P.Ł.K. and M.J.; Writing—review and editing, M.M.A.N.R., P.Ł.K. and M.J. All authors have read and agreed to the published version of the manuscript.

Funding: This research was financially supported by China Agriculture Research System (grant 438 no. CARS-45-27), Jiangsu Agricultural Industry Technology System (grant no. JATS [2019] 467), and the National first-class discipline program of Food Science and Technology (grant no. JUFSTR20180201).

Institutional Review Board Statement: Not applicable.

Informed Consent Statement: Not applicable.

Data Availability Statement: All data generated or analyzed during this study are included in this published article.

Conflicts of Interest: The authors declare no conflict of interest. The funders had no role in the design of the study, in the collection, analyses, or interpretation of data, in the writing of the manuscript, or in the decision to publish the results.

References

1. Shehzad, Q.; Rehman, A.; Jafari, S.M.; Zuo, M.; Khan, M.A.; Ali, A.; Khan, S.; Karim, A.; Usman, M.; Hussain, A. Improving the Oxidative Stability of Fish Oil Nanoemulsions by co-Encapsulation with Curcumin and Resveratrol. *Colloids Surf. B Biointerfaces* **2021**, *199*, 111481. [CrossRef] [PubMed]
2. Comunian, T.; Babazadeh, A.; Rehman, A.; Shaddel, R.; Akbari-Alavijeh, S.; Boostani, S.; Jafari, S. Protection and Controlled Release of Vitamin C by Different Micro/Nanocarriers. *Crit. Rev. Food Sci. Nutr.* **2020**, 1–22. [CrossRef] [PubMed]
3. Batool, M.; Kauser, S.; Nadeem, H.R.; Perveen, R.; Irfan, S.; Siddiqa, A.; Shafique, B.; Zahra, S.M.; Waseem, M.; Khalid, W.; et al. A Critical Review on Alpha Tocopherol: Sources, RDA and Health Benefits. *J. Appl. Pharm.* **2020**, *12*, 39–56. [CrossRef]
4. Song, H.Y.; Moon, T.W.; Choi, S.J. Impact of Antioxidant on the Stability of β-Carotene in Model Beverage Emulsions: Role of Emulsion Interfacial Membrane. *Food Chem.* **2019**, *279*, 194–201. [CrossRef]
5. Fang, Z.; Xu, X.; Cheng, H.; Li, J.; Liang, L. Comparison of Whey Protein Particles and Emulsions for the Encapsulation and Protection of α-Tocopherol. *J. Food Eng.* **2019**, *247*, 56–63. [CrossRef]
6. Rehman, A.; Tong, Q.; Jafari, S.M.; Assadpour, E.; Shehzad, Q.; Aadil, R.M.; Iqbal, M.W.; Rashed, M.M.; Mushtaq, B.S.; Ashraf, W. Carotenoid-Loaded Nanocarriers: A Comprehensive Review. *Adv. Colloid Interface Sci.* **2020**, *275*, 102048. [CrossRef] [PubMed]
7. Katouzian, I.; Jafari, S.M. Nano-Encapsulation as a Promising Approach for Targeted Delivery and Controlled Release of Vitamins. *Trends Food Sci. Technol.* **2016**, *53*, 34–48. [CrossRef]
8. Khan, S.; Rehman, A.; Shah, H.; Aadil, R.M.; Ali, A.; Shehzad, Q.; Ashraf, W.; Yang, F.; Karim, A.; Khaliq, A. Fish Protein and Its Derivatives: The Novel Applications, Bioactivities, and Their Functional Significance in Food Products. *Food Rev. Int.* **2020**, 1–28. [CrossRef]
9. Rehman, A.; Jafari, S.M.; Tong, Q.; Karim, A.; Mahdi, A.A.; Iqbal, M.W.; Aadil, R.M.; Ali, A.; Manzoor, M.F. Role of Peppermint Oil in Improving the Oxidative Stability and Antioxidant Capacity of Borage Seed Oil-Loaded Nanoemulsions Fabricated by Modified Starch. *Int. J. Biol. Macromol.* **2020**, *153*, 697–707. [CrossRef]
10. Nejadmansouri, M.; Hosseini, S.M.H.; Niakosari, M.; Yousefi, G.H.; Golmakani, M.T. Physicochemical Properties and Oxidative Stability of Fish Oil Nanoemulsions as Affected by Hydrophilic Lipophilic Balance, Surfactant to Oil Ratio and Storage Temperature. *Colloids Surf. A Physicochem. Eng. Asp.* **2016**, *506*, 821–832. [CrossRef]
11. Salvia-Trujillo, L.; Decker, E.A.; McClements, D.J. Influence of an Anionic Polysaccharide on the Physical and Oxidative Stability of Omega-3 Nanoemulsions: Antioxidant Effects of Alginate. *Food Hydrocoll.* **2016**, *52*, 690–698. [CrossRef]
12. Rashidinejad, A.; Bahrami, A.; Rehman, A.; Rezaei, A.; Babazadeh, A.; Singh, H.; Jafari, S.M. Co-Encapsulation of Probiotics with Prebiotics and Their Application in Functional/Synbiotic Dairy Products. *Crit. Rev. Food Sci. Nutr.* **2020**, 1–25.
13. Rehman, A.; Jafari, S.M.; Tong, Q.; Riaz, T.; Assadpour, E.; Aadil, R.M.; Niazi, S.; Khan, I.M.; Shehzad, Q.; Ali, A.; et al. Drug Nanodelivery Systems Based on Natural Polysaccharides against Different Diseases. *Adv. Colloid Interface Sci.* **2020**, *284*, 102251. [CrossRef]
14. Qian, C.; Decker, E.A.; Xiao, H.; McClements, D.J. Physical and Chemical Stability of β-Carotene-Enriched Nanoemulsions: Influence of pH, Ionic Strength, Temperature, and Emulsifier Type. *Food Chem.* **2012**, *132*, 1221–1229. [CrossRef] [PubMed]
15. Rehman, A.; Tong, Q.; Shehzad, Q.; Aadil, R.M.; Khan, I.M.; Riaz, T.; Jafari, S.M. Rheological Analysis of Solid-Like Nanoencapsulated Food Ingredients by Rheometers. In *Characterization of Nanoencapsulated Food Ingredien*; Elsevier: Amsterdam, The Netherlands, 2020; pp. 547–583.
16. Garavand, F.; Cacciotti, I.; Vahedikia, N.; Rehman, A.; Tarhan, Ö.; Akbari-Alavijeh, S.; Shaddel, R.; Rashidinejad, A.; Nejatian, M.; Jafarzadeh, S.; et al. A Comprehensive Review on the Nanocomposites Loaded with Chitosan Nanoparticles for Food Packaging. *Crit. Rev. Food Sci. Nutr.* **2020**, 1–34. [CrossRef]
17. Rehman, A.; Jafari, S.M.; Aadil, R.M.; Assadpour, E.; Randhawa, M.A.; Mahmood, S. Development of Active Food Packaging Via Incorporation of Biopolymeric Nanocarriers Containing Essential Oils. *Trends Food Sci. Technol.* **2020**, *101*, 106–121. [CrossRef]
18. Chaiyasit, W.; McClements, D.J.; Decker, E.A. The Relationship between the Physicochemical Properties of Antioxidants and Their Ability to Inhibit Lipid Oxidation in Bulk Oil and Oil-in-Water Emulsions. *J. Agric. Food Chem.* **2005**, *53*, 4982–4988. [CrossRef]
19. Serfert, Y.; Drusch, S.; Schwarz, K. Chemical Stabilisation of Oils Rich in Long-Chain Polyunsaturated Fatty Acids during Homogenisation, Microencapsulation and Storage. *Food Chem.* **2009**, *113*, 1106–1112. [CrossRef]
20. Nadeem, M.; Ghaffar, A.; Hashim, M.M.; Murtaza, M.A.; Ranjha, M.M.A.N.; Mehmood, A.; Riaz, M.N. Sonication and Microwave Processing of Phalsa Drink: A Synergistic Approach. *Int. J. Fruit Sci.* **2021**, *21*, 993–1007. [CrossRef]
21. Ranjha, M.M.A.N.; Amjad, S.; Ashraf, S.; Khawar, L.; Safdar, M.N.; Jabbar, S.; Nadeem, M.; Mahmood, S.; Murtaza, M.A. Extraction of Polyphenols from Apple and Pomegranate Peels Employing Different Extraction Techniques for the Development of Functional Date Bars. *Int. J. Fruit Sci.* **2020**, *20*, S1201–S1221. [CrossRef]
22. Ranjha, M.M.A.N.; Irfan, S.; Lorenzo, J.M.; Shafique, B.; Kanwal, R.; Pateiro, M.; Arshad, R.N.; Wang, L.; Nayik, G.A.; Roobab, U.; et al. Sonication, a Potential Technique for Extraction of Phytoconstituents: A Systematic Review. *Processes* **2021**, *9*, 1406. [CrossRef]
23. Ranjha, M.M.A.N.; Kanwal, R.; Shafique, B.; Arshad, R.N.; Irfan, S.; Kieliszek, M.; Kowalczewski, P.Ł.; Irfan, M.; Khalid, M.Z.; Roobab, U.; et al. A Critical Review on Pulsed Electric Field: A Novel Technology for the Extraction of Phytoconstituents. *Molecules* **2021**, *26*, 4893. [CrossRef]

24. Nadeem, H.R.; Akhtar, S.; Ismail, T.; Sestili, P.; Lorenzo, J.M.; Ranjha, M.M.A.N.; Jooste, L.; Hano, C.; Aadil, R.M. Heterocyclic Aromatic Amines in Meat: Formation, Isolation, Risk Assessment, and Inhibitory Effect of Plant Extracts. *Foods* **2021**, *10*, 1466. [CrossRef]

25. Ranjha, M.M.A.N.; Irfan, S.; Nadeem, M.; Mahmood, S. A Comprehensive Review on Nutritional Value, Medicinal Uses, and Processing of Banana. *Food Rev. Int.* **2020**, 1–27. [CrossRef]

26. Ranjha, M.M.A.N.; Shafique, B.; Wang, L.; Irfan, S.; Safdar, M.N.; Murtaza, M.A.; Nadeem, M.; Mahmood, S.; Mueen-ud-Din, G.; Nadeem, H.R. A Comprehensive Review on Phytochemistry, Bioactivity and Medicinal Value of Bioactive Compounds of Pomegranate (*Punica granatum*). *Adv. Tradit. Med.* **2021**, 1–21. [CrossRef]

27. Faikoh, E.N.; Hong, Y.-H.; Hu, S.-Y. Liposome-Encapsulated Cinnamaldehyde Enhances Zebrafish (*Danio rerio*) Immunity and Survival when Challenged with *Vibrio vulnificus* and *Streptococcus agalactiae*. *Fish Shellfish. Immunol.* **2014**, *38*, 15–24. [CrossRef]

28. Muhoza, B.; Xia, S.; Cai, J.; Zhang, X.; Duhoranimana, E.; Su, J. Gelatin and pectin complex coacervates as carriers for cinnamalde-hyde: Effect of pectin esterification degree on coacervate formation, and enhanced thermal stability. *Food Hydrocoll.* **2019**, *87*, 712–722. [CrossRef]

29. Liu, Q.; Cui, H.; Muhoza, B.; Duhoranimana, E.; Xia, S.; Hayat, K.; Hussain, S.; Tahir, M.U.; Zhang, X. Fabrication of Low Environment-Sensitive Nanoparticles for Cinnamaldehyde Encapsulation by Heat-Induced Gelation Method. *Food Hydrocoll.* **2020**, *105*, 105789. [CrossRef]

30. Rashed, M.M.; Ghaleb, A.D.; Li, J.; Al-Hashedi, S.A.; Rehman, A. Functional-Characteristics of *Zanthoxylum schinifolium* (Siebold & Zucc.) Essential Oil Nanoparticles. *Ind. Crops Prod.* **2021**, *161*, 113192.

31. Rehman, A.; Ahmad, T.; Aadil, R.M.; Spotti, M.J.; Bakry, A.M.; Khan, I.M.; Zhao, L.; Riaz, T.; Tong, Q. Pectin Polymers as Wall Materials for the Nano-Encapsulation of Bioactive Compounds. *Trends Food Sci. Technol.* **2019**, *90*, 35–46. [CrossRef]

32. Mohsin, A.; Zaman, W.Q.; Guo, M.; Ahmed, W.; Khan, I.M.; Niazi, S.; Rehman, A.; Hang, H.; Zhuang, Y. Xanthan-Curdlan Nexus for Synthesizing Edible Food Packaging Films. *Int. J. Biol. Macromol.* **2020**, *162*, 43–49. [CrossRef]

33. Ashraf, W.; Latif, A.; Lianfu, Z.; Jian, Z.; Chenqiang, W.; Rehman, A.; Hussain, A.; Siddiquy, M.; Karim, A. Technological Advancement in the Processing of Lycopene: A Review. *Food Rev. Int.* **2020**, 1–27. [CrossRef]

34. Zhao, L.; Tong, Q.; Wang, H.; Liu, Y.; Xu, J.; Rehman, A. Emulsifying Properties and Structure Characteristics of Octenyl Succinic Anhydride-Modified Pullulans with Different Degree of Substitution. *Carbohydr. Polym.* **2020**, *250*, 116844. [CrossRef]

35. Rashidinejad, A.; Boostani, S.; Babazadeh, A.; Rehman, A.; Rezaei, A.; Akbari, S.; Shaddel, R.; Jafari, S. Opportunities and Challenges for the Nanodelivery of Green Tea Catechins in Functional Foods. *Food Res. Int.* **2021**, *142*, 110186. [CrossRef] [PubMed]

36. Walker, R.M.; Gumus, C.E.; Decker, E.A.; McClements, D.J. Improvements in the Formation and Stability of Fish Oil-in-Water Nanoemulsions Using Carrier Oils: MCT, Thyme Oil, & Lemon Oil. *J. Food Eng.* **2017**, *211*, 60–68.

37. Qiu, C.; Zhao, M.; Decker, E.A.; McClements, D.J. Influence of Protein Type on Oxidation and Digestibility of Fish Oil-in-Water Emulsions: Gliadin, Caseinate, and Whey Protein. *Food Chem.* **2015**, *175*, 249–257. [CrossRef] [PubMed]

38. Yi, J.; Lam, T.I.; Yokoyama, W.; Cheng, L.W.; Zhong, F. Beta-Carotene Encapsulated in Food Protein Nanoparticles Reduces Peroxyl Radical Oxidation in Caco-2 Cells. *Food Hydrocoll.* **2015**, *43*, 31–40. [CrossRef]

39. Hategekimana, J.; Masamba, K.G.; Ma, J.; Zhong, F. Encapsulation of Vitamin E: Effect of Physicochemical Properties of Wall Material on Retention and Stability. *Carbohydr. Polym.* **2015**, *124*, 172–179. [CrossRef] [PubMed]

40. Sharif, H.R.; Williams, P.A.; Sharif, M.K.; Khan, M.A.; Majeed, H.; Safdar, W.; Shamoon, M.; Shoaib, M.; Haider, J.; Zhong, F. Influence of OSA-Starch on the Physico Chemical Characteristics of Flax Seed Oil-Eugenol Nanoemulsions. *Food Hydrocoll.* **2017**, *66*, 365–377. [CrossRef]

41. Yang, Y.; McClements, D.J. Encapsulation of Vitamin E in Edible Emulsions Fabricated Using a Natural Surfactant. *Food Hydrocoll.* **2013**, *30*, 712–720. [CrossRef]

42. Abbas, S.; Bashari, M.; Akhtar, W.; Li, W.W.; Zhang, X. Process Optimization of Ultrasound-Assisted Curcumin Nanoemulsions Stabilized by OSA-Modified Starch. *Ultrason. Sonochemistry* **2014**, *21*, 1265–1274. [CrossRef]

43. Bonilla, J.; Atarés, L.; Vargas, M.; Chiralt, A. Effect of Essential Oils and Homogenization Conditions on Properties of Chitosan-Based Films. *Food Hydrocoll.* **2012**, *26*, 9–16. [CrossRef]

44. Liang, R.; Shoemaker, C.F.; Yang, X.; Zhong, F.; Huang, Q. Stability and Bioaccessibility of β-Carotene in Nanoemulsions Stabilized by Modified Starches. *J. Agric. Food Chem.* **2013**, *61*, 1249–1257. [CrossRef] [PubMed]

45. Zhao, X.; Liu, F.; Ma, C.; Yuan, F.; Gao, Y. Effect of Carrier Oils on the Physicochemical Properties of Orange Oil Beverage Emulsions. *Food Res. Int.* **2015**, *74*, 260–268. [CrossRef]

46. Nagy, K.; Kerrihard, A.L.; Beggio, M.; Craft, B.D.; Pegg, R.B. Modeling the Impact of Residual Fat-Soluble Vitamin (FSV) Contents on the Oxidative Stability of Commercially Refined Vegetable Oils. *Food Res. Int.* **2016**, *84*, 26–32. [CrossRef]

47. Sánchez, M.d.R.H.; Cuvelier, M.-E.; Turchiuli, C. Effect of α-Tocopherol on Oxidative Stability of Oil during Spray Drying and Storage of Dried Emulsions. *Food Res. Int.* **2016**, *88*, 32–41. [CrossRef]

48. Poyato, C.; Navarro-Blasco, I.; Calvo, M.I.; Cavero, R.Y.; Astiasarán, I.; Ansorena, D. Oxidative Stability of O/W and W/O/W Emulsions: Effect of Lipid Composition and Antioxidant Polarity. *Food Res. Int.* **2013**, *51*, 132–140. [CrossRef]

49. Shahidi, F.; Zhong, Y. Revisiting the Polar Paradox Theory: A Critical Overview. *J. Agric. Food Chem.* **2011**, *59*, 3499–3504. [CrossRef] [PubMed]

50. Jacobsen, C.; Let, M.B.; Nielsen, N.S.; Meyer, A.S. Antioxidant Strategies for Preventing Oxidative Flavour Deterioration of Foods Enriched with n-3 Polyunsaturated Lipids: A Comparative Evaluation. *Trends Food Sci. Technol.* **2008**, *19*, 76–93. [CrossRef]
51. Carocho, M.; Ferreira, I.C. A Review on Antioxidants, Prooxidants and Related Controversy: Natural and Synthetic Compounds, Screening and Analysis Methodologies and Future Perspectives. *Food Chem. Toxicol.* **2013**, *51*, 15–25. [CrossRef]
52. Bakır, T.; Sönmezoğlu, İ.; İmer, F.; Apak, R. Polar Paradox Revisited: Analogous Pairs of Hydrophilic and Lipophilic Antioxidants in Linoleic Acid Emulsion Containing Cu (II). *J. Sci. Food Agric.* **2013**, *93*, 2478–2485. [CrossRef]
53. Liu, Y.; Hou, Z.; Yang, J.; Gao, Y. Effects of Antioxidants on the Stability of β-Carotene in O/W Emulsions Stabilized by Gum Arabic. *J. Food Sci. Technol.* **2015**, *52*, 3300–3311. [CrossRef]
54. Hategekimana, J.; Chamba, M.V.; Shoemaker, C.F.; Majeed, H.; Zhong, F. Vitamin E Nanoemulsions by Emulsion Phase Inversion: Effect of Environmental Stress and Long-Term Storage on Stability and Degradation in Different Carrier Oil Types. *Colloids Surf. A Physicochem. Eng. Asp.* **2015**, *483*, 70–80. [CrossRef]
55. Matalanis, A.; Decker, E.A.; McClements, D.J. Inhibition of Lipid Oxidation by Encapsulation of Emulsion Droplets within Hydrogel Microspheres. *Food Chem.* **2012**, *132*, 766–772. [CrossRef]
56. Majeed, H.; Liu, F.; Hategekimana, J.; Sharif, H.R.; Qi, J.; Ali, B.; Bian, Y.-Y.; Ma, J.; Yokoyama, W.; Zhong, F. Bactericidal Action Mechanism of Negatively Charged Food Grade Clove Oil Nanoemulsions. *Food Chem.* **2016**, *197*, 75–83. [CrossRef] [PubMed]
57. Laguerre, M.; Bayrasy, C.; Panya, A.; Weiss, J.; McClements, D.J.; Lecomte, J.; Decker, E.A.; Villeneuve, P. What Makes Good Antioxidants in Lipid-Based Systems? The next Theories beyond the Polar Paradox. *Crit. Rev. Food Sci. Nutr.* **2015**, *55*, 183–201. [CrossRef] [PubMed]
58. Berton, C.; Ropers, M.-H.; Bertrand, D.; Viau, M.; Genot, C. Oxidative Stability of Oil-in-Water Emulsions Sabilised with Protein or Surfactant Emulsifiers in Various Oxidation Conditions. *Food Chem.* **2012**, *131*, 1360–1369. [CrossRef]
59. Kargar, M.; Spyropoulos, F.; Norton, I.T. Microstructural Design to Reduce Lipid Oxidation in Oil-Inwater Emulsions. *Procedia Food Sci.* **2011**, *1*, 104–108. [CrossRef]
60. Wang, B.; Vongsvivut, J.; Adhikari, B.; Barrow, C.J. Microencapsulation of Tuna Oil Fortified with the Multiple Lipophilic Ingredients Vitamins A, D3, E, K2, Curcumin and Coenzyme Q10. *J. Funct. Foods* **2015**, *19*, 893–901. [CrossRef]
61. Qian, C.; Decker, E.A.; Xiao, H.; McClements, D.J. Nanoemulsion Delivery Systems: Influence of Carrier Oil on β-Carotene Bioaccessibility. *Food Chem.* **2012**, *135*, 1440–1447. [CrossRef]
62. Mao, L.; Xu, D.; Yang, J.; Yuan, F.; Gao, Y.; Zhao, J. Effects of Small and Large Molecule Emulsifiers on the Characteristics of β-Carotene Nanoemulsions Prepared by High Pressure Homogenization. *Food Technol. Biotechnol.* **2009**, *47*, 336–342.
63. Yi, J.; Li, Y.; Zhong, F.; Yokoyama, W. The Physicochemical Stability and *In Vitro* Bioaccessibility of Beta-Carotene in Oil-In-water Sodium Caseinate Emulsions. *Food Hydrocoll.* **2014**, *35*, 19–27. [CrossRef]

applied
sciences

Article

Extraction of Galactolipids from Waste By-Products: The Feasibility of Green Chemistry Methods

Łukasz Woźniak [1,*], Monika Wojciechowska [2], Krystian Marszałek [1] and Sylwia Skąpska [1]

[1] Department of Fruit and Vegetable Product Technology, Institute of Agricultural and Food Biotechnology–State Research Institute, 36 Rakowiecka Street, 02532 Warsaw, Poland; krystian.marszalek@ibprs.pl (K.M.); sylwia.skapska@ibprs.pl (S.S.)
[2] Department of Bacterial Physiology, Faculty of Biology, University of Warsaw, 1 Miecznikowa Street, 02096 Warsaw, Poland; monwoj45@wp.com
[*] Correspondence: lukasz.wozniak@ibprs.pl

Abstract: Galactolipids are a class of lipids present, inter alia, in the plastid membranes of plant cells. Apart from their biological significance, they are recognized as an important group of bioactive agents, especially in the treatment of osteoarthritis. The aim of this research was to evaluate the usefulness of the green chemistry approach in the extraction of these compounds. Waste products of food processing were selected as a raw material to improve the sustainability of the process even further, and their galactolipid content was investigated using an LC-MS analysis. The rosehip pomace, which was recognized as the most promising amongst materials used in this study, was subjected to supercritical fluid extraction (SFE) and ultrasound-assisted extraction (UAE). It transpired that SFE using pure CO_2 was not an effective method for the extraction of galactolipids, although the use of ethanol as a cosolvent favored the separation. The results of UAE were also very promising—the improvement of the extraction yield up to 74% was observed. The green chemistry approaches used for galactolipid isolation were compared with a conventional processing method and proved to be an interesting alternative.

Keywords: supercritical fluid extraction; ultrasound-assisted extraction; rosa canina; green chemistry; sustainability

Citation: Woźniak, Ł.; Wojciechowska, M.; Marszałek, K.; Skąpska, S. Extraction of Galactolipids from Waste By-Products: The Feasibility of Green Chemistry Methods. *Appl. Sci.* **2021**, *11*, 12088. https://doi.org/10.3390/app112412088

Academic Editors: Marek Kieliszek and Przemyslaw Lukasz Kowalczewski

Received: 7 October 2021
Accepted: 16 December 2021
Published: 18 December 2021

Publisher's Note: MDPI stays neutral with regard to jurisdictional claims in published maps and institutional affiliations.

1. Introduction

Galactolipids are a group of lipids that occur mainly in the thylakoids of higher plants and algae [1]. It is the main class of lipids found in plastid membranes, and their presence is crucial for the synthesis of the photosynthetic apparatus. Two main groups of galactolipids are distinguished: monogalactosyldiacylglycerols (MGDGs), which account for 50% of all lipids in thylakoids, and digalactosyldiacylglycerols (DGDGs), which in turn account for 30% [2]; their structures are presented in Figure 1. Their synthesis takes place in the plastid envelope membranes by galactosyltransferase enzymes [3] An impact of a balanced diet rich in vegetables and fruits on consumers' health due to the presence of bioactive compounds has been known for decades. These compounds include galactolipids, which are considered to be an important class of nutraceuticals. So far, several positive effects have been attributed to galactolipids, e.g., MGDGs isolated from fresh spinach leaves inhibit the activation of the Epstein–Barr virus (EBV) [4]. Another study by Kuriyama et al. showed that the glycolipid fraction of spinach, containing, among other things, MGDGs and DGDGs, shows potential anti-cancerogenic features [5]. Moreover, the presence of galactolipids along with other compounds (such as carotenoids or terpenoids) present in rosehip (*Rosa canina* L.) counteract osteoarthritis [6]. Its anti-inflammatory as well as anti-rheumatic properties were attributed to a galactolipid named GOPO. So far, many studies have been carried out that clearly show that in patients suffering from chronic inflammation of the bones and joints, there was a reduction in pain and an improvement in joint mobility after the administration of powdered rosehips [7]. More importantly, no side effects of the administered substances have been observed so far [8]. Products containing GOPO are commercially available on the pharmaceutical market; producers claim that thanks to the presence of this galactolipid, they can delay arthritic changes and slow down the degradation of cartilage tissue. It must, however, be noted, that the EFSA has so far not made any health claims connected with galactolipids. The details of the methods used by the manufacturers of galactolipid-based diet supplements are confidential; however, based on chemical properties of galactolipids, it can be assumed that extraction with organic solvents is the method of choice.

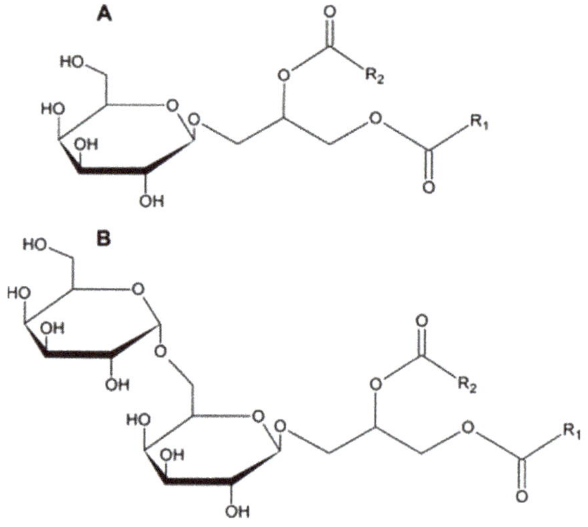

Figure 1. Structure of MGDGs (**A**) and DGDGs (**B**) [9]. Reproduced from [9] with permission. Copyright 2012 Elsevier.

In recent decades, there has been an increasing awareness of the environmental impact of production and processing, which has led to a global tendency to ensure the sustainability of such operations. The term "green chemistry" was coined to describe an alternative approach to the operations that were developed to be more environmentally friendly, often at the price of their profitability. The main principles of green extraction include the use of alternative solvents, innovative technologies, and the selection of renewable resources [10]. Some of the techniques have already found a place as the methods of choice in industrial applications, while others are still in the first testing stage. A broader description of the various methods of green extraction can be found in excellent reviews on the subject [11,12].

Ultrasound-assisted extraction (UAE) refers to the modification of a typical solid–liquid extraction by applying high-intensity ultrasonic waves to the sample. Typical implementations of the process include the use of baths, with generators built in the wall of the vessel, and horns, which are immersed in the sample. Compared to other novel processing techniques, UAE is very inexpensive. The main driving force of the process is acoustic cavitation—the creation, expansion, and implosive collapse of microbubbles in ultrasound-irradiated liquids. The use of ultrasound also causes agitation, vibrations, shockwaves, microjets, and radical formation. The UAE benefits from enhanced mass transfer and an increase of the surface area due to the disintegration of the solid matrix [13].

Supercritical fluid extraction (SFE) takes advantage of the properties of substances in a supercritical state and uses them instead of typical liquid solvents. Many compounds can be used in supercritical processes, although the solvent of choice in the vast majority of applications is carbon dioxide, which is characterized by mild critical-point conditions, chemical inertia, low price, and negligible toxicity. Apart from that, the low viscosity of supercritical fluids improves mass transfer during extraction, while the transition to gas after the pressure decrease results in obtaining solvent-free extracts [14]. The advantages of SFE led to its industrial implementation in the production of, among other things, decaffeinated coffee, hop extracts, alcohol-free wine and beer, and spice extracts [15].

The aim of the work was to evaluate the possibility of using green extraction technology for the extraction of galactolipids from plant material. SFE and UAE were the two techniques selected for the experiments. The techniques were selected according to their status as the method that is already implemented in industrial practice. The methods differ in principle of the operation: SFE uses an alternative solvent, while UAE improves currently existing techniques by an intensification of the mass transfer. Additionally, waste products from fruit and vegetable processing were selected as a material to further improve the sustainability of the process. This paper is the first report on the extraction of galactolipids using green chemistry methods, and the authors believe that it can provide valuable information for prospective industrial applications.

2. Materials and Methods

2.1. Materials

Certified standards of MGDGs and DGDGs were acquired from Avanti Polar Lipids (Alabaster, AL, USA). Chromatographic solvents of appropriate purity were acquired from Merck (Darmstadt, Germany), while other chemicals came from either Merck or Avantor (Radnor, PA, USA).

As significant amounts of galactolipids are present exclusively in plant material, the selection of matrices to be analyzed was limited to the waste products of fruit and vegetable processing. The selected material included pomace from apple (*Malus domestica* Borkh.), carrot (*Daucus carota* Hoffm.), and rosehip (*Rosa canina* L.), as well as wastes from the processing of broccoli (*Brassica oleracea* var. *italica*), fava beans (*Vicia faba* L.), and potato (*Solanum tuberosum* L.). The material was produced on a laboratory scale using typical processing methods used in the industry. Samples were kept at a temperature of $-18\,^{\circ}\mathrm{C}$ prior to analysis and extraction.

2.2. Supercritical Fluid Extraction (SFE)

The SFE process was conducted on a SFE Spe-ed 4 apparatus (Applied Separations, Allentown, PA, USA) using a 24 mL extraction vessel. The portions of material were thermostated for 15 min prior to extraction; in the experiments using cosolvent, the cosolvent was added along with the sample at this point. Each extraction run was conducted with a constant flow rate of approx. $100\,\mathrm{g\,h^{-1}}$ for 1 h; our previous publications using similar parameters proved that such a duration is sufficient for extraction [16–19].

The solvating strength of supercritical fluids is strongly connected with their parameters of state; therefore various temperatures and pressures were evaluated. Their lower limits were set by critical parameters of carbon dioxide ($31.0\,^{\circ}\mathrm{C}$, 7.38 MPa), while the upper limit was set by the thermal stability of the galactolipids ($75\,^{\circ}\mathrm{C}$) and technical limitations of the equipment (40 MPa). A 515 HPLC Pump (Waters, Milford, MA, USA) was used to volumetrically deliver the co-solvent to the vessel. Ethanol and carbon dioxide were mixed in a small chamber prior to entering the vessel. The experiments included extraction runs at combinations of three levels of the temperature (35, 55, and $75\,^{\circ}\mathrm{C}$) and the levels of the pressure (10, 20, and 40 MPa). Next, the three levels of cosolvent flow (0.0, 0.2 and $0.4\,\mathrm{mL\,min^{-1}}$) were evaluated in combinations with the temperatures.

2.3. Ultrasound-Assisted Extraction (UAE)

The UAE process was conducted using an ultrasound bath MKD-6 (MKD Ultrasonic, Warsaw, Poland) with a sonication power of 300 W and an ambient temperature. The plant material was placed inside a Falcon vessel (50 mL) with solvent, sonicated for a fixed amount of time and shaken for 1 min at $300\,\mathrm{min^{-1}}$ (SK-O330 Pro, DragonLab Instruments,

Beijing, China). The control samples were prepared using the same matrix–solvent ratio and kept for a fixed amount of time. and shaken for 1 min.

The first stage of the experiment was to compare the efficiency of twelve solvents for the extraction of galactolipids—the yield of both sonicated and control samples was compared. The results along with properties of the solvents were used to select the most appropriate solvent for further experiments. The second part was an analysis of an impact of process parameters (sonication time, mass–solvent ratio) on the efficiency of the process.

2.4. Analysis of Galactolipid Content

The method presented by Zábranská and co-workers was used for the qualitative and quantitative analysis of galactolipids [9]. The ground plant sample (6 g) was placed in a flask, 15 mL of isopropanol was added, and the mixture was kept at a temperature of 75 °C for 15 min. Subsequently, 7.5 mL of chloroform and 3 mL of water were added to the flask and then shaken for 1 h at 300 min^{-1}. The extract was collected and a sample was reextracted with 10 mL of chloroform–methanol mixture (2:1) three times. The extracts were collected and washed with 20 mL of 1M potassium chloride. The organic phase was dehydrated with 5 g of anhydrous magnesium sulfate, filtered through cotton wool, and concentrated on a rotary evaporator (R-300 Rotavapor, Büchi, Flawil, Switzerland) to approximately 20% of the initial volume. The extracts obtained using SFE and UAE methods were only subjected to the last part of the aforementioned protocol (from dehydration with $MgSO_4$).

Semi-preparative thin-layer chromatography was used to isolate the main groups of galactolipids. The extract was transferred to a 10×20 cm TLC silica gel 60G (0.2 mm) plate (Merck, Darmstadt, Germany) and eluted with a mixture of chloroform–methanol–water (80:18:2). The zones corresponding to MGDGs and DGDGs typically had retention factors of 0.7 and 0.4, respectively. Nevertheless, standards were also placed on the plate and visualized under UV light after spraying with a primuline solution (0.05% in methanol-water (8:2) mixture). The zones corresponding to galactolipid groups were scraped, extracted with a mixture of chloroform-methanol (2:1), and filtered through cotton glass.

Liquid chromatography analyses were performed on an H-class liquid chromatograph coupled to a mass spectrometer with a time-of-flight analyzer (UPLC-TOF-HRMS; Waters, Milford, MA, USA). The samples (5 µL) were separated using a UPLC C18 Cortecs (2.1 × 100 mm, 1.6 µm) column (Waters). Despite using the same equipment, separate gradients were employed for the galactolipid groups—the details are presented below in Table 1. The mass spectrometer operated in the positive electrospray ionization mode. The ion source temperature was 150 °C, while the desolvation temperature was 350 °C. The nebulizing gas (N_2) flow rate was 750 L min^{-1}, the cone gas flow rate was 40 L min^{-1}, and the capillary bias was 3.2 kV.

Table 1. Gradients used for the LC-MS analysis of the galactolipids.

	MGDGs		
Time (min)	Water		Methanol
0	25%		75%
80	0%		100%
90	0%		100%
95	25%		75%

	DGDGs		
Time (min)	Water	Acetonitrile	Methanol
0	35%	65%	0%
50	0%	80%	20%
60	0%	0%	100%
80	0%	0%	100%
85	35%	65%	0%

The method implemented allowed the separation of galactolipids and their partial identification. As reported by the authors of the method, galactolipids in plant samples are present as a mixture of compounds with various fatty acids attached to the glycerol scaffold (Figure 1) [9]. The implementation of MS detection made it possible to distinguish between the compounds based on their m/z ratio, as the elongation of the carbon chain or saturation of the double bond can be easily identified. GOPO, the compound that, according to the literature, has the highest pharmacological potential is MGDG, with two α-linoleic acids (18:3) attached; due to its structure, the identification is certain as no lipids of the same molecular mass could occur. As MGDG analysis was used to a greater extent during the research, the separation of these galactolipids is presented in Figure S1 in the Supplementary Materials.

2.5. Statistical Analysis

Data were analyzed using Statistica 7.1 software (StatSoft, Tulsa, OK, USA). An ANOVA test at $\alpha = 0.05$ was used to determine the statistical significance of the differences in mean values; if such differences existed, a post hoc Tukey test was performed. All of the results were obtained from analyses of three independent samples.

3. Results and Discussion

3.1. Galactolipids in Processing By-Products

The waste materials were analyzed for their galactolipid content, with a special emphasis on GOPO, which has the greatest pharmacological potential, as was highlighted earlier. The content of the analytes is presented in Table 2.

Table 2. Content of galactolipids and dry mass in the waste materials analyzed.

Material	Total Content of MGDGs (mg kg^{-1})	Content of GOPO (mg kg^{-1})	Total Content of DGDGs (mg kg^{-1})	Dry Mass Content (%)
apple pomace	1.0 ± 0.4	≤ 0.1	0.5 ± 0.2	27.4 ± 0.3
carrot pomace	≤ 0.1	≤ 0.1	≤ 0.1	22.3 ± 0.4
rosehip pomace	56.4 ± 1.1	41.0 ± 1.5	3.2 ± 0.3	24.6 ± 0.4
broccoli wastes	24.2 ± 1.2	1.2 ± 0.2	14.3 ± 0.7	13.5 ± 0.7
fava bean wastes	1.9 ± 0.5	≤ 0.1	1.2 ± 0.2	43.2 ± 0.2
potato wastes	≤ 0.1	≤ 0.1	≤ 0.1	31.1 ± 0.2

The content of analytes varied significantly between the samples, which was partially the effect of using various morphological parts of plants. The subterranean parts of plants did not contain measurable levels of galactolipids, apples and beans were characterized by low levels, and broccoli had a high content of these compounds, as its green florets are abundant in chloroplasts. The rosehip was the richest source of galactolipids among the matrices analyzed, having an MGDGs/DGDGs ratio distinct from other resources.

The rosehip was also characterized by a high content of GOPO, which accounted for almost 70% of galactolipids in the sample. This compound was also detected in broccoli, although its levels were much lower. Additionally, the broccoli wastes were characterized by a high water content, which would hinder the extraction with supercritical carbon dioxide and organic solvents of low polarity. The matrix could be dried prior to the process, although, depending on the technique selected, this would lead to the thermal decomposition of analytes (convection-drying) or generate substantial costs (freeze-drying). Therefore, only the rosehip pomace was selected for further experiments. Additionally, as levels of DGDGs in the material are much lower than those of MGDGs and require additional chromatographic analysis, only levels of MGDGs were evaluated in the later parts of the work.

The galactolipid content in plants can be related to their physiological function because they are essential components of chromoplast membranes [1]. The review paper by Christensen et al. confirms this assumption. The authors reported that the levels of both MGDGs and DGDGs in leafy vegetables are by far the highest [2]. Such results were confirmed by other authors such as Lee et al., who reported a total galactolipid content of 100–200 mg kg^{-1} (of dry weight) in green leafy vegetables (parsley, spinach, and perilla) [20]. The only two literature reports on GOPO levels in rosehip are by Larsen and coworkers, who recorded a level of 270 mg kg^{-1} in dried fruits [9], and Chrubasik at al., who reported 300 mg kg^{-1} in a diet supplement based on dried rosehip [21], which is comparable to green vegetables. Our data provide a total galactolipid content of approximately 60 mg kg^{-1} in rosehip and less than 40 mg kg^{-1} in broccoli; the results are consistent with the literature considering the humidity of the samples and interspecies variations.

3.2. Supercritical Fluid Extraction (SFE)

In the first series of the extraction runs, nine sets of process parameters were evaluated—temperatures of 35, 55, and 75 °C were combined with pressures of 10, 20, and 40 MPa. None of the extraction variants was able to obtain a measurable yield of galactolipids. The possible explanation was an insufficient polarity of pure carbon dioxide; therefore, another set of experiments was conducted using ethanol as a cosolvent. The literature data highlight that higher pressures strongly improve the solvating properties of carbon dioxide; therefore, only 40 MPa was applied to limit the number of experiments [14]. Table 3 presents tested parameters and summarizes the results of this part of the work. The cosolvent flow values was selected to make up 0%, 10%, and 20% of total flow.

Table 3. Impact of supercritical fluid extraction parameters on the efficiency of MGDGs extraction from rosehip pomace. All the experiments were performed at the pressure of 40 MPa. The letters denote statistical significance of the differences ($p \leq 0.05$).

Temperature (°C)	Cosolvent (mL min^{-1})	MGDGs Yield (mg kg^{-1} of Sample)
35	0.0	≤0.1 a
35	0.2	22.7 ± 3.1 b
35	0.4	48.8 ± 2.5 c
55	0.0	≤0.1 a
55	0.2	20.9 ± 2.5 b
55	0.4	49.2 ± 2.0 c
75	0.0	≤0.1 a
75	0.2	21.4 ± 1.3 b
75	0.4	47.9 ± 1.6 c

The use of a cosolvent allowed the extraction of significant amounts of galactolipids and the yields obtained were in the range from 37 to 87%. The flow rate of ethanol appeared to be the crucial parameter for the process efficiency: the best results were observed for the 0.4 mL min^{-1} share of the co-solvent, while pure carbon dioxide was unable to extract any galactolipids. The two-fold increase of ethanol content (0.2 vs. 0.4 mL min^{-1}) resulted in improvement of the yield obtained higher than 100%; this may suggest that obtaining proper polarity of CO_2–ethanol mixture may be more significant than simply adding ethanol to the system.

The possibility of implementation of supercritical fluids for isolation of galactolipids from other matrices was reported by few teams. The paper by Ijima et al. described experiments on extraction of galactolipids from spinach (*Spinacia oleracea*) leaves by SFE with and without a modifier [22]. Similarly to our results, the use of pure CO_2 did not allow MGDG isolation, while ethanol as a cosolvent gave good results. The authors used a different matrix and tested only one parameter set; thus, full comparison of results is not possible. The work by Yang et al. showed that supercritical CO_2 and ethanol mixture was able to extract over 80% of lipids from *Cyanobacterium Arthrospira*, while pure carbon dioxide gave negligible yield [23]. The possibility of application of less orthodox solvent was reported by Nekrasov and coworkers, who successfully extracted galactolipids from fern fronds with dimethyl ether–water–ethanol mixture in the supercritical state [24].

3.3. Ultrasound-Assisted Extraction (UAE)

The experiments on the possible use of UAE in the isolation of galactolipids began with the evaluation of twelve commonly used solvents of various polarities. The rosehip pomace was mixed with them in a ratio of 1:10 (m/v) and subsequently extracted in two variants: with and without sonication. The yields of MGDGs obtained are listed in Table 4.

Table 4. Impact of sonication on the efficiency of MGDG extraction from rosehip pomace using various solvents. All yields in mg per kg of pomace; sonication time was 5 min. An asterisk next to the solvent name denotes a significant difference between sonicated and control samples, while letters stand for differences between solvents (both at $p \leq 0.05$).

Solvent	Relative Polarity [25]	Sonicated	Control
hexane	0.009	$\leq$0.1 a	$\leq$0.1 a
toluene *	0.099	2.2 ± 0.3 b	1.5 ± 0.2
methyl t-butyl ether *	0.124	8.5 ± 0.5 d	7.2 ± 0.2
ethyl acetate *	0.228	11.5 ± 0.8 e	9.5 ± 0.4
chloroform *	0.265	34.1 ± 1.6 h	28.6 ± 0.7
dichloromethane *	0.309	32.9 ± 1.0 h	27.8 ± 0.5
acetone *	0.355	14.3 ± 0.1 f	11.7 ± 0.6
acetonitrile *	0.460	9.0 ± 0.4 d	7.1 ± 0.3
isopropanol *	0.546	7.2 ± 0.1 c	6.2 ± 0.3
ethanol *	0.654	15.2 ± 0.4 g	13.0 ± 0.2
methanol *	0.762	14.6 ± 0.7 fg	11.9 ± 0.6
water	1.000	$\leq$0.1 a	$\leq$0.1 a

Apart from the most and the least polar (water and hexane, respectively), all the solvents were able to extract MGDGs to some extent. The best results were obtained for chlorinated hydrocarbons, yet even dichloromethane, which is much safer for people and the environment than chloroform, can be hardly considered a green solvent. Therefore, the solvent characterized by a lower extraction efficiency, but of lesser environmental and health impact, must be chosen. Ethanol combines low toxicity with a low cost; thus it was chosen for the further experiments. Additionally, its use as a cosolvent during SFE allows a systematic comparison of these two techniques. The sonication made it possible to significantly increase the yield of extraction in all the solvents evaluated, which indicated the potential of the technique.

The next part of the experiments was the investigation of the impact of the sonication time and ratio of pomace and ethanol on the efficiency of the extraction—the results are presented in Figure 2. Two solid–liquid ratios and three sonication times were investigated. The control samples were extracted with ethanol for the same amount of time without any means of a process intensification. As was stated before, the samples were shaken for 1 min prior to removal of the solvent.

The amphiphilic character of galactolipids was visible during the selection of the appropriate solvent. Both highly polar (water) and non-polar (hexane) solvents were unsuitable for the extraction. The impact of the sonication time on the yield presented in Figure 2 is in accordance with the extraction curves reported by other authors [26–28]. Sonication leads to the disintegration of cells on the surface of the matrix particles, which at the beginning of the process leads to a significant increase in yield, although longer treatment does not have such an effect, as the majority of the cells are already broken. The comparison of the extraction curves suggests that differences between the sonicated and the control samples are mainly the result of improvement of the mass transfer. The bigger gap between sonicated and control yields in the case of 1:25 ratio is supposedly the result of a limited solubility of galactolipids in ethanol. The sonication of the rosehip samples led to a substantial improvement in the yields obtained (up to 74%) with a relatively low cost of implementation. The abovementioned facts indicate that the method could be suitable to improve technological aspects as well as the sustainability of the process. Evidently, the implementation of UAE for an industrial-scale operation would require optimization as such processes are not easily scaled up.

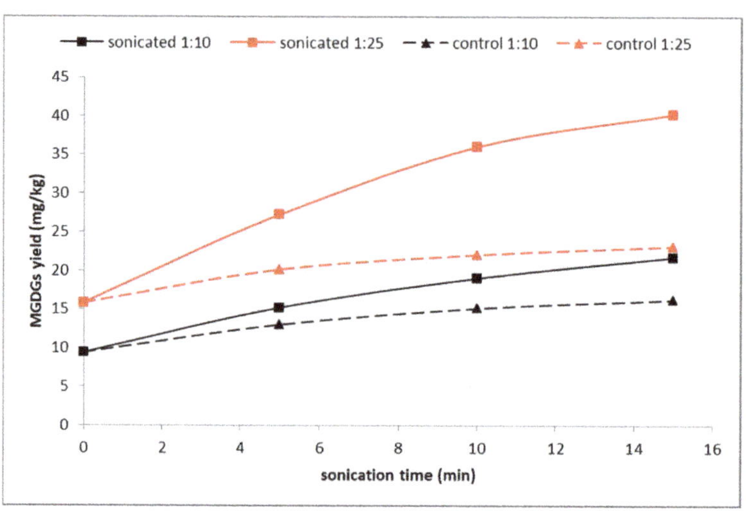

Figure 2. Impact of the sonication time and the solid–liquid ratio (m/V) on the efficiency of MGDGs extraction from rosehip pomace with ethanol.

3.4. Comparison of the Techniques and General Remarks

In order to compare efficiency of the extraction methods, an additional UAE run using the same processing time (1 h) and the solid–liquid ratio (4.8 m/V) was performed along with control sample. The results are presented in Figure 3.

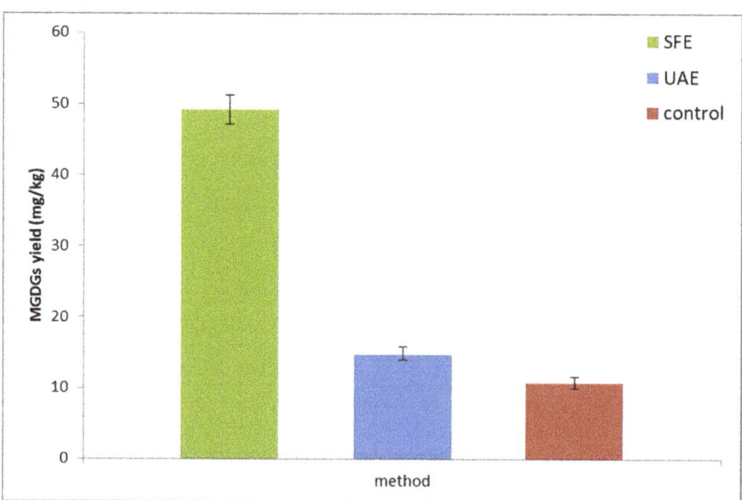

Figure 3. The comparison of MGDGs extraction efficiency using different techniques with the same processing time (1 h) and solid–liquid ratio (4.8, m/V). SFE was conducted at a temperature of 55 °C and at 40 MPa. There were significant differences ($p \leq 0.05$) between all the samples tested.

The use of such low solid–liquid ratio significantly hindered the extraction process, as the limited solubility of MGDGs played more crucial role. Nevertheless, about 36% improvement in efficiency was observed after use of the sonication. SFE yields were approx. 5-fold higher than the control sample, proving the superiority of this technique. On the other hand, the use of SFE is connected with high initial costs of the machinery, while UAE can be considered as a very low-priced and easy-to-implement technology. Nevertheless, we believe that the presented data are a satisfactory proof of suitability of green extraction techniques for isolation of galactolipids from waste materials of fruit and vegetable processing.

Rosehip preparations are often sold in the form of capsules filled with milled fruits instead of extract. Such an approach is undoubtedly a much cheaper alternative; however, it is disadvantageous from a bioavailability point of view. The highest intestinal absorption of galactolipids would probably be achieved after ingestion of encapsulated or micronized extracts, while absorption after administration of the milled tissue would probably be very low [29,30].

Supplementary Materials: The following are available online at https://www.mdpi.com/article/10.3390/app112412088/s1. Figure S1: The chromatograms of MGDGs standard mixture on various m/z values.

Author Contributions: Conceptualization, Ł.W. and S.S.; methodology, Ł.W. and K.M.; formal analysis, Ł.W. and M.W.; investigation, Ł.W. and M.W.; writing—original draft preparation, Ł.W. and M.W.; writing—review and editing, K.M. and S.S.; visualization, Ł.W.; supervision, K.M. and S.S. All authors have read and agreed to the published version of the manuscript.

Funding: This research received no external funding.

Institutional Review Board Statement: Not applicable.

Informed Consent Statement: Not applicable.

Data Availability Statement: Not applicable.

Conflicts of Interest: The authors declare no conflict of interest.

References

1. Hölzl, G.; Dörmann, P. Structure and function of glycoglycerolipids in plants and bacteria. *Prog. Lipid Res.* **2007**, *46*, 225–243. [CrossRef] [PubMed]
2. Christensen, L.P. Galactolipids as potential health promoting compounds in vegetable foods. *Recent Pat. Food Nutr. Agric.* **2009**, *1*, 50–58. [CrossRef]
3. Weatherby, K.; Carter, D. *Chromera velia*: The missing link in the evolution of parasitism. *Adv. Appl. Microbiol.* **2013**, *85*, 119–144. [PubMed]
4. Wang, R.; Furomoto, T.; Motoyama, K.; Okazaki, K.; Kondo, A.; Fukui, H. Possible anti-tumor promoters in *Spinacia oleracea* (spinach) and comparison of their contents among cultivars. *Biosci. Biotechnol. Biochem.* **2002**, *66*, 248–254. [CrossRef] [PubMed]
5. Kuriyama, I.; Musumi, K.; Yonezawa, Y.; Takemura, M.; Maeda, N.; Iijima, H.; Hada, T.; Yoshida, H.; Mizushina, Y. Inhibitory effects of glycolipids fraction from spinach on mammalian DNA polymerase activity and human cancer cell proliferation. *J. Nutr. Biochem.* **2005**, *16*, 594–601. [CrossRef]
6. Gruenwald, J.; Uebelhack, R.; Moré, M.I. *Rosa canina*—Rose hip pharmacological ingredients and molecular mechanics counteracting osteoarthritis—A systematic review. *Phytomedicine* **2019**, *60*, 152958. [CrossRef]
7. Warholm, O.; Skaar, S.; Hedman, E. The effects of a standardized herbal remedy made from a subtype of *Rosa canina* in patients with osteoarthritis: A double-blind, randomized, placebo—controlled clinical trial. *Curr. Therap. Res.* **2003**, *64*, 21–31. [CrossRef]
8. Larsen, E.; Kharazmi, A.; Christensen, L.P.; Christensen, S.B. An antiinflammatory galactolipid from Rose hip (*Rosa canina*) that inhibits chemotaxis of human peripheral blood neutrophils in vitro. *J. Nat. Prod.* **2003**, *66*, 994–995. [CrossRef] [PubMed]
9. Zábranská, M.; Vrkoslav, V.; Sobotníková, J.; Cvačka, J. Analysis of plant galactolipids by reversed-phase high-performance liquid chromatography/mass spectrometry with accurate mass measurement. *Chem. Phys. Lipids* **2012**, *165*, 601–607. [CrossRef] [PubMed]
10. Chemat, F.; Vian, M.A.; Cravotto, G. Green extraction of natural products: Concept and principles. *Int. J. Mol. Sci.* **2012**, *13*, 8615–8627. [CrossRef] [PubMed]
11. Chemat, F.; Rombaut, N.; Meullemiestre, A.; Turk, M.; Perino, S.; Fabiano-Tixier, A.S.; Abert-Vian, M. Review of green food processing techniques. Preservation, transformation, and extraction. *Innov. Food Sci. Emerg. Technol.* **2017**, *41*, 357–377. [CrossRef]

12. Mena-García, A.; Ruiz-Matute, A.I.; Soria, A.C.; Sanz, M.L. Green techniques for extraction of bioactive carbohydrates. *Trends Anal. Chem.* **2019**, *119*, 115612. [CrossRef]

13. Tiwari, B.K. Ultrasound: A clean, green extraction technology. *Trends Anal. Chem.* **2015**, *71*, 100–109. [CrossRef]

14. de Melo, M.M.R.; Silvestre, A.J.D.; Silva, C.M. Supercritical fluid extraction of vegetable matrices: Applications, trends and future perspectives of a convincing green technology. *J. Supercrit. Fluids* **2014**, *92*, 115–176. [CrossRef]

15. Brunner, G. Supercritical fluids: Technology and application to food processing. *J. Food Eng.* **2005**, *67*, 21–33. [CrossRef]

16. Woźniak, Ł.; Marszałek, K.; Skąpska, S. Extraction of phenolic compounds from sour cherry pomace with supercritical carbon dioxide: Impact of process parameters on the composition and antioxidant properties of extracts. *Sep. Sci. Technol.* **2016**, *51*, 1472–1479. [CrossRef]

17. Woźniak, Ł.; Marszałek, K.; Skąpska, S.; Jędrzejczak, R. The application of supercritical carbon dioxide and ethanol for the extraction of phenolic compounds from chokeberry pomace. *Appl. Sci.* **2017**, *7*, 322. [CrossRef]

18. Woźniak, Ł.; Szakiel, A.; Pączkowski, C.; Marszałek, K.; Skąpska, S.; Kowalska, H.; Jędrzejczak, R. Extraction of triterpenic acids and phytosterols from apple pomace with supercritical carbon dioxide: Impact of process parameters, modelling of kinetics, and scaling-up study. *Molecules* **2018**, *23*, 2790. [CrossRef]

19. Woźniak, Ł.; Połaska, M.; Marszałek, K.; Skąpska, S. Photosensitizing furocoumarins: Content in plant matrices and kinetics of supercritical carbon dioxide extraction. *Molecules* **2020**, *25*, 3805. [CrossRef]

20. Lee, S.J.; Song, Y.; Chung, M.Y.; Kim, I.H.; Kim, B.H. Isolation and compositional analysis of galactoglycerolipids from perilla [*Perilla fructescens* (L.) Britton] leaves and comparison to the galactoglycerolipids from spinach and parsley. *J. Food Sci.* **2020**, *85*, 4271–4280. [CrossRef]

21. Chrubasik, C.; Wiesner, L.; Black, A.; Müller-Ladner, U.; Chrubasik, S. A one-year survey on the use of a powder from *Rosa canina lito* in acute exacerbations of chronic pain. *Phytother. Res.* **2008**, *22*, 1141–1148. [CrossRef]

22. Iijima, H.; Musumi, K.; Hada, T.; Maeda, N.; Yonezawa, Y.; Yoshida, H.; Mizushina, Y. Inhibitory effect of monogalactosyldiacylglycerol, extracted from spinach using supercritical CO_2, on mammalian DNA Polymerase activity. *J. Agric. Food Chem.* **2006**, *54*, 1627–1632. [CrossRef] [PubMed]

23. Yang, X.; Li, Y.; Li, Y.; Ye, D.; Yuan, L.; Sun, Y.; Han, D.; Hu, Q. Solid Matrix-Supported supercritical CO_2 enhances extraction of γ-linolenic acid from the cyanobacterium Arthrospira (Spirulina) platensis and bioactivity evaluation of the molecule in zebrafish. *Mar. Drugs* **2019**, *17*, 203. [CrossRef] [PubMed]

24. Nekrasov, E.V.; Tallon, S.J.; Vyssotski, M.V.; Catchpole, O.J. Extraction of lipids from New Zealand fern fronds using near-critical dimethyl ether and dimethyl ether-water-ethanol mixtures. *J. Supercrit. Fluids* **2021**, *170*, 105137. [CrossRef]

25. Welton, T.; Reinchardt, C. *Solvents and Solvent Effects in Organic Chemistry*, 3rd ed.; Wiley-VCH: Wienheim, Germany, 2003.

26. Da Porto, C.; Natolino, A. Extraction kinetic modelling of total polyphenols and total anthocyanins from saffron floral bio-residues: Comparison of extraction methods. *Food Chem.* **2018**, 137–143. [CrossRef]

27. Dias, A.L.B.; Sergio, C.S.A.; Santos, P.; Barbero, G.F.; Rezende, C.A.; Martínez, J. Ultrasound-assisted extraction of bioactive compounds from dedo de moça pepper (*Capsicum baccatum* L.): Effects on the vegetable matrix and mathematical modeling. *J. Food Eng.* **2017**, *198*, 36–44. [CrossRef]

28. Natolino, A.; Da Porto, C. Kinetic models for conventional and ultrasound assistant extraction of polyphenols from defatted fresh and distilled grape marc and its main components skins and seeds. *Chem. Eng. Res. Des.* **2020**, *156*, 1–12. [CrossRef]

29. McClements, D.J.; Decker, E.A.; Park, Y. Controlling lipid bioavailability through physiochemical and structural approaches. *Crit. Rev. Food Sci. Nutr.* **2009**, *49*, 48–67. [CrossRef] [PubMed]

30. Capuano, E.; Pellegrini, N. An integrated look at the effect of structure on nutrients bioavailability in plant foods. *J. Sci. Food Agric.* **2019**, *99*, 492–498. [CrossRef] [PubMed]

applied sciences

Article

The Effects of Plasma-Activated Water on Heavy Metals Accumulation in Water Spinach

Chih-Yao Hou [1], Ting-Khai Kong [2], Chia-Min Lin [1,†] and Hsiu-Ling Chen [2,*,†]

[1] Department of Seafood Science, National Kaohsiung University of Science and Technology, Kaohsiung 811, Taiwan; chihyaohou@nkust.edu.tw (C.-Y.H.); cmlin@mail.naku.edu.tw (C.-M.L.)

[2] Department of Food Safety/Hygiene and Risk Management, College of Medicine, National Cheng Kung University, Tainan 701, Taiwan; 10912015@gs.ncku.edu.tw

* Correspondence: hsiulinchen@mail.ncku.edu.tw

† The authors has the same contribution.

Abstract: Toxic heavy metals accumulate in crops from the environment through different routes and may interfere with biochemical reactions in humans, causing serious health consequences. Plasma technology has been assessed for the promotion of seed germination and plant growth in several past studies. Therefore, the aim of the present study was to evaluate whether the growth rate of plants can be increased with the application of non-thermal plasma, as well as to reduce the accumulation of heavy metals in leafy vegetables (water spinach). In this study, several kinds of plasma treatments were applied, such as treatment on the seeds (PTS + NTW), irrigation water (NTS + PAW) or both (PTS + PAW). The results of the study showed that the heavy metals accumulated in water spinach were affected by the heavy metals available in the soil. The bioconcentration factor (BCF) of Cd in water spinach decreased from 0.864 to 0.543 after plasma treatment in seed or irrigating water, while the BCF of Pb was low and did not show any significant changes. Therefore, the results suggest that plasma treatment may suppress Cd absorption, but not for Pb. In this study, plasma treatment did not help to improve the product yield of water spinach planted in Cd-added soil. In the future, fertilizers can be used to supply nutrients that are not provided by plasma-activated water to support the growth of water spinach.

Keywords: plasma-activated water; heavy metals; water spinach; plant growth

Citation: Hou, C.-Y.; Kong, T.-K.; Lin, C.-M.; Chen, H.-L. The Effects of Plasma-Activated Water on Heavy Metals Accumulation in Water Spinach. *Appl. Sci.* **2021**, *11*, 5304. https://doi.org/10.3390/app11115304

Academic Editors: Marek Kieliszek and Przemyslaw Lukasz Kowalczewski

Received: 14 May 2021
Accepted: 28 May 2021
Published: 7 June 2021

Publisher's Note: MDPI stays neutral with regard to jurisdictional claims in published maps and institutional affiliations.

1. Introduction

Advancements in technology, industrialization, and urbanization have sped up the pollution of the environment. Potential toxic heavy metals, including lead (Pb), cadmium (Cd), mercury (Hg), arsenic (As), nickel (Ni), chromium (Cr), copper (Cu), zinc (Zn), cobalt (Co), and other metal elements in the environment are emitted into water and soil through industrial wastewater discharge, commercial waste disposal and burial, exhaust emissions, and the improper use of agricultural materials. Toxic heavy metals are less mobile in the environment, and they possess a high residual capacity, and hence are easily accumulated in the environment. Heavy metal uptake can occur through mechanisms such as phytoextraction, phytostabilization, rhizofiltration, and phytovolatilization, which is the uptake, absorption, translocation, and/or transpiration process of metals by the plant into the above-ground portion of the roots [1]. Most of the heavy metals in soil can accumulate in crops, and consumption of crops that are contaminated with heavy metals can pose serious human health issues and present some adverse effects [2]. In Taiwan, the content of the heavy metals Cd and Pb in leafy vegetables should not exceed 0.2 mg/kg F.W. and 0.3 mg/kg F.W., respectively, according to the Taiwan Food and Drug Administration's regulations.

Plasma is considered to be a state of matter other than solid, liquid, and gas. Basically, plasma can be generated by subjecting different types of gases to an electric or electromagnetic field, which provides sufficient energy for ionization, or dissociation or exciting collisions. The increased number of collisions between the electrons due to the magnetic field causes an increase in the energy in the ionized gas. As a result, a quasi-neutral gas is formed that contains different species, such as electrons, ions, radicals, atoms, and molecules in their fundamental or excited states, with a net neutral charge [3]. Plasma can be categorized into high-temperature plasma (equilibrium plasma), thermal plasma (quasi-equilibrium plasma), and non-thermal plasma (non-equilibrium plasma) [4]. If the plasma ions and neutral components remain at or near room temperature, it is considered to be low-temperature or non-equilibrium plasma. The special characteristics of plasma with a strong thermodynamic non-equilibrium nature and low gas temperature include containing a lot of reactive chemical species, which makes non-thermal plasma applicable to several different industries [5]. Therefore, the application of plasma in agriculture is one of the most promising applications being studied in recent years. The main research direction in agricultural applications and food safety are as follows: inducing changes in the plant characteristics, eliminating pathogens or inhibiting microorganisms, promoting seed germination, increasing crop quantity, and improving crop preservation, and pesticide elimination [6–9].

Pawłat et al. [10] applied an atmospheric pressure plasma jet that operated with dielectric barrier discharge (DBD) on *Lavatera thuringiaca* L. seeds' germination at different exposure times (1, 2, 5, 10, and 15 min). The experimental results revealed that the germination capacity and germination energy of the seeds in the experimental group were improved as a result of plasma stimulation before sowing. However, excessive exposure can cause damage to the seeds, resulting in decreased germination. The germination rate, germination vigor, relative conductivity, and water uptake of the oilseed rape were improved, while the plant yield was increased by 28.2% after the seeds were treated by a low-vacuum helium cold plasma at a parameter of 100 W, 13.56 MHz, and 150 Pa for 15 s [11]. Research on direct and indirect plasma treatments of several vegetable species with DBD was conducted by Liu et al., who found that the germination of seeds was improved, although the efficiency depended on the species [12].

Ma et al. proved that the polysaccharides of a mutant strain of *Ganoderma lingzhi mycelia* increased by 25.6% after treatment with atmospheric DBD non-thermal plasma [7]. Another similar study by Takaki et al. included the application of a pulsed high-voltage electrical stimulation to cultivation mushrooms at 50–130 kV, where the fruiting body formation increased by 1.3–2 times [9].

Plasma-treated water or plasma-activated water (PAW) is produced by applying plasma to water in order to generate active oxygen molecules and excited OH radicals. Atomic oxygen and nitrogen generated in gaseous plasma could transfer into liquid and be easily converted into other reactive oxygen species (ROS) and reactive nitrogen species (RNS), such as nitrates (NO_3^-), nitrites (NO_2^-), and hydrogen peroxides (H_2O_2) in the solution. Air is considered to be the most abundant processing gas at low cost, and acts as a great source of oxygen (O_2) and nitrogen (N_2) with ultraviolet radiation and other chemical species in liquids. PAW is a mixture of high biochemically reactive solution of an acidic condition due to the changes in oxidation-reduction potential (ORP), electrical conductivity (EC), and the formation of RON and RNS [13,14]. Judée et al. evaluated the potential application of plasma treatment of tap-water (PATW) produced by cold atmospheric pressure DBD for improving agricultural quality [15]. The study revealed that the formation of nitrite, nitrate, ammonium ions (NH_4^+), and H_2O_2 in the water could induce and enhance crop growth, and that the consumption of bicarbonate ions could promote seedling growth. The length of the seedlings increased by 34% and 128.4% after 3 and 6 days of PATW treatment. The production of reactive species using plasma-activated water could change or affect the seeds both physically and chemically. Thirumdas et al. reviewed the relevant research on PAW and indicated that the physical and chemical properties of PAW are related to the acidity, conductivity, redox potential, and concentration of ROS and RNS in the water, which affects microorganisms [14,16]. In addition, Thirumdas et al. suggested that PAW has a synergistic effect on food disinfection and can also promote the growth of seed seedlings [14]. The increment in the proportion of nitrate and nitrite ions in PAW may be the actual reason for the improvement in plant growth. Soaking seeds in PAW can play an antibacterial role and can also promote seed germination and plant growth. PAW may be used to increase crop yields and confront drought environmental conditions [12,14].

Several previous studies in the literature have proven the efficiency of plasma treatment for improving the germination and growth rate of plant species. Hence, this research focuses on assessing the application of non-thermal plasma in decreasing the heavy metal accumulation of vegetables planted in contaminated soil by initiating the growth of the vegetables.

2. Materials and Methods

2.1. Chemicals

Cadmium (II) acetate anhydrous (Sigma-Aldrich, India, 99.995%), lead acetate trihydrate (Sigma-Aldrich, India, 99–102.5%), ICP multi-element standard solution VI (Merck, Germany, 1000 mg/L), rhodium ICP standard (Merck, Germany, 1000 mg/L), nitric acid (JT Baker, Canada, 70%), nitric acid 65% for analysis (Merck, Germany), and hydrochloric acid fuming 37% for analysis (Merck, Germany) were used in the current study.

2.2. Man-Made Contaminated Soil

For preparing Cd-contaminated soil and Pb-contaminated soil, we dissolved 12.81 g of lead (II) acetate in 1 L of water and mixed them into 10 kg of cultivation soil, purchased from the gardening market (SINON CORPORATION, Tainan, Taiwan) to prepare 700 mg/kg wet weight (W.W.) man-made contaminated soil. For Cd-contaminated soil, 0.14 g of cadmium acetate was weighed and dissolved in 1 L of water, followed by mixing with 10 kg of cultivated soil. The concentration was then expected to be 7 mg/kg W.W. The man-made contaminated soil was continuously mixed well for 1 week to ensure that the metals were stable and uniformly distributed.

2.3. Selection of Vegetables

Water spinach (*Ipomoea aquatica*) seeds were selected and purchased (Known-You Seed Co., LTD, Kaohsiung, Taiwan), as leafy vegetables are relatively more frequently consumed by the Taiwanese population, on the basis of data released by the National Food Consumption Database.

2.4. The Plasma Device and Parameters

The non-thermal atmospheric plasma generator was designed and produced by the Aerothermal and Plasma Physics Lab (APPL), Department of Mechanical Engineering, National Chiao Tung University (Hsinchu, Taiwan). This system included a high-voltage power supply, an air pump, and an atmospheric pressure plasma jet (APPJ) (patent US10,121,638B1) generator [17]. This system was specifically designed to generate plasma beneath the water surface. Air was pumped into the electrode by the air pump under a regular atmosphere at a flow rate of 10 standard liters per minute (slm). The input air was ionized by the high-voltage electrode to generate plasma, which was then delivered through a quartz capillary into the water. Reverse osmotic (RO) water was used as the water source. The input power and frequency for the power supply were AV 110 V and 60 Hz, respectively. The output voltage, frequency, and power were set at 3.0 kV, 16 kHz, and 60 Watts, respectively.

2.5. Plasma Treatment of Seeds and Irrigation Water

The first step in the plasma application was to prepare PAW. A total of 500 mL of reverse osmosis (RO) water was prepared in a 1000-mL beaker. Two plasma jets were immersed in water, and the beaker was covered with a plastic wrap. The power was set to 60 W for plasma generation and the water was treated for 20 min. For the control group, RO water without plasma treatment was used as the irrigation water for plantation purposes. For seed treatment, 15 seeds were selected randomly and placed in a beaker containing 200 mL of RO water. One plasma jet was immersed in water, and the power was set to 60 W for plasma generation. The seeds were treated for 7 min after a preliminary study revealed that this was optimal for achieving high efficiency in germination. For the control group, the seeds were placed in the water for 7 min without any plasma application.

2.6. Plantation of Water Spinach

Planting pots (15 cm diameter, 13 cm height) were used for plantation purposes. Clean cultivation soil (control group) and man-made contamination soil with added metals were weighed. A total of 550 g of soil was placed into each plantation pot. For seedling preparation, the treated and untreated seeds were then placed on wet filter paper in a Petri dish for a germination period of 5 days. Six of the best-grown seedlings were selected and moved to each pot for growing. The plasma treatment was applied to the seeds and the water, in combinations as follows: Group 1, no plasma treatment on seeds or irrigation water (NTS + NTW); Group 2, no plasma treatment on seeds, but with the use of PAW as irrigation water (NTS + PAW); Group 3, plasma treatment on seeds, but no plasma treatment was used on irrigation water (PTS + NTW); and Group 4, plasma treatment on seeds, and using PAW as the irrigation water (PTS + PAW). Each group was planted in three pots (triplicate), and each pot contained six seedlings. The procedure was shown in Figure 1. The plantation was carried out for 5 weeks under neutral day conditions (photoperiod: 13-h light between 7:00 a.m. to 8:00 p.m.) with the aid of a T9 fluorescent lamp (6500 K, 18 W). Each pot of water spinach was watered with 150 mL PAW or RO water thrice each week, while the plant height measurement and photoshoot were performed on the 4th day of each week.

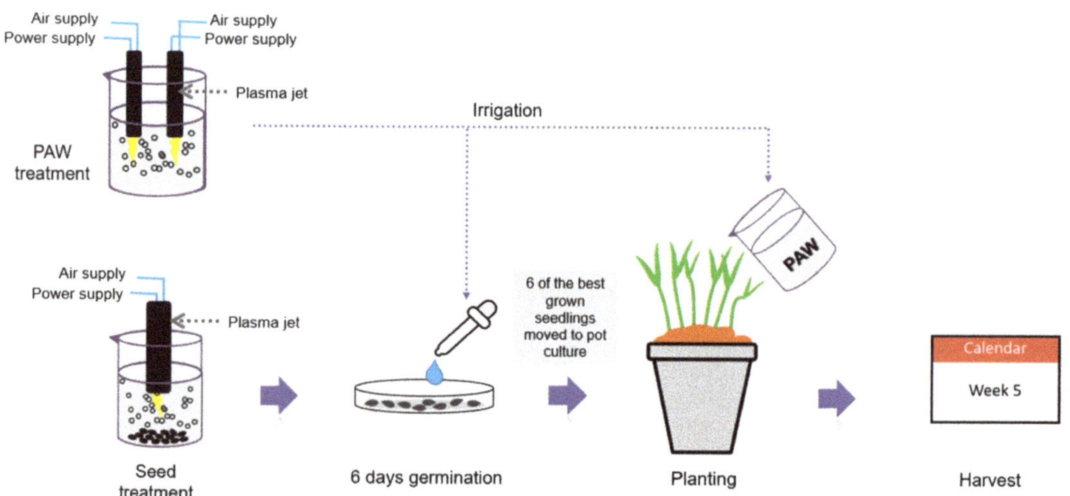

Figure 1. The procedure of the plantation.

2.7. Physiochemical Properties Analysis

2.7.1. Determination of Nitrates and Nitrites of PAW

The NO_3^- and NO_2^- contents of water samples were determined by NitraVer® 4 Nitrate Reagent Powder Pillow and NitraVer® 2 Nitrite Reagent Powder Pillow, respectively, with the aid of DR900 Multiparameter Portable Colorimeter, according to the instructions provided by the manufacturer (HACH, Loveland, CO, USA).

2.7.2. Determination of pH, Oxidation-Reduction Potential (ORP) and Electrical Conductivity (EC), and H_2O_2 of PAW

Physicochemical properties such as pH, ORP, and EC were determined immediately after the generation of PAW. The pH values and conductivity of the PAW were evaluated using a pH electrode and electro-conductivity probe (Cole-Parmer PC200-2 pH/Conductivity Meter Kit). On the other hand, the redox potential was measured using an SP-2100 Laboratory pH/ORP meter with the corresponding ORP probe. H_2O_2 was measured using a Hydrogen Peroxide Test Kit (Model HYP-1, product no.2291700) (HACH, USA).

2.8. Metals Analysis

2.8.1. Preparation of Soil Samples

The cultivated soil samples were collected at the beginning and the end of the plantation from each pot. A PET bottle was used to store the samples. Before the freeze-drying process, the samples were stored at $-18\ °C$ in a freezer for 24 h. The freeze-drying process of the vegetable samples and soil samples lasted for between 48 and 72 h. For the soil samples, a mortar and pestle were used to grind the samples into a fine powder, and the samples were then filtered through a sifting screen.

2.8.2. Determination of Heavy Metals in Soils

This sample pretreatment refers to Microwave-Assisted Aqua Regia Digestion Method for the Analysis of Heavy Metals in Soil (NIEA S301.61B). First, 0.25 g of the sample was placed into microwave digestion vessels. Three milliliters of 37% HCl and 1 mL 65% HNO_3 were then added to the vessels, and the mixture was allowed to react for 30 min. The mixtures were then digested with a microwave oven (MarsXpress microwave system, CEM) at 1800 W. The digested mixtures were then filtered with the Hydrophilic PTFE Filter (30 mm × 0.45 µm) and additional deionized water was added to make the volume of the mixtures 50 mL. The determination was performed in triplicate. The samples, blank samples, and standard solutions with different concentrations were analyzed with Inductively Coupled Plasma Optical Emission Spectrometer (ICP-OEs, PerkinElmer Avio 200, San Jose, CA, USA). Detected concentrations in samples that were lower than the method detection limit (MDL) are expressed as ND, and <LOQ is stated for detected concentrations that were lower than the limit of quantification.

2.8.3. Preparation of Vegetable Samples

The edible portions of the water spinach were cut at a position around 1 cm above the soil surface for chemicals analysis. The harvested samples were weighed and then cleaned using RO water to remove the dirt. The water spinach samples were dried with paper towels and placed in zipper bags with the corresponding labels. Samples collected from the same treatment combination were maintained in the same zipper bag to make it a mixed sample from three pots of a plant. Before the freeze-drying process, the samples were stored at $-18\ °C$ in the freezer for 24 h. The freeze-drying process of the vegetable samples and soil samples lasted for between 48 and 72 h. A homogenized blender was used to chop the vegetable samples into smaller pieces.

2.8.4. Determination of Heavy Metals in Vegetables

First, 0.3 g of the samples were weighed and placed into microwave digestion vessels. Then, 3 mL of 67% HNO_3 was added to the vessels and mixed with the samples.

The mixtures were then digested with a microwave oven (MarsXpress microwave system, CEM) at 1800 W. The digested mixtures were then filtered with the Hydrophilic PTFE Filter (30 mm $\times$ 0.45 μm) and additional deionized water was added to make the volume of the mixtures 25 mL for further analysis. The samples, blank samples, and standard solutions with different concentrations were then analyzed with Inductively Coupled Plasma Mass Spectrometry (ICP-MS; PerkinElmer NexION 2000, San Jose, CA, USA).

2.8.5. Quality Assurance/Quality Control for Metal Analysis

The recovery efficiency tests for Pb and Cd were conducted using the same sample analysis procedure with the addition of a standard solution. The recovery rates ranged from 103% and 89% for Pb and Cd in the soil samples and 108% and 93% for Pb and Cd in vegetables in the present study. The method detection limit (MDL) was used with a concentration that was slightly lower than the lowest concentration of the calibration curve. Measurements at this concentration were repeated seven times to estimate the standard deviation, and MDL was set as three times the standard deviation. The MDL ranged from 0.047 μg/g and 0.082 μg/g for Pb and Cd g in soil and 0.230 μg/g and 0.132 μg/g for Pb and Cd in vegetables.

2.9. Bioconcentration Factor

Bioconcentration factor (BCF) is described as the ability of plants to absorb heavy metals from contaminated soils. The BCF was calculated as suggested by previous studies [18–20]:

$$\mathrm{BCF} = \frac{\text{Heavy metal concentration in vegetables}}{\text{Heavy metal concentration in soil}} \tag{1}$$

2.10. Statistical Analysis

The data were analyzed by the IBM SPSS Statistics 23. All determinations in this study were performed in triplicate. The significant differences in the concentration of heavy metals in soil were analyzed by Kruskal–Wallis test. The results were presented as mean $\pm$ standard deviation. $p = 0.05$ indicated statistical significance.

3. Results and Discussion

3.1. Physiochemical Properties of Irrigation Water

The generation of PAW can be confirmed by analyzing the acidity, conductivity, ORP, and concentration of ROS and RNS in the water [14]. Thirumdas et al. mentioned that the increase in the NO_3^- and NO_2^- contents in PAW could be the main reason behind enhanced plant growth and seed germination [14]. In this study, RO water was used as the water source and treated for 5, 10, 15, and 20 min to generate PAW for the selection of the most favorable PAW as irrigation water. The physiochemical characteristics of PAW under different treatment times are summarized in Table S1. In a nutshell, the NO_3^- and NO_2^- concentrations, as well as the EC, ORP and H_2O_2 values, increased with increasing treatment time, except that the pH values of PAW decreased with increasing treatment time (Figures S1–S4).

Figure 2 illustrates the concentration of NO_3^- generated by plasma in this study. The concentration of NO_3^- increased from untreated (control) to 20-min treatment. The concentration of NO_3^- decreased in the order 20 min > 15 min > 10 min > 5 min > control. The increasing trend nitrate concentration with increasing treatment time was similar to results reported in some other studies [21,22]. The findings of Zhou et al. [21] indicated that NO_2^- and NO_3^- are formed in water, with the constant rate following zero-order kinetics, indicating a direct effect of the plasma. Meanwhile, they also reported that air microplasma treatment resulted in the highest concentration of nitrites and nitrates, followed by nitrogen plasma and oxygen plasma. Li et al. [22] also mentioned that the concentrations of NO_2^-, NO_3^-, and H_2O_2 increased with increasing discharge current or activation time [22]. Figure 3 depicts the concentrations of NO_2^- in PAW treated for different lengths of time in this study. However, the NO_2^- concentration of PAW in this study do not follow the time-dependent trend with treatment suggested by other researchers. NO_2^- was not detected in RO water, but it was present when water dissolved the ions resulting from the breakdown of nitrogen and oxygen in the plasma treatment. The concentration of NO_2^- was highest after 10-min PAW treatment (PAW10), followed by PAW20 > PAW15 > PAW5. Furthermore, some studies in the literature have reported that no NO_2^- was detected after 1 and 5 min of plasma activation [23]. NO_3^- plays important roles as the nutrient that leads to the production of amino acids and nitrogen compounds and as signal molecules in plant development and metabolism. Hence, PAW treatment for 20 min was selected for the irrigation water throughout the whole plantation period of water spinach.

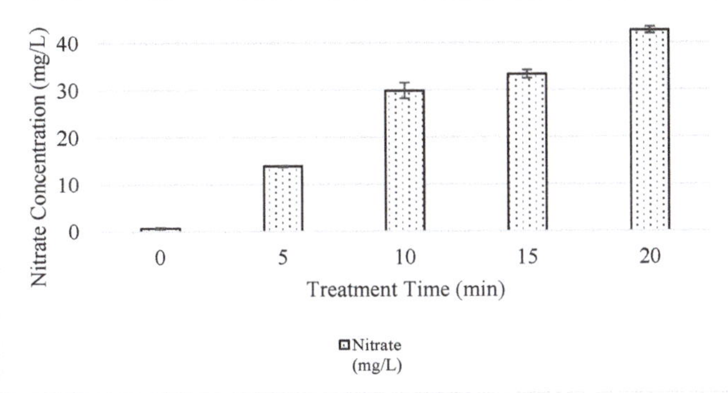

Figure 2. Concentration of nitrates (NO_3^-) produced in PAW under different treatment times ($n = 3$).

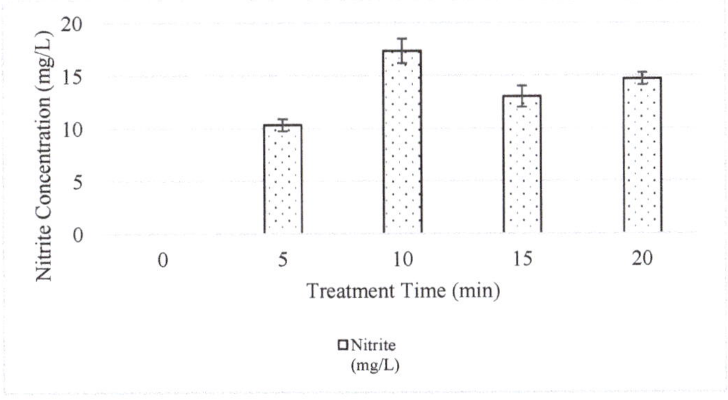

Figure 3. Concentration of nitrites (NO_2^-) produced in PAW under different treatment times ($n = 3$).

3.2. Concentration of Heavy Metals in Man-Made Contamination Soil

The initial heavy metal concentration was determined before plantation. The Cd-added cultivation soil was expected to be 7 mg/kg W.W., while Pb-added cultivation soil was expected to be 700 mg/kg W.W. The expected heavy metal concentrations in the soil were selected to be 3-times higher than the reference of the monitored and controlled standard for metals in soil stated by the Taiwan EPA; therefore, the added Cd concentration was 18.18–22.28 mg/kg dry weight (D.W.), and the Pb concentration was 2064–2492 mg/kg D.W (Table 1).

Table 1. The metal concentrations of control soil, Cd-added soil and Pb-added soil (mg/kg D.W.).

Treatment	*n*	Cd Concentration		Pb Concentration	
		Control Soil	Cd-Added Soil	Control Soil	Pb-Added Soil
NTS + NTW	3	ND-<LOQ	18.18–20.43	2.60–3.17	2222–2492
NTS + PAW	3	<LOQ	19.06–19.53	2.36–3.77	2064–2470
PTS + NTW	3	ND-<LOQ	20.93–22.28	1.84–2.32	2182–2429
PTS + PAW	3	ND-<LOQ	21.17–21.92	1.77–2.80	2086–2448

Note: The results are represented as min–max value. ND: Not detected; LOQ: Limit of quantification.

3.3. Harvested Water Spinach

The total weight of harvested water spinach edible portions was higher in water spinach planted in soil with Pb contaminants, but the total quantity of harvested water spinach decreased with either plasma treatment of the seeds or when PAW was used as the irrigation water (Table 2). The sequence for the height of the water spinach planted in Cd-added soil was as follows (Figure S5): PTS + NTW > Control > NTS + PAW > PTS + PAW. As the average heights of water spinach in NTS + PAW and PTS + PAW treatment groups were lower than those in the control group in Cd-added soil, it can be concluded that PAW usage and combined treatments of PTS and PAW inhibited plant growth in terms of plant height in this study. The sequence for the height of water spinach planted in Pb-added soil was as follows (Figure S6): PTS + PAW > PTS + NTW > Control > NTS + PAW. Overall, PAW usage may inhibit growth in terms of plant height when Pb was present in the soil.

Table 2. The total weight of harvested edible parts of water spinach (in term of fresh weight).

Treatment	*n*	Total Weight (g) of Water Spinach Grown in		
		Control Soil	Cd-Added Soil	Pb-Added Soil
NTS + NTW	3	9.24	9.43	7.28
NTS + PAW	3	7.72	7.40	6.38
PTS + NTW	3	9.29	6.48	7.38
PTS + PAW	3	5.91	7.43	8.39

Note: The data shown are the sum values of three repeats.

3.4. Concentration of Heavy Metals in Water Spinach

To investigate the effectiveness of plasma treatment for reducing the accumulation of heavy metals in vegetables, the heavy metal concentration in water spinach was analyzed. The Cd concentration of water spinach planted in the control and Cd-added soil group are presented in Table 3. When the water spinach was planted in soil with a Cd contaminant, the Cd accumulated by the water spinach was reduced by either plasma treatment on seeds or by using PAW as the irrigation water, but not by both. The Cd concentration of the water spinach in NTS + PAW, PTS + NTW, and PTS + PAW declined by 17.8%, 22.9%, and 4.5%, respectively, when compared with the group without any treatment, especially PTS + NTW group showed the best outcome on declining Cd accumulation in water spinach. Hence, the plasma treatment may demonstrate positive effects for reducing Cd accumulation in water spinach, which is consistent with the results of Kabir et al., who reported that Ar/O$_2$- and Ar/Air plasma-treated plants exhibited significantly reduced Cd concentrations in the shoot and root in wheat crop when reducing the expression of Cd transporters (*TaLCT1* and *TaHMA2*) in the root [24]. Meanwhile, Table 3 depicts the results of water spinach planted in soil with Pb contaminant. The Pb concentration of water spinach in NTS + PAW and PTS + NTW decreased by 0.57% and 1.42%, respectively. Moreover, PTS + PAW treatment on water spinach increased the Pb accumulation in the water spinach by 64.8%. Hence, individual plasma treatment of seeds or water did not demonstrate any obvious positive effects for reducing Pb accumulation in water spinach.

From a different perspective, although the concentration of Cd in the Cd-added soil (18.18–20.43 mg/kg D.W.) was approximately 100 times lower than the concentration of Pb in the Pb-added cultivation soil (2064–2492 mg/kg D.W.), the Cd concentration in the water spinach (12.1–15.7 mg/kg D.W.) planted in the Cd-added soil was greater than the Pb concentration in the water spinach (6.94–11.6 mg/kg D.W.) planted in the Pb-added soil. This result was consistent with that of Intawongse and Dean [25], who reported increasing uptake of Cd, Cu, Mn, and Zn by plants correlating with increasing concentration in soil, while the uptake of Pb was poor.

Table 3. The metal concentrations of water spinach planted in control, Cd-added soil and Pb-added soil (in terms of dry weight).

Treatment	Cd Concentration		Pb Concentration	
	Control Soil (mg/kg D.W.)	Cd-Added Soil (mg/kg D.W.)	Control Soil (mg/kg D.W.)	Pb-Added Soil (mg/kg D.W.)
NTS + NTW	0.265	15.7	0.703	7.04
NTS + PAW	0.182	12.9	0.246	7.00
PTS + NTW	0.215	12.1	0.284	6.94
PTS + PAW	0.207	15.0	0.697	11.6

Note: Samples used in the analysis were the mixtures of 3 pots of plants.

The combined effect of plasma-treated seeds and water may depend on the nature of the seeds [26]. Ahn et al. [27] conducted an experiment to compare various plasma conditions with different plasma generators on corn seeds grown in six separated locations in Illinois. The effect of plasma treatment on corn yield was affected by weather and soil quality. The Ar/O_2 plasma affected the wheat plant growth and development by increasing the ascorbate peroxidase (APX) activity in shoots and significantly increased in superoxide dismutase (SOD) activity and the TaSOD expression in the roots of wheat [28]. Safari et al. [29] reported that plasma treatment of 1 min resulted in an improved effect on the total leaf area, shoot, and root lengths, although plasma treatment for 2 min significantly impaired the growth, reducing the total biomass. The high intensity of plasma treatment could cause some changes in the gene expression, inhibiting the growth rates and inducing morphological and physiological differences [29]. According to some studies in the literature, and based on our results, the non-significant results with respect to plant growth following watering with PAW may be affected by metals in the soil, the cultivation environment, and plasma treatment of the seeds.

With respect to heavy metals, the presence of Pb in soil reduced the yield of water spinach, while Cd in soil did not affect the yield relative to that in the untreated group (NTS + NTW). The group with the PTS + PAW combination only showed a positive effect in terms of the total weight of the water spinach planted in Pb-soil.

In a nutshell, the average height and total weight of the water spinach were affected by the presence of heavy metals in soil. The total weight and the average plant height were restricted by the high concentration of Pb. On the other hand, these restricted parameters can be overcome or improved through plasma treatment. In this study, plasma treatment did not help to improve the product yield of water spinach planted in Cd-added soil, but it was able to increase the average plant height of water spinach after seed treatment. Finally, in the PTS + NTW group, the growth of the water spinach was increased in terms of both total weight and average height compared to that in the control soil.

Heavy metals enter the plants through soil uptake and accumulates in different organs, which may reduce plant growth and productivity, while some metabolic processes are also associated with Cd toxicity and its tolerance [30]. A high concentration of Pb can cause several toxic symptoms in plants and may lead to growth retardation in plants, the inhibition of photosynthesis, root blackening, changes in hormonal status, disturbance of mineral and water balance, and other symptoms [31]. Hence, the unsatisfactory plant growth observed in this study may be an effect of heavy metal toxicity.

As previously mentioned, the concentration of heavy metals in Cd-added and Pb-added cultivation soil was extremely high, which may have led to the restriction of plant growth. Since the heavy metal concentration used in this study was high and the overall results of plant growth were not consistent with those in some other works in the literature (Table 4), in the future, the plantation environment could be changed and fertilizer could be used to supply some nutrients that could not be provided by PAW in order to support the growth of water spinach.

Table 4. Summary of plasma treatment applications in seed germination and plant growth.

| | | **A. Plasma Treatment on Seeds** | | | |
| | | | | **Outcomes** | |
Author	**Target**	**Plasma Generation**	**Parameters**	**Seed Germination**	**Plant Growth**
Ling et al. [11]	Oilseed rape (*Brassica napus* L. cv Zhongshuang 9)	Low-vacuum helium cold plasma (Radio frequency discharge)	Helium gas 100 W, 15 s, 13.56 MHz, 150 Pa	Positive	Positive
Jiang et al. [32]	Tomato (*Solanum lycopersicum* L. cv. Shanghai 906)	Inductive helium capacitive coupled plasma (CCP), Computer-controlled plasma treatment apparatus HD-2N (Radiofrequency)	Helium gas 80 W, 15 s, 13.56 MHz, 150 Pa, 3.5 eV (electron temperature)	Positive	Positive
Ling et al. [33]	Peanut (*Arachis hypogaea* L. cv. Eyou 7)	Inductive helium discharge with HD-2N units (Radiofrequency generator)	Helium gas, 0–120 W, 15 s, 13.56 MHz, 150 Pa	Positive	Positive
Li et al. [34]	Soybean (*Glycine max* L. Merr cv. Zhongdou 40)	Inductive helium discharge with computer-controlled plasma treatment apparatus HD-2N units (Radiofrequency generator)	Helium gas 0–120 W, 15 s, 13.56 MHz, 150 Pa	Positive	Positive
Zhang et al. [35]	Maize, peppers, wheat, soybeans, tomatoes, eggplants, pumpkins etc.	Electromagnetic shielding and suspension electrode technology; High (glow) radiofrequency discharger produces plasma	Air, 13.56 MHz, 80–180 W, 30–200 Pa, 5–90 s	Positive	Positive
Saberi et al. [36]	Wheat (*Triticum aestivum* L.)	Non-thermal Radio Frequency Plasma	Air, 80 W, 13.56 MHz, 0.1 mbar, 0–240 s	-	Positive
Jiang et al. [37]	Wheat (*Triticum* spp.)	Cold plasma generator	Helium gas, 60–100 W, 15 s, 3×10^9 MHz, 13 eV, 150 Pa	Positive	Positive
Safari et al. [29]	*Capsicum annum* PP805 Godiva	DBD plasma	Argon gas, 23 kHz, 11 kV, 80 W, 94.98 cm^2 plasma treatment areas, 0.84 W/cm power density, 0–2 min	-	Positive

Table 4. *Cont.*

A. Plasma Treatment on Seeds					
				Outcomes	
Author	**Target**	**Plasma Generation**	**Parameters**	**Seed Germination**	**Plant Growth**
Mihai et al. [38]	Radish seed	Non-thermal plasma-surface discharge	Air, 15 kV, 2.7 W, 20 min, Gas flow rate = 1 L/min	No Change	No Change
de Groot et al. [39]	Cotton seeds variety Sicot 74BRF	Cold atmospheric-pressure plasma (CAP)	Air/Argon gas; Flow rate = 1 L/min, AC power supply, 1 kHz, sine wave with 38 kVpp for air and 11 kVpp for argon, treatment time: 0, 3, 27 min with dry air and 81 min with argon gas	Positive	Not significant but positive for Air-27 min and Ar-81 min, and negative for Air-3 min
Iranbakhsh et al. [40]	Wheat (*Triticum aestivum* L. cv. Parsi)	Dielectric barrier discharge (DBD)	Nitrogen and Helium gas, 20 kHz, 15 kV, 100 W, 254.3 cm^2; 0.4 W/cm; 100 Pa, 15, 30, 60, 120 s, repetition with 1, 2, 4 times with 24 h intervals	-	Positive
Pawlat et al. [41]	*Lavatera thuringiaca* L.	Gliding arc reactor	Nitrogen gas (8 L/min), 680 V, 50 Hz, 33 mA, 40 W, 1, 2, 5, 10, 15 min	Positive	-
Pawlat et al. [10]	*Lavatera thuringiaca* L.	Dielectric barrier discharge (DBD) plasma jet	Helium: 1.6 dm^3/min, Nitrogen: 0.03 dm^3/min, 3.7 kV; 17 kHz; mean of 6 W, 1, 2, 5, 10, 15 min	Positive	-
Rahman et al. [28]	Wheat (BARI Gom 22)	Low pressure dielectric barrier discharge (DBD)	Ar60%/Air40%; Ar60%/Oxygen40%, 5 kV, 4.5 kHz, ~45 W, 90 s	Positive	-

B. Plasma Treatment on Seeds through Aqueous Media					
				Outcomes	
Author	**Target**	**Plasma Generation**	**Parameters**	**Seed Germination**	**Plant Growth**
Zhou et al. [21]	Mung bean seeds (*Vigna radiata* Linn. Wilczek)	Atmospheric pressure microplasma array	He, N_2, artificial Air, O_2, (2 standard liters per min), 36 microplasma jet units, 4.5 kV, 9.0 kHz, 25 W, 10 min	Positive	Positive
Liu et al. [12]	Tomato, Lettuce, Mung bean, Sticky bean, Radish, Dianthus, Mustard, Wheat	Dielectric barrier discharge (DBD)	N_2, O_2, Synthetic Air, 1.5 L/min, 0–18 kV, 500 Hz, Power consumption: ~2.5 W, Vpp = 20 kV, 2, 4, 6 min	Positive	Positive

Table 4. *Cont.*

	C. Combination: Plasma Treatment on Seeds and Water				
Author	**Target**	**Plasma Generation**	**Parameters**	**Outcomes**	
				Seed Germination	**Plant Growth**
Bafoil et al. [42]	*Arabidopsis thaliana* (Early stage)	Floating electrode dielectric-barrier discharge (FE-DBD)	Ambient air, 10 kV for the voltage, 9.7 kHz for the frequency and 1 us for the pulse duration, 15 min	Positive	-
	Distilled water and tap water (Later stage)	Plasma jet, the floating electrode-dielectric barrier discharges (FE-DBD)	Helium gas, 10 kV, 9.7 kHz, 1 μs pulse duration, 3 L/min, 15 min, 30 mL water used in each treatment	-	Positive
Sivachandiran et al. [26]	Water	Cylindrical double DBD reactor in air under atmospheric pressure and room temp	Synthetic Air (Air Liquide), Flow rate = 1 L/min, Pulse width: 120 ns; 21 kV; 2.4 A; 400 Hz; Max energy: 7 mJ, 250 mL DI water activated for 15 min and 30 min	Positive when treat on dry seeds, no significant influences on wet seeds	Negative for stem length on P-10 min seeds and P-20 min seed + PAW-30 min
	Radish, Tomato, Sweet Pepper seeds	Plate-to-plate double DBD reactor in air under atmospheric pressure and room temp	Synthetic Air (Air Liquide), Flow rate = 1 L/min, Pulse width: 120 ns; 21 kV; 18 A; 200 Hz; Max energy: 57 mJ, 10 min; Plasma discharge volume: 130 W/cm		

3.5. BCF of Vegetables Planted in Contaminated Soils

Bioaccumulation or bioconcentration refers to an increase in particular chemicals or elements in organisms over a period of time, which may harm the organism itself or its predators [43]. The BCF of Cd and Pb were calculated, and the results are depicted in Table 5 and Figure S7. Overall, BCF of Cd in water spinach (0.543–0.864) was higher when compared to that of Pb (0.003–0.006). The treatment of NTS + NTW indicated the highest BCF of Cd in water spinach, followed by PTS + PAW and NTS + PAW, while PTS + NTW showed the lowest BCF of Cd in water spinach. The BCF of Cd in water spinach decreased from 0.864 to 0.543 following the plasma treatment, while the BCF of Pb was low and did not show any significant changes. These results are similar to those reported by Huang et al. [44], which showed that the BCF of Cd in two water spinach cultivars was much greater than that of BCFs of Pb, irrespective of whether it was in the leaf, stem, or root.

Saad et al. [45] showed that the bioaccumulation factor (BAF) of Cd in water spinach was slightly higher than that of Pb, albeit not significant.

Table 5. Bioconcentration factors (BCF) of Cd and Pb in water spinach.

Treatment	n	Cd	Pb
NTS + NTW	3	0.819 ± 0.048	0.003 ± 0.000
NTS + PAW	3	0.669 ± 0.008	0.003 ± 0.000
PTS + NTW	3	0.564 ± 0.019	0.003 ± 0.000
PTS + PAW	3	0.695 ± 0.012	0.005 ± 0.000
p-value		0.016 *	0.092

Note: * $p < 0.05$, Kruskal–Wallis Test. Samples used in the analysis were the mixtures of three pots of plants.

Sengar et al. [46] mentioned that several factors may affect the amount and speed of uptake and transpiration of Pb into plants, such as soil particle size, soil cation-exchange capacity, root surface area, organic matter content, root exudation, mycorrhizal transpiration rate, and soil pH values. In addition, Pb absorption by plants may also be affected by the life stage and passive environmental factors. At the lethal concentration of Pb, the barriers in the root endodermis will be broken and cause lead to enter the vascular tissues. This point may further affect the mineral nutrition and water balance, inactivate the enzyme activities, change the hormonal status, and affect the membrane structure and permeability [46]. Pourrut et al. [47] reported that once Pb enters the root system, approximately 95% or more of Pb is accumulated in the roots and only a small fraction is translocated to the aerial plant section. This phenomenon might be different for each kind of heavy metal, and is specific to Pb. Pb is blocked in the endodermis by the Casparian strip, which is followed by symplastic transport. In addition, the majority of the Pb entering the root is removed by the plant's detoxification systems. Some plant species are able to tolerate higher concentrations of Pb ions due to various detoxification mechanisms, including selective metal uptake, excretion, complexation by specific ligands, and compartmentalization.

According to a report by Wang et al. [48], lower pH influences the metal uptake of plants because Cd and Zn are more soluble at lower pH; this may be an effective strategy for enhancing the phyto-extraction of metals from soil. Ye et al. [49] also reported that soil pH and organic matter content are important in affecting the Cd availability in soil. Both of these elements determined the adsorption of Cd at soil binding sites, as well as its solubility and mobility in soil. The determination of soil pH was not performed in this study; hence, the mechanisms of low pH of PAW are expected to increase the metal solubility, while increased uptake by plants needs to be determined in the future.

In this study, the accumulation of Cd in water spinach was reduced, but the Pb concentration in water spinach in the PTS + PAW treated group was enhanced. Our results for Pb concentration increment in water spinach were similar to those of Kabir et al. [24]. Meanwhile, our study reported a reduction in Cd concentration in water spinach through the inhibition of expression of Cd transporter in the root by plasma treatment, but an increase in Pb concentration may be related to the activation of Pb transporter in the root of water spinach by plasma treatment on seeds and PAW irrigation.

4. Conclusions

The accumulation of heavy metals by water spinach was found to be dependent on the type of heavy metals and their concentration in soil. A higher concentration of Cd was accumulated in water spinach compared to that of Pb. The PTS + NTW treatment group demonstrated the highest reduction in Cd accumulation in water spinach. However, plasma treatment of seeds or PAW irrigation was able to reduce the accumulation of heavy metals in water spinach, except for water spinach in the group of PTS + PAW grown in Pb-added soils. The limitations of this study, such as the insufficiency of the planting space, the fixed condition of the PAW generator, and the small sample size, could have led to imperfections in the study. An increased sample size in order to achieve a more powerful analysis, modification of the optimal parameters (plasma power, treatment time, etc.) of the PAW treatments, and the selection of a more suitable PAW generator will all be taken into account in future studies. Meanwhile, since the heavy metal concentration used in this study was high and the plant growth in the current study was not consistent with that described in some other studies in the literature, this study on the effects of various concentrations of Cd and Pb on the growth and development of water spinach is a prerequisite for research into the combined influence of PAW and heavy metals on this plant species.

Supplementary Materials: The following are available online at https://www.mdpi.com/article/10.3390/app11115304/s1. Table S1 The physiochemical properties of PAW under different treatment time, Figure S1 The pH values of PAW under different treatment time ($n = 3$), Figure S2 The ORP values of PAW under different treatment time ($n = 3$), Figure S3 The electrical conductivity of PAW under different treatment time ($n = 3$), Figure S4 Concentration of H_2O_2 produced in PAW at different treatment time ($n = 3$), Figure S5 The average height of water spinach planted in Cd-added soil under different treatments, Figure S6 The average height of water spinach planted in Pb-added soil under different treatments, Figure S7 Bioconcentration factors of Cd and Pb in water spinach.

Author Contributions: Conceptualization, H.-L.C.; methodology and processing, T.-K.K., C.-Y.H., C.-M.L. and H.-L.C.; chemical analysis, T.-K.K.; resources, C.-Y.H., C.-M.L. and H.-L.C.; writing-original draft preparation, T.-K.K. and C.-Y.H.; writing-review and editing, C.-M.L. and H.-L.C.; supervision, C.-Y.H.; project administration, C.-M.L. and H.-L.C.; funding acquisition, H.-L.C. All authors have read and agreed to the published version of the manuscript.

Funding: This work has been supported by Lab of Hsiu-Ling Chen.

Institutional Review Board Statement: Not applicable.

Data Availability Statement: This study did not report any data.

Acknowledgments: The authors acknowledge that the plasma jet system was designed and developed by the Mechanical and Mechatronics Systems Research Lab, Industrial Technology Research Institute (ITRI) (Hsinchu, Taiwan).

Conflicts of Interest: The authors declare no conflict of interest.

References

1. Tangahu, B.V.; Sheikh Abdullah, S.R.; Basri, H.; Idris, M.; Anuar, N.; Mukhlisin, M. A Review on Heavy Metals (As, Pb, and Hg) Uptake by Plants through Phytoremediation. *Int. J. Chem. Eng.* **2011**, *2011*, 1–31. [CrossRef]
2. Rai, P.K.; Lee, S.S.; Zhang, M.; Tsang, Y.F.; Kim, K.H. Heavy metals in food crops: Health risks, fate, mechanisms, and management. *Environ. Int.* **2019**, *125*, 365–385. [CrossRef] [PubMed]
3. Pankaj, S.K.; Keener, K.M. Cold plasma: Background, applications and current trends. *Curr. Opin. Food Sci.* **2017**, *16*, 49–52. [CrossRef]
4. Yan, D.; Sherman, J.H.; Keidar, M. Cold atmospheric plasma, a novel promising anti-cancer treatment modality. *Oncotarget* **2017**, *8*, 15977–15995. [CrossRef] [PubMed]
5. Nehra, V.; Kumar, A.; Dwivedi, H. Atmospheric non-thermal plasma sources. *Int. J. Eng.* **2008**, *2*, 53–68.
6. Brun, P.; Bernabè, G.; Marchiori, C.; Scarpa, M.; Zuin, M.; Cavazzana, R.; Zaniol, B.; Martines, E. Antibacterial efficacy and mechanisms of action of low power atmospheric pressure cold plasma: Membrane permeability, biofilm penetration and antimicrobial sensitization. *J. Appl. Microbiol.* **2018**, *125*, 398–408. [CrossRef]
7. Ma, Y.; Zhang, Q.; Zhang, Q.; He, H.; Chen, Z.; Zhao, Y.; Wei, D.; Kong, M.; Huang, Q. Improved production of polysaccharides in Ganoderma lingzhi mycelia by plasma mutagenesis and rapid screening of mutated strains through infrared spectroscopy. *PLoS ONE* **2018**, *13*, e0204266. [CrossRef]
8. Sarangapani, C.; O'Toole, G.; Cullen, P.; Bourke, P. Atmospheric cold plasma dissipation efficiency of agrochemicals on blueberries. *Innov. Food Sci. Emerg. Technol.* **2017**, *44*, 235–241. [CrossRef]
9. Takaki, K.; Yoshida, K.; Saito, T.; Kusaka, T.; Yamaguchi, R.; Takahashi, K.; Sakamoto, Y. Effect of Electrical Stimulation on Fruit Body Formation in Cultivating Mushrooms. *Microorganisms* **2014**, *2*, 58–72. [CrossRef]
10. Pawłat, J.; Starek, A.; Sujak, A.; Terebun, P.; Kwiatkowski, M.; Budzeń, M.; Andrejko, D. Effects of atmospheric pressure plasma jet operating with DBD on Lavatera thuringiaca L. seeds' germination. *PLoS ONE* **2018**, *13*, e0194349. [CrossRef]
11. Ling, L.; Jiangang, L.; Hanliang, S.; Yuanhua, D. Effects of low-vacuum helium cold plasma treatment on seed germination, plant growth and yield of oilseed rape. *Plasma Sci. Technol.* **2018**, *20*, 095502.

12. Liu, B.; Honnorat, B.; Yang, H.; Arancibia, J.; Rajjou, L.; Rousseau, A. Non-thermal DBD plasma array on seed germination of different plant species. *J. Phys. D Appl. Phys.* **2018**, *52*, 025401. [CrossRef]

13. Zhou, R.; Zhou, R.; Prasad, K.; Fang, Z.; Speight, R.; Bazaka, K.; Ostrikov, K.K. Cold atmospheric plasma activated water as a prospective disinfectant: The crucial role of peroxynitrite. *Green Chem.* **2018**, *20*, 5276–5284. [CrossRef]

14. Thirumdas, R.; Kothakota, A.; Annapure, U.; Siliveru, K.; Blundell, R.; Gatt, R.; Valdramidis, V.P. Plasma activated water (PAW): Chemistry, physico-chemical properties, applications in food and agriculture. *Trends Food Sci. Technol.* **2018**, *77*, 21–31. [CrossRef]

15. Judée, F.; Simon, S.; Bailly, C.; Dufour, T. Plasma-activation of tap water using DBD for agronomy applications: Identification and quantification of long lifetime chemical species and production/consumption mechanisms. *Water Res.* **2018**, *133*, 47–59. [CrossRef] [PubMed]

16. Herianto, S.; Hou, C.; Lin, C.; Chen, H. Nonthermal plasma-activated water: A comprehensive review of this new tool for enhanced food safety and quality. *Compr. Rev. Food Sci. Food Saf.* **2021**, *20*, 583–626. [CrossRef]

17. Lin, C.-M.; Chu, Y.-C.; Hsiao, C.-P.; Wu, J.-S.; Hsieh, C.-W.; Hou, C.-Y. The Optimization of Plasma-Activated Water Treatments to Inactivate Salmonella Enteritidis (ATCC 13076) on Shell Eggs. *Foods* **2019**, *8*, 520. [CrossRef]

18. Hamzah, A.; Hapsari, R.I.; Wisnubroto, E.I. Phytoremediation of cadmium-contaminated agricultural land using indigenous plants. *Int. J. Environ. Agric. Res.* **2016**, *2*, 8–14.

19. Ndeda, L.; Manohar, S. Bio concentration factor and translocation ability of heavy metals within different habitats of hydrophytes in Nairobi Dam, Kenya. *J. Environ. Sci. Toxicol. Food Technol.* **2014**, *8*, 42–45.

20. Takarina, N.D.; Pin, T.G. Bioconcentration Factor (BCF) and Translocation Factor (TF) of Heavy Metals in Mangrove Trees of Blanakan Fish Farm. *Makara J. Sci.* **2017**, *21*, 77–81. [CrossRef]

21. Zhou, R.; Zhou, R.; Zhang, X.; Zhuang, J.; Yang, S.; Bazaka, K.; Ostrikov, K. Effect Atmospheric-Pressure N_2, He, Air, and O_2 Microplasmas on Mung Bean Seed Germination and Seedling Growth. *Sci. Rep.* **2016**, *6*, 32603. [CrossRef]

22. Li, X.; Li, X.; Gao, K.; Liu, R.; Liu, R.; Yao, X.; Gong, D.; Su, Z.; Jia, P. Comparison of deionized and tap water activated with an atmospheric pressure glow discharge. *Phys. Plasmas* **2019**, *26*, 033507. [CrossRef]

23. Porto, C.L.; Ziuzina, D.; Los, A.; Boehm, D.; Palumbo, F.; Favia, P.; Tiwari, B.; Bourke, P.; Cullen, P.J. Plasma activated water and airborne ultrasound treatments for enhanced germination and growth of soybean. *Innov. Food Sci. Emerg. Technol.* **2018**, *49*, 13–19. [CrossRef]

24. Kabir, A.H.; Rahman, M.; Das, U.; Sarkar, U.; Roy, N.C.; Reza, A.; Talukder, M.R.; Uddin, A. Reduction of cadmium toxicity in wheat through plasma technology. *PLoS ONE* **2019**, *14*, e0214509. [CrossRef] [PubMed]

25. Intawongse, M.; Dean, J.R. Uptake of heavy metals by vegetable plants grown on contaminated soil and their bioavailability in the human gastrointestinal tract. *Food Addit. Contam.* **2006**, *23*, 36–48. [CrossRef] [PubMed]

26. Sivachandiran, L.; Khacef, A. Enhanced seed germination and plant growth by atmospheric pressure cold air plasma: Combined effect of seed and water treatment. *RSC Adv.* **2017**, *7*, 1822–1832. [CrossRef]

27. Ahn, C.; Gill, J.; Ruzic, D.N. Growth of Plasma-Treated Corn Seeds under Realistic Conditions. *Sci. Rep.* **2019**, *9*, 1–7. [CrossRef] [PubMed]

28. Rahman, M.; Sajib, S.A.; Rahi, S.; Tahura, S.; Roy, N.C.; Parvez, S.; Reza, A.; Talukder, M.R.; Kabir, A.H. Mechanisms and Signaling Associated with LPDBD Plasma Mediated Growth Improvement in Wheat. *Sci. Rep.* **2018**, *8*, 1–11. [CrossRef]

29. Safari, N.; Iranbakhsh, A.; Ardebili, Z.O. Non-thermal plasma modified growth and differentiation process of Capsicum annuum PP805 Godiva in in vitro conditions. *Plasma Sci. Technol.* **2017**, *19*, 055501. [CrossRef]

30. Shanmugaraj, B.M.; Malla, A.; Ramalingam, S. Cadmium Stress and Toxicity in Plants: An Overview. In *Cadmium Toxicity and Tolerance in Plants*; Elsevier: Amsterdam, The Netherlands, 2019; pp. 1–17.

31. Nas, F.; Ali, M. The effect of lead on plants in terms of growing and biochemical parameters: A review. *MOJ Eco. Environ. Sci.* **2018**, *3*, 265–268.

32. Jiang, J.; Jiangang, L.; Yuanhua, D. Effect of cold plasma treatment on seedling growth and nutrient absorption of tomato. *Plasma Sci. Technol.* **2018**, *20*, 044007. [CrossRef]

33. Ling, L.; Jiangang, L.; Minchong, S.; Jinfeng, H.; Hanliang, S.; Yuanhua, D.; Jiafeng, J. Improving seed germination and peanut yields by cold plasma treatment. *Plasma Sci. Technol.* **2016**, *18*, 1027.

34. Li, L.; Jiang, J.F.; Li, J.; Shen, M.; He, X.; Shao, H.; Dong, Y. Effects of cold plasma treatment on seed germination and seedling growth of soybean. *Sci. Rep.* **2015**, *4*, 5859.

35. Zhang, B.; Li, R.; Yan, J. Study on activation and improvement of crop seeds by the application of plasma treating seeds equipment. *Arch. Biochem. Biophys.* **2018**, *655*, 37–42. [CrossRef] [PubMed]

36. Saberi, M.; Modarres-Sanavy, S.A.M.; Zare, R.; Ghomi, H. Amelioration of Photosynthesis and Quality of Wheat under Non-thermal Radio Frequency Plasma Treatment. *Sci. Rep.* **2018**, *8*, 1–8. [CrossRef]

37. Jiang, J.; He, X.; Li, L.; Li, J.; Shao, H.; Xu, Q.; Ye, R.; Dong, Y. Effect of Cold Plasma Treatment on Seed Germination and Growth of Wheat. *Plasma Sci. Technol.* **2014**, *16*, 54–58. [CrossRef]

38. Mihai, A.; Dobrin, D.; Magureanu, M.; Popa, M. Positive effect of non-thermal plasma treatment on radish seeds. *Rom. Rep. Phys.* **2014**, *66*, 1110–1117.

39. de Groot, G.J.; Hundt, A.; Murphy, A.B.; Bange, M.P.; Mai-Prochnow, A. Cold plasma treatment for cotton seed germination improvement. *Sci. Rep.* **2018**, *8*, 1–10. [CrossRef]

40. Iranbakhsh, A.; Ghoranneviss, M.; Ardebili, Z.O.; Tackallou, S.H.; Nikmaram, H. Non-thermal plasma modified growth and physiology in Triticum aestivum via generated signaling molecules and UV radiation. *Biol. Plant.* **2017**, *61*, 702–708. [CrossRef]
41. Pawlat, J.; Starek, A.; Sujak, A.; Kwiatkowski, M.; Terebun, P.; Budzeń, M. Effects of atmospheric pressure plasma generated in GlidArc reactor on Lavatera thuringiaca L. seeds' germination. *Plasma Process. Polym.* **2018**, *15*, 1700064. [CrossRef]
42. Bafoil, M.; Jemmat, A.; Martinez, Y.; Merbahi, N.; Eichwald, O.; Dunand, C.; Yousfi, M. Effects of low temperature plasmas and plasma activated waters on Arabidopsis thaliana germination and growth. *PLoS ONE* **2018**, *13*, e0195512. [CrossRef]
43. Eid, E.M.; Shaltout, K.H.; Alamri, S.A.M.; Sewelam, N.A.; Galal, T.M. Uptake prediction of ten heavy metals by Corchorus olitorius L. cultivated in soil mixed with sewage sludge. *Food Energy Secur.* **2020**, *9*, e203. [CrossRef]
44. Huang, B.; Xin, J.; Dai, H.; Liu, A.; Zhou, W.; Liao, K. Translocation analysis and safety assessment in two water spinach cultivars with distinctive shoot Cd and Pb concentrations. *Environ. Sci. Pollut. Res.* **2014**, *21*, 11565–11571. [CrossRef]
45. Saad, F.N.M.; Lim, F.J.; Izhar, T.N.T.; Odli, Z.S.M. Evaluation of phytoremediation in removing Pb, Cd and Zn from contaminated soil using Ipomoea Aquatica and Spinacia Oleracea. *IOP Conf. Ser. Earth Environ. Sci.* **2020**, *476*, 012142. [CrossRef]
46. Sengar, R.; Gautam, M.; Garg, S.K.; Sengar, K.; Chaudhary, R. Lead Stress Effects on Physiobiochemical Activities of Higher Plants. *Rev. Environ. Contam. Toxicol.* **2008**, *196*, 73–93. [PubMed]
47. Pourrut, B.; Shahid, M.; Dumat, C.; Winterton, P.; Pinelli, E. Lead Uptake, Toxicity, and Detoxification in Plants. *Rev. Environ. Contam. Toxicol.* **2011**, *213*, 113–136. [PubMed]
48. Wang, A.S.; Angle, J.S.; Chaney, R.L.; Delorme, T.A.; Reeves, R.D. Soil pH Effects on Uptake of Cd and Zn by Thlaspi caerulescens. *Plant Soil* **2006**, *281*, 325–337. [CrossRef]
49. Ye, X.; Ma, Y.; Sun, B. Influence of soil type and genotype on Cd bioavailability and uptake by rice and implications for food safety. *J. Environ. Sci.* **2012**, *24*, 1647–1654. [CrossRef]

MDPI
St. Alban-Anlage 66
4052 Basel
Switzerland
Tel. +41 61 683 77 34
Fax +41 61 302 89 18
www.mdpi.com

Applied Sciences Editorial Office
E-mail: applsci@mdpi.com
www.mdpi.com/journal/applsci